NUTRITION TECHNOLOGY OF PROCESSED FOODS

NUTRITION TECHNOLOGY OF PROCESSED FOODS

Nicholas D. Pintauro

NOYES DATA CORPORATION
Park Ridge, New Jersey London, England
1975

Library of Congress Catalog Card Number: 75-14912
ISBN: 0-8155-0586-8
Printed in the United States

Published in the United States of America by
Noyes Data Corporation
Noyes Building, Park Ridge, New Jersey 07656

FOREWORD

The detailed, descriptive information in this book is based on U.S. patents issued since 1961 that deal with the nutrition technology of processed foods. Where it was necessary to round out the complete technological picture, even earlier, but very relevant patents were included.

This book serves a double purpose in that it supplies detailed technical information and can be used as a guide to the U.S. patent literature in this field. By indicating all the information that is significant, and eliminating legal jargon and juristic phraseology, this book presents an advanced, commercially oriented review of modern methods of modifying and enhancing the nutritional content of foods as depicted in U.S. patents.

The U.S. patent literature is the largest and most comprehensive collection of technical information in the world. There is more practical, commercial, timely process information assembled here than is available from any other source. The technical information obtained from a patent is extremely reliable and comprehensive; sufficient information must be included to avoid rejection for "insufficient disclosure." These patents include practically all of those issued on the subject in the United States during the period under review; there has been no bias in the selection of patents for inclusion.

The patent literature covers a substantial amount of information not available in the journal literature. The patent literature is a prime source of basic commercially useful information. This information is overlooked by those who rely primarily on the periodical journal literature. It is realized that there is a lag between a patent application on a new process development and the granting of a patent, but it is felt that this may roughly parallel or even anticipate the lag in putting that development into commercial practice.

Many of these patents are being utilized commercially. Whether used or not, they offer opportunities for technological transfer. Also, a major purpose of this book is to describe the number of technical possibilities available, which may open up profitable areas of research and development. The information contained in this book will allow you to establish a sound background before launching into research in this field.

Advanced composition and production methods developed by Noyes Data are employed to bring our new durably bound books to you in a minimum of time. Special techniques are used to close the gap between "manuscript" and "completed book." Industrial technology is progressing so rapidly that time-honored, conventional typesetting, binding and shipping methods are no longer suitable. We have bypassed the delays in the conventional book publishing cycle and provide the user with an effective and convenient means of reviewing up-to-date information in depth.

The Table of Contents is organized in such a way as to serve as a subject index. Other indexes by company, inventor and patent number help in providing easy access to the information contained in this book.

15 Reasons Why the U.S. Patent Office Literature Is Important to You –

1. The U.S. patent literature is the largest and most comprehensive collection of technical information in the world. There is more practical commercial process information assembled here than is available from any other source.

2. The technical information obtained from the patent literature is extremely comprehensive; sufficient information must be included to avoid rejection for "insufficient disclosure."

3. The patent literature is a prime source of basic commercially utilizable information. This information is overlooked by those who rely primarily on the periodical journal literature.

4. An important feature of the patent literature is that it can serve to avoid duplication of research and development.

5. Patents, unlike periodical literature, are bound by definition to contain new information, data and ideas.

6. It can serve as a source of new ideas in a different but related field, and may be outside the patent protection offered the original invention.

7. Since claims are narrowly defined, much valuable information is included that may be outside the legal protection afforded by the claims.

8. Patents discuss the difficulties associated with previous research, development or production techniques, and offer a specific method of overcoming problems. This gives clues to current process information that has not been published in periodicals or books.

9. Can aid in process design by providing a selection of alternate techniques. A powerful research and engineering tool.

10. Obtain licenses — many U.S. chemical patents have not been developed commercially.

11. Patents provide an excellent starting point for the next investigator.

12. Frequently, innovations derived from research are first disclosed in the patent literature, prior to coverage in the periodical literature.

13. Patents offer a most valuable method of keeping abreast of latest technologies, serving an individual's own "current awareness" program.

14. Copies of U.S. patents are easily obtained from the U.S. Patent Office at 50¢ a copy.

15. It is a creative source of ideas for those with imagination.

CONTENTS AND SUBJECT INDEX

INTRODUCTION

In the years immediately following World War II, the thrust of the R&D programs conducted by food companies was toward the development of convenience foods. These products were not generally accepted unless in addition to the convenience they provided, they also were highly acceptable in flavor and texture. Examples of these products are "ready-to-eat" cereals, "instant" foods, and snack foods.

At the White House Conference on Food, Nutrition and Health held in 1969, reports of the unsatisfactory state of health and nutrition of some segments of the American population were presented. These reports stimulated an interest in publicizing the nutritional value (or lack of value) of convenience foods. It was hoped that this information would ultimately stimulate food manufacturers and processors to improve the nutritional quality of their products.

On January 19, 1973 the Food and Drug Administration issued guidelines for participation by the food companies in a program for nutritional labeling designed to assist the consumer in choosing foods based on their nutritive value. In implementing this program, it was not only found that there was a lack of reliable information on the content and variability of food nutrients but also a lack of reliable methods for nutritional analysis. Apparently, nutritional labeling was beyond the capabilities of most food manufacturers.

Despite the fact that most foods, obviously, are consumed for nutritional purposes (the exceptions being coffee, tea, spices, etc.) food labels do not carry information on caloric, protein, fat, carbohydrate and other nutritional component content. Even under the proposed regulations labels for most foods are not required to contain such information. Only in cases where any mention or claim is made for nutritional content or quality, is it necessary for the manufacturer to comply with the labeling guidelines.

There is much information and knowledge available on nutritional studies with animals and humans. Most of the available information on human nutrition, however, concerns malnutrition in underdeveloped nations, special diets, and studies on the relationship between nutrition and disease. The main source of nutritional information was to be found for the most part in the patent literature which discusses both the ingredients and technology of food processing of products specifically designed for their nutritive benefits. This book is based on that source of information. In setting it up patents were selected and abstracted when information was given on: processes designed to preserve the nutritional quality of foods; formulations, ingredients, supplements or other additives intended to improve the nutritional quality of foods; or where nutritional values were given which might then be used for labeling purposes designed to assist the consumer in food selection and planning.

It will be obvious to the reader that many existing manufacturing processes are similar to those included in this publication. However, the value of including them here resides in the relationships that are provided for raw materials, formulated products, processes and nutritional compositions. These relationships can be used as guides to the food technologist in the development of new products or modification of existing products to comply with nutritional label statements.

The chapters are organized with regard to: product groups (bread and baked goods, dairy products, rice and potatoes); special dietary products; and processes to prepare stable supplements containing vitamins, minerals and amino acids.

To add to the usefulness of the book, the reader is advised to examine the most recent releases publicized in the Federal Register and other official sources for updated information on regulations pertaining to the use of ingredients in formulating products to be marketed with nutritional labeling.

CEREAL PRODUCTS

LYSINE FORTIFICATION

Special Fractions from Corn

As cereal grain kernels develop in the field, protein remains essentially loose and free from the starch granules, but when the ripened kernel dehydrates, protein and starch agglomerate. These agglomerates can be finely ground but the protein remains attached to the starch. Users of cereal flours until recently, had to depend on varietal differences for flours having desired protein concentrations. Air classification has now provided a means for obtaining enriched protein fractions that can be added back to control the composition of flour.

Dry milling methods, the major method of producing edible cereal products, generally start with tempering the kernels with water, after which the hydrated kernel is dehulled (usually abrasively), the germ is removed, and the endosperm collected and dried giving the three major cereal endosperm materials: grits, meal, and flour. If any separation of protein and starch is desired, the above cereal endosperm materials are fine ground and air classified.

Air classification separation of these finely ground products has been successful in obtaining high-protein concentrates from corn and sorghum. High-protein flours are desirable for specialty baked goods such as breads, crackers, and snack foods. High-protein flours can also be used to supplement carbohydrate food products. Dry milling and air classification procedures for producing specialty wheat flours are expensive in that the fractions obtained having the desired amount of protein are small. Only a small portion of the total wheat protein can be shifted to high-protein fractions. The remaining fractions have slightly less protein.

In the process by *D.D. Christianson, A.C. Stringfellow and J.S. Wall; U.S. Patent 3,661,593; May 9, 1972; assigned to U.S. Secretary of Agriculture* a variety of protein concentrates can be obtained from cereal grains. A variety of cereal endosperm materials were subjected to an isotonic buffer. The buffer used for the examples was composed of 0.1 M potassium phosphate buffer (pH 7.5) containing 0.006 M magnesium chloride.

The process has essentially only three parameters: amount of buffer used in relation to the cereal endosperm product; time of soaking; and temperature of soaking. A quantity of buffer sufficient to cover each cereal endosperm particle at from 4°C to room temperature will slough the protein sheath from the starch granules as seen by the scanning electron microscope. Amounts of protein found in the various fractions obtained from classification procedures such as sieving or air classification are also used as criteria for showing this starch-protein agglomerate disruption. Since endosperm materials from normal milling give only a small separation when subjected to classification procedures, particle size weight distribution of the treated materials subjected to the same procedures provides a good qualitative measure of disruption.

When the buffer is filtered off and the remaining material weighed, it was found that an amount of buffer equal to 45% of the cereal endosperm product had been absorbed. This was felt to be the minimum amount of buffer that would achieve the desired disruption of protein from starch. When a slurry of buffer and high-lysine corn flour was stirred for 1 hr at 24°C, the weight distribution of the sieved fractions was very nearly the same as the weight distribution obtained from the product of the preferred treatment (i.e., 0.5 liter buffer per 50 g flour, stirred 24 hr at 4°C). The percent protein in the fractions obtained from the 3-hr, 45% buffer treatment are essentially the same as the comparable fractions obtained from the 24-hr, 4°C stirred slurry treatment.

Percent protein analysis and the weight distribution in the sieved fraction obtained from water treated-hammer milled grits and those having no treatment show that little or no disruption of the starch-protein agglomerates has taken place. The similarity between these data indicates that the distribution of particle size is due to hammer milling and not the water treatment. The fact that there is little difference in percent protein in any of the fractions from these two samples means that the protein has not sloughed off the starch particles.

The protein and starch are essentially evenly distributed among all fractions. Data from the buffer treated samples show considerable distribution in both percent protein and in particle size. There are fractions that vary from 5.6 to 18% protein. Since the original grits contain 12.4% protein, it is evident that the protein has been separated from the starch by the buffer treatment. The similarity in the data from the 24-hr, 4°C and the 1-hr, 22°C buffer treated-hammer milled grits substantiates the fact that the parameters of time and temperature are not critical.

The various untreated cereal endosperm materials do give some distribution when subjected to the sieving procedure (see Example 1), but in each example the major portion stays in the 90μ sieve. In all the buffer treated samples the major portion of the weight is distributed among the smaller mesh sizes. In Examples 2, 3, and 4, the largest weight fraction is in the smallest mesh size. Protein distribution in the fractions obtained from buffer treated materials, in every example, is highest in the largest mesh size and progressively decreases as the mesh size decreases. The result is a number of cereal endosperm products containing protein concentrations that vary from as little as one third to almost three times the amount of protein in the original dry milled material.

Salt solutions other than those in the isotonic buffer were used to treat wheat flour under the conditions described in Example 1. Weight distributions obtained from sieving these products are compared to weight distributions obtained from untreated control, water treated, and buffer treated wheat flour samples. It has been shown that untreated and water treated cereal endosperm products do not give the desired distribution of protein, and the following table shows that Na_2CO_3 and Na_3PO_4 solutions do not give the desired weight distribution indicating no starch-protein disruption.

Furthermore, scanning electron microscopy shows little, if any, starch-protein disruption from the Na_2CO_3 and Na_3PO_4 solution treatments. Only the isotonic buffer will slough off protein sheaths that naturally surround the starch particles of dry milled cereal endosperm materials as shown by scanning electron microscopy and normal classification procedures.

Example 1: 50 g of Ponca hard red winter wheat flour analyzing 10.3% protein was stirred with 0.5l of isotonic buffer for 24 hr at 4°C. The isotonic buffer is composed of 0.1 M KH_2PO_4 and 0.006 M $MgCl_2 \cdot 6H_2O$ at pH 7.5. At the end of the 24-hr soaking period, the resulting mixture of starch and protein was dried. A portion of the dried material was sieved through a series of BMC micromesh screens (30, 45, 60, 75, and 90μ) in a Ro-Tap sieve shaker for 10 minutes.

	pH	Percent Weight 90μ	75μ	60μ	45μ	30μ
Control		36	22	16	17	9
Water	6.0	71	23	5	1	0
0.5% Na_2CO_3	8.5	55	24	9	6	3
0.5% Na_3PO_4	6.9	87	9	3	1	0
Buffer	7.2	11	25	24	26	15

Example 2: 50 g of red endosperm sorghum flour obtained by dry milling having 8.6% protein were buffer treated, sieved and analyzed for protein in the same manner as described in Example 1.

Example 3: 50 g lots of experimentally milled high-lysine corn (opaque-2) flour having 6.3% protein were buffer treated as in Example 1. A portion of the dried material was sieved and analyzed and compared with nontreated and sieved original flour. A second portion was pin milled three times at 14,000 rpm in an Alpine 16 Kolloplex pin mill and air classified by weight into five fractions in a Pillsbury laboratory model air classifier by a uniform standardized procedure. A sample of the original flour was also pin milled and air classified as above. All air classified fractions were analyzed for protein content, as shown in the following table.

Fraction number	Weight (%) A[1]	B[2]	Protein (%) A[1]	B[2]
1	6.9	10.2	13.5	12.3
2	9.3	19.5	7.4	7.3
3	22.7	36.3	4.1	2.7
4	22.6	13.9	5.3	2.6
5	38.2	20.2	5.1	6.7

[1] A = Before treatment, original flour pin milled 3 times at 14,000 r.p.m.
[2] B = After treatment, pin milled 3 times at 14,000 r.p.m.

Example 4: 50 g of commercially obtained corn bread flour having 5.8% protein was buffer treated and dried as in Example 1. One portion was sieved and analyzed for protein and compared to nontreated and sieved original flour. A second portion of the buffer treated material and a sample of the original corn flour were pin milled, air classified, and analyzed for protein as in Example 3, as shown in the following table.

Fraction number	Weight (%) A[1]	B[2]	Protein (%) A[1]	B[2]
1	10.8	13.7	13.0	14.6
2	11.3	21.5	8.2	7.6
3	44.8	47.0	5.7	3.5
4	14.9	14.1	5.8	3.5
5	18.1	3.7	7.3	3.8

[1] A = Before treatment, original flour pin milled 3 times at 14,000 r.p.m.
[2] B = After treatment, pin milled 3 times at 14,000 r.p.m.

Lysine-Phosphate Combinations

H.C. White; U.S. Patent 3,231,388; January 25, 1966; assigned to The Dow Chemical Company found that a L-lysine·H_3PO_4 complex overcomes the disadvantages of using other L-lysine salts in the fortification of cereals and baked goods in that (1) cereals and foodstuffs fortified with L-lysine·H_3PO_4 in the indicated amounts are not objectionably bitter; (2) L-lysine·H_3PO_4 is relatively nonhygroscopic and is, therefore, more easily handled than other salts, e.g., L-lysine·HCl; (3) nutritionally significant amounts of L-lysine·H_3PO_4 can be added to bread dough without causing an undesirable decrease in baked loaf volume;

and (4) the addition of nutritionally significant amounts of L-lysine·H_3PO_4 to doughs does not cause an objectionably dark browning on baking.

L-lysine·H_3PO_4 is prepared by reacting a mixture of substantially equimolar proportions of L-lysine and H_3PO_4, i.e., from 1 to 1.1 mols of phosphoric acid per mol of L-lysine, as concentrated phosphoric acid, in an aqueous reaction medium and precipitating the formed L-lysine·H_3PO_4 at room temperature with a water-miscible organic liquid precipitant, i.e, any lower monohydric alcohol or acetone, in amount sufficient to precipitate said salt, preferably with added methanol to provide at least 50 volume percent of methanol. The resulting precipitated L-lysine·H_3PO_4 is separated from the reaction medium and is washed with methanol or other precipitant.

At a relative humidity less than 80%, L-lysine·H_3PO_4 is essentially nonhygroscopic, since it gains only up to 0.2% of its anhydrous weight on 120 hr exposure to air at relative humidities up to 80%. In contrast, L-lysine·HCl is essentially nonhygroscopic only at a relative humidity below 57%. As a result of their greater hygroscopicity, L-lysine·HCl and other hygroscopic L-lysine salts are not easily used as a means of supplementation of cereals such as wheat, corn, rice, barley and oats, especially in flour form, and foodstuffs containing them. In practice, the L-lysine·H_3PO_4 is added with mixing at any convenient stage in the preparation of the cereal-containing foodstuff.

Example 1: Two loaves of sponge dough process white bread were baked from the same dough, except that to one was added L-lysine·HCl and to the other was added L-lysine·H_3PO_4, each at a level of 2 g of lysine salt per pound of bread, equivalent to 0.44 weight percent, bread basis. A visual diminution in loaf volume was evident with the L-lysine·HCl fortified bread, but not with the L-lysine·H_3PO_4 fortified bread.

Example 2: L-lysine salts, as indicated below, were incorporated at the remix stage of preparation of sponge dough process bread to the extent of 0.25 weight percent, with observations as noted below.

Loaf No.	L-lysine salt	Crust color
1	Control	Normal brown.
2	L-lysine/H_3PO_4	Do.
3	(L-lysine)$_2$-tartrate	Considerably darker.
4	(L-lysine)$_3$-citrate	Do.
5	L-lysine-sorbate	Do.
6	L-lysine/HCl	Do.

Loaf No. 6 was extremely bitter in taste. Loaf No. 2 had no objectionable bitterness compared with the control. Loaf No. 2 was most acceptable from the standpoint of crust color. Loaves 3, 4 and 5 would not be acceptable because of excessive crust coloration. Similar advantageous results are obtained when 0.1 to 1 weight percent of L-lysine·H_3PO_4 is added to wheat cereal, oatmeal and barley, or to flours and mixes containing cereal products, whether for use in cooking or baking.

Protein Coating

Nutritionists do not regard wheat products as a complete answer to mass feeding programs because their protein content is only about 10 to 12%. Thus, nutritionists advocate that a single food for mass feeding should contain a higher protein content, on the order of 20%, whereby it may serve as a balanced source of carhohydrate and protein nutrients. Another item is that wheat products are relatively deficient in one of the essential amino acids, lysine.

A process to fortify wheat products with protein is described by *R.P. Graham, M.R. Hart and A.I. Morgan, Jr.; U.S. Patent 3,450,540; June 17, 1969; assigned to the U.S. Secretary of Agriculture* as follows.

(1) The cereal—wheat, for instance—is first moistened. This is done by soaking it in water

until the moisture content of the grain is 20%. Products with the bran retained will require a longer soak than those with the bran removed. In typical operations with whole raw wheat or gelatinized whole-kernel wheat, the soaking time is 10 minutes; with cracked wheat products the soaking time may be as short as 15 to 20 seconds. Usually, the soaking is done in water at room temperature (75°F). After application of water, the grain is drained and while still wet is subjected to the next step of the procedure.

This moistening step also provides a convenient time at which to absorb water-soluble supplements into the grain. Thus, one may add to the moistening water such substances as water-soluble vitamins (thiamine, riboflavin, ascorbic acid, etc.) and/or mineral salts, such as calcium chloride, whereby these substances will be absorbed by the grain and the product will be thus enriched in these factors.

(2) The moistened wheat is coated by tumbling it with the proteinous additive, this latter being in a powder form. The tumbling is readily achieved by placing the moistened wheat and protein powder in a drum and rotating the drum to uniformly coat the grains. The wet condition of the grains is important as it enables the powder to stick to the grains. The relative proportion of grain and protein powder may be varied depending on what degree of protein fortification is desired for the final product and depending on the protein content of the powder to be applied.

In typical operations where the protein is soybean flour, for example, a preferred ratio is 15 to 25 parts per 100 parts of wheat. At the completion of the tumbling operations, the product consists of wheat grains coated with the protein powder. However, at this point the coating is not durable and will readily slough off by rubbing action. Accordingly, further treatment is required to stabilize the coating.

(3) In this step the coated grain is subjected to steaming. This is conventionally done in an ordinary food blancher. The coated grain is placed on metal screens and passed through the device where it is exposed to live steam (at 212°F). This steaming is continued for a period of 2 to 5 minutes, depending on the depth of the grain on the screens. A thinner bed of grain will be heated quicker and hence will require a shorter sojourn in the blancher than will a thicker bed of grain. If desired, the grain may be subjected to shaking or vibrating while being steamed but this is not essential. The steaming operation provides several important results.

For one thing, it causes a setting or consolidation of the coating—i.e., it consolidates the individual particles of protein in the coating into a homogeneous material. It also adheres the coating to the grain so that the tendency of the coating to rub off is greatly diminished. Also, the steaming has the effect of decreasing the flavor of the added protein. This is important, for example, where the protein is soybean flour which naturally has a rather undesirable beany flavor. The steaming markedly suppresses this beany flavor so that it does not detract from the desired taste of the grain itself. Another item is that the steaming destroys substances—such as antitrypsin or other enzyme inhibitors—which may be naturally present in the added protein and which substances interfere with digestion of the protein.

It is thus evident that the steaming accomplishes valuable results. However, at this point the coating is not stable to cooking. If the product is cooked—as by boiling in water—at least part of the coating will slough off with the result that the product has an undesirable mealy texture. In the next step the product is toasted to stabilize the coating to the end that the product can be cooked with retention of the coating.

(4) The toasting can be readily accomplished in a device which includes a treating chamber provided with a top and bottom—both of screening or perforated metal. Also provided is a duct for directing a draft of hot air upwardly through the chamber. In operation, the coating grain, still moist from the previous steaming, is placed in the chamber where it is tumbled about and suspended in the steam of hot air (350°F) passing through the chamber.

Usually, the exposure to the hot air current is continued for 1 to 2 minutes depending on the size of the grains. In any event, the time of treatment should not be so extended as to cause the grains to be excessively brown. In the preferred treatment, the toasting operation is discontinued when the grains assume a light tan color.

(5) (Optional) If the product after toasting is not dry enough for keeping, it is subjected to a conventional drying step, for example, by contacting it with warm air (150°F) until its moisture content is 10 to 14%. By applying the procedure described above, one obtains cereal particles each enveloped with an abrasion-resistant coating of the proteinous material. This product constitutes a valuable food as it contains the carbohydrates, proteins and vitamins of the cereal plus the added protein of the coating.

The products are not only valuable nutritionally but have an attractive appearance as the coating is smooth and covers the grain uniformly. Also, the products have an attractive light tan color. A further item is that the individual coated grains are not sticky but remain separate on storage and can be readily dispensed from containers by pouring just as can be, say, raw wheat. The adventitious rubbing of the grains against one another in packaging, shipping, etc. does not detach the coating. Thus the coating may well be described as rugged and abrasion-resistant. The coating does not interfere in any way with cooking of the products—they are cooked in the same way as the uncoated grain would be cooked.

For example, if the grain is in a precooked condition prior to coating, the products would be heated in boiling water 10 to 15 minutes. A somewhat longer time would be used if the grain were not precooked. In any case, the coating remains on the grain during the cooking so the consumer has the benefit of the protein fortification as he consumes the grain. Another item is that the products form a dish of excellent nutty flavor even in cases where the protein supplement is derived from soybeans, which normally impart an undesirable beany taste to food products.

The process may be applied to other cereals, for example, rice, oats, barley corn, rye, sesame, milo, etc. The cereal used as the starting material for the coating operation may be in a condition retaining the bran, it may be partly or completely debranned, it may retain the germ or may be partly or completely degermed. Moreover, the cereal may be in a raw or precooked state and it may be in the form of whole kernels or cracked into kernel fragments.

The protein powder used for the coating is preferably low-fat soybean flour since it has a high protein content—50%—and is low in price and readily available on the market. Also, it contains a substantial proportion of lysine and thus remedies the deficiency of this amino acid in wheat. The material is referred to as low-fat since it is prepared from soybeans which have been defatted by solvent extraction or equivalent procedures. Other protein materials which may be used are: defatted fish flour; casein; gluten and other proteins or protein concentrates derived from wheat; zinc; nonfat milk solids; and flours prepared from beans of the genus Phaseolus, i.e., lima beans, navy beans, etc.

Example 1: 500 g of precooked, peeled wheat was placed in a perforated metal basket and immersed in a tank of cold water for 10 minutes. The basket was then removed from the tank and swung vigorously several times to drain the wheat. The moistened wheat was then poured into a drum together with 125 g of low-fat soybean flour. The drum was closed, shaken several times to mix the contents, then rolled for about 4½ minutes to coat the flour uniformly on the wheat grains. The drum was then opened and the coated grain spread on a screen which was placed in a conventional food blancher wherein the grain was contacted with steam (212°F) for 2 minutes.

The steam coated grain was then introduced into a toasting device. This piece of equipment includes a receptacle having a screened top and bottom, and means for directing a stream of hot air upwardly through the receptacle at a velocity high enough to suspend and tumble about the grain contained therein. In this instance the grain was contacted with the hot air stream (150°) for 1.75 minutes. The toasted grain was then placed in a conventional dehydrator where it was contacted with warm (140°F) air until its moisture content was 12%.

The product was observed to consist of light tan colored oval particles of uniform size. Its protein content was 20%. The coating was smooth and strongly adherent. If the product was subjected to cracking forces, the coating was not separated. The entire grains broke and the pieces still retained their proportion of adherent coating. A sample of the product was simmered in an equal volume of water for 15 minutes and examined and tasted. It was found that the grains remained separated, the coating was retained on the grains, and the product had an excellent nutty flavor—no bean flavor could be detected.

Preparation of Starting Material – The precooked peeled wheat used as a base for the coating described above in Example 1 was prepared as follows:

(a) Soaking: Raw wheat was fed at the rate of 250 lb/min into a trough, U-shaped in cross-section, provided with a screw, and filled with hot water. Heating jackets were provided to maintain the mixture at 135°F at the feed end, 180°F at the exit end. The time of sojourn of the wheat in the trough was one hour.

(b) Tempering: The soaked wheat was fed through a bin wherein it was held at 180°F. Sojourn time was 30 minutes.

(c) Cooking: The soaked and tempered wheat was passed through a reel wherein it was contacted with steam (212°F) for 12 minutes (to gelatinize the starch in the grains).

(d) Lye Treatment: The cooked wheat, still in the reel was sprayed with hot (180°F), 25% aqueous NaOH, using an amount of the solution equal to 20% the weight of the wheat. The lye-coated wheat was held in the reel 3 minutes while steam was applied to it to keep it hot (180°F). The lye-treated wheat was then rinsed on a rotating screen with cold water to quench it and remove at least part of the NaOH.

(e) Peeling: The lye-treated wheat was slurried with excess water and the slurry was pumped under 40 psi pressure and at a rate of 25 gal/min through a hydrocyclone. This device has a diameter (at the top) of 3" and a length of 24". The discharge tip had a diameter of ¼", the inlet and outlet were standard ½" pipe. Pressure drop through the device was 40 psi. The peeled grain issuing from the discharge tip was washed thoroughly with water, using counterflow through a trough equipped with a screw conveyor.

(f) Acid Treatment: The washed, peeled grain was passed through another trough equipped with a screw conveyor wherein it was contacted with 1% aqueous acetic acid at 120°F for 5 minutes. In running this step, about 1 lb of acid per 100 lb of grain was metered into the trough to make up for the acid neutralized by the alkali in the grain. The acid-treated grain was then washed with water on a screen.

(g) Drying: The grain, which at this point contained 50% H_2O was dried. Drying was in a belt-trough dryer wherein a gently-mixing bed of the grain was contacted with a stream of warm air. Other types of dryers such as rotary kilns, tunnel dryers, etc. can, of course, be used. In any case, the drying is continued until the grain contains 10 to 14% moisture. Yield of the product, which has been named WURLD wheat, was 85 lb/100 lb of raw wheat. Properties of the product are listed below:

Color	Yellow-white
Colored bran content	0%
Cooking time	15 min
Density	40 lb/bu
Protein	11%
Fat	1.0%
Thiamine	0.12 mg/100 g
Riboflavin	0.10 mg/100 g
Germ	Absent

Another phase of the process concerns a flaked product of cereal and the protein-fortifying material. This is a product especially adapted for feeding of infants and its production is described in detail below. In preparing the flaked product, a selected cereal is subjected to the steps of moistening, tumbling with the protein additive, and steaming, all as described above. The coated product, while still warm from the steaming, is then pressed between rolls (preferably heated to 200° to 225°F) to form flakes of a thickness of 5 to 20 thousandths of an inch. Optionally, the flakes may be then toasted as described. Ordinarily, the flakes so produced have a low moisture so that they will keep safely without spoilage. If, however, it is necessary to further dry them this can be done in conventional manner as by contacting them with warm (150°F) air. Usually, a final moisture content of 10 to 14% is preferred.

Example 2: Raw, cracked wheat (100 parts) was placed in a perforated metal basket and immersed in a tank of cold water for about ¼ minute. The basket was then removed from the tank and swung vigorously several times to drain the wheat. The moistened wheat was poured into a drum together with low-fat soybean flour (25 parts). The drum was closed, shaken several times, then rolled for 1½ minutes to coat the flour uniformly on the cracked kernels. The drum was then opened and the coated material spread on a screen which was placed in a food blancher where it was contacted with steam (212°F) for 1.5 minutes.

The steam material, while still hot, was fed into the nip between two rotating rolls maintained at 225°F. The distance between the rolls was adjusted to press the coated particles into flakes 0.010" thick. It was observed that the soybean materials adhered strongly to each flake so that it did not slough off when the flakes were subjected to rubbing action. Protein content of the product was 20% (original wheat 12%). A sample of the product was cooked by adding an amount of boiling for 2 minutes. This produced a mush-like dish eminently suitable for infant feeding and which had an excellent flavor in which the beany flavor of soybeans could not be detected.

Solution Absorption Method

Treatment of cereals with amino acids to increase the nutritional value is described by *G.A. Geurts; U.S. Patent 3,617,303; November 2, 1971; assigned to Stamicarbon N.V. Heelen; Netherlands*. It would be most desirable to further increase the amino acid content of cereals, especially where high amino acid contents are important in special dietary problems. Further, where a cereal with a high amino acid content can be obtained, these cereals may be blended in relatively small proportions with untreated natural cereals to provide any desired final amino acid content without the necessity of treating all or substantially all of the cereal in order to obtain a relatively high average amino acid content.

Briefly stated, it has been found that in the treatment of a cereal with an aqueous solution of at least one amino acid, a considerably higher amino-acid content may be obtained with preservation of the granular character of the cereal, by contacting the cereal with water for a period of time sufficient to increase the water content of the cereal and thereafter contacting the cereal having an increased water content with the solution of at least one amino acid.

The increase of the water content of the cereal is not altogether critical, but in order to effect a substantial increase in the amount of amino acid absorbed, the increase in water content, expressed as weight percent increase over the weight of the natural cereal, should be at least 4% and especially at least 10%. The increase in water content of the cereal is important, since the amount of amino acid absorbed by the cereal is dependent, as well as upon the cereal. For example, in the treatment of wheat with an aqueous solution containing L-lysine monohydrochloride, the L-lysine monohydrochloride content in the wheat may be increased to 8% by weight, if the wheat contains an additional 30% by weight of water (calculated with respect to the weight of wheat containing its natural water content).

It has been found also that the amount of amino acid absorbed can be further increased if the cereal which has an increased water content is broken before being treated with the solution of an amino acid. For example, if the same wheat mentioned above (natural water content of 14% is increased in water content by 30% with respect to the weight of the wheat and the wheat grains are then broken, the treatment with an aqueous solution of L-lysine monohydrochloride, will produce a L-lysine monohydrochloride content of the wheat of 12% by weight, as opposed to the 8% by weight of the wheat treated with unbroken grains. Preferably, the cereal is first increased in water content and subsequently broken, because this facilitates the breaking procedure, but the reserve of the steps may be used.

The term broken grains is defined as grains which have had the husks cracked, but with substantial preservation of the granular character of the cereal. The husk may be broken by subjecting the grains to a light pressure, e.g., between two rubber-lined rollers rotating in opposite directions, between rotating discs, by fluid blasts and by vigorous tumbling. From a theoretical point of view, it would be assumed that an increased water content of the cereal to be treated with the amino acid solution would have an adverse effect or at least no effect at all on the amount of amino acids absorbed by the cereal. If cereal with the natural water content is subjected to a prolonged treatment, e.g., 10 hr or more, the amount of amino acids absorbed by the cereal will be much smaller than that absorbed by the cereal with an increased water content in a much shorter time.

Generally speaking, a sufficient increase in water content of the cereal to be treated may be obtained by contacting, e.g., soaking, spraying, dipping, etc., the water with the cereal for a relatively short time such as from ¼ to 6 hours, e.g., for 1 to 4 hours. While the temperature of the water for increasing the water content of the cereal is not critical, elevated temperatures are preferred since the absorption of water by the cereal takes place more rapidly at higher temperatures. The water temperature is therefore suitably up to the boiling point of water, e.g., between 40° and 70°C. However, if desired, the cereal may also be contacted with steam (preferably low pressure, e.g., up to 20 psig, saturated steam) instead of with water.

Since as noted above, a higher amino acid content of the cereal can be obtained with increasing amount of water which has been absorbed by the cereal, by controlling the amount of water absorbed by the cereal, i.e., controlling the time and temperature of the water contacting step, the amount of amino acid absorbed may also be controlled. However, valuable results may be obtained over a relatively wide range of percent increases in water content of the cereal over cereal with the water content of the natural cereal. The particular amount of water absorbed will depend upon the particular cereal being treated. For example, the water content of wheat is increased by an amount corresponding to a 4 to 30% by weight increase of water content with respect to the wheat with a natural water content.

In the case of maize, the percentage increase in water content preferably varies from 10 to 45% by weight. Of course, after the cereal of increased water content is treated with the aqueous solution of one or more amino acids, the cereal may be dried, e.g., at up to 85°C, especially between 40° to 70°C, to reduce the moisture content to any desired level, e.g., the natural level.

Example 1: 100 g of wheat grains (natural water content 14% by weight; L-lysine content 0.2% by weight) were contacted by soaking with 200 g of water at 60°C for 3½ hours. The grains absorbed 30 g of water during that period (the increased water content percentage of the soaked wheat with respect to the wheat with a natural water content amounts to 30% by weight). The grains were separated off by filtration and subsequently contacted with an aqueous solution containing 30% by weight of L-lysine monohydrochloride at 60°C for 4 hours.

The grains were filtered off, washed with flowing water, and dried in an oven at 60°C to a water content of 15% by weight. The L-lysine monohydrochloride content of the dried wheat grains was determined to be 8% by weight. As a comparison, wheat grains (natural water content of 14% by weight) were treated in the same manner as above but without the water soaking step. The L-lysine monohydrochloride content of the dried wheat grains was determined to be only 3.8% by weight.

Example 2: 100 g of wheat grains were contacted by soaking with 220 g of water at 60°C for 3 hours. The water which had not been absorbed was removed by filtration, after which the grains were broken by passing the wheat grains between two rubber-lined rollers rotating in opposite directions (shortest distance between the rollers: 2 mm; diameter of the rollers: 8 cm). The amount of water absorbed by the grains amounted to 28 g.

Subsequently, the broken grains were contacted by soaking with an aqueous solution containing 25% by weight of L-lysine monohydrochloride at 55°C for 3½ hours. The grains were filtered off, washed with flowing water, and dried at 55°C in an oven. The dried wheat grains contain 11.8% by weight of L-lysine monohydrochloride and 14% by weight of water.

Example 3: 100 g of maize grains (natural water content 12.5% by weight; L-lysine content 0.3% by weight), were contacted by soaking with 195 g of water at 55°C for 3½ hours. The maize grains, which have absorbed 40 g of water in the treatment, were separated off by filtration and subsequently treated with an aqueous solution containing 30% of L-lysine monohydrochloride at 60°C for 3 hours. Thereafter, the grains were filtered off, washed with flowing water, and dried at 60°C in an oven. The L-lysine monohydrochloride content of the dried maize grains was 9.5% by weight, while the water content was 13% by weight.

As can be seen from the above examples, cereals with exceptional high increases in the amino acid content may be obtained. For example a content of lysine (which is an essential amino acid) of over 1,100% of the original lysine content may be obtained in the treating of unbroken cereals. In the treating of broken cereals a lysine content may be obtained which amounts to over 3,400% of the original lysine content.

PROTEIN BASE FROM CEREALS

Fractions from Corn Germ

It is known that corn germ is potentially a good source of protein. The protein content in cereal germs is of much better nutritional quality than that protein which may be obtained from the endosperm portion of the cereal grain. However, to date, there is no simple, low cost yet efficient method of upgrading the protein content in germ to produce a protein rich fraction relatively low in fiber content such that it will be useful for human consumption.

Cereal endosperm proteins are in general low in the essential amino acids, lysine, tryptophan, and to some extent methionine. Protein from corn germ, on the other hand, has been reported to have both good amino acid balance and high biological value. Thus, because of its low cost and nutritionally good protein, spent corn germ has frequently been suggested as having potential use in food products. A major obstacle to the use of corn germ in food products is, however, its relatively high content of fibrous materials and its characteristic bitter flavor. Attempts to obtain corn germ protein in a form more suitable for food use have generally been unsatisfactory because of low product yields and/or high processing costs. These procedures have generally utilized classical procedures of dissolving protein by chemical (alkaline) means and precipitating with acid.

Corn germ is also a potential source of a fibrous product which can easily be fractionated into cellulose and hemicellulose components. The hemicellulose component can be used for the same end-applications as the commercial vegetable gums are now utilized or as a source material for furfural production. Cellulose from this source could be used in any of the many applications that cellulose is now used such as in the manufacture of viscose rayon, cellophane, lacquers, etc. However, starch which is normally present in the spent germ flake is solubilized under the alkaline conditions needed to dissolve the hemicellulose and cannot readily be separated.

A protein rich product from a cereal grain is prepared by *J.E. Freeman and R.M. Olson; U.S. Patent 3,615,655; October 26, 1971; assigned to CPC International Inc.* In general, the process includes the steps of rupturing the cells of a cereal germ, abrading the ruptured germ to free protein particles which adhere to the ruptured germ and separating the freed protein particles from the remainder of the ruptured germ cell fragments. A protein rich germ fraction is thus produced which has a relatively low content of fibrous material. Also, a fibrous germ residue fraction is produced which is relatively low in protein content. Both fractions are useful, with the high protein content portion being particularly suitable for human consumption in a number of food materials such as bread, and the fibrous fraction being useful as a source of cellulose and hemicellulose.

The first step involves rupturing the germ cell walls in order to partially free the proteinaceous fraction. The starting material may be either a spent grain germ such as a spent corn germ or whole previously unprocessed grain germ. Spent germ, that is deoiled grain germ, such as corn or grain sorghum germ may be produced by either dry or wet milling schemes. In the usual wet milling system cleaned whole grain is softened by steeping for two days in warm water that is acidified with a small amount of sulfur dioxide. The temperature of the steepwater is normally maintained between 116° and 132°F. After steeping, the grain, usually containing 45 to 55% moisture, is coarsely ground in degermination mills to liberate the germ. The germ is then separated from the remainder of the constituents of the whole grain by flotation on a starch slurry or by mechanical separation in hydroclones.

Separation by either method is possible because of the lower density of the steeped germ which contains about 45 to 55% oil on a dry basis. The germ is then washed, dried, heated and put through continuous mechanical screw presses. Here under stress of the high pressure most of the oil is pressed out while the solid part of the germ, still containing some residual oil is usually sold for animal feed. The solid residue of the germ is usually designated as spent germ.

Spent germ may be also derived from a dry milling technique. After moisture adjustment, the grain may be passed through a series of cracking and reducing rolls, or break rolls, to break the kernels apart and give a mixture of bran, germ and endosperm particles. In more modern corn dry milling plants, the germ may be freed from the residual of the tempered kernel by such mechanical means as Entoleter mills or Beall degermination mills. In either case the mixture of germ, bran, and endosperm is sifted to effect a separation between hulls and germ on the one hand and grit or endosperm particles on the other. Most of the bran can be separated from the germ by aspiration. The germ from the dry milling process is also expelled and may be treated by essentially the same procedure as used with wet-milled germ.

However, in either case the spent germ residue from these procedures has too much fiber and not enough protein to be used in human foodstuffs. The next step involves freeing the protein particles from the remainder of the germ cell. This is best accomplished by abrading in some manner the ruptured germ substance to release the protein rich fraction from the surface of the cell fragments. Conventional impact and attrition mills are most suitable here. For best results, this step should be carried out when the already ruptured germ is in a liquid slurry form. Thus, a slurry of ruptured germ in a nonpolar solvent such as hexane, heptane or other aliphatic solvents may be prepared and the slurry agitated to the desired degree to free the protein fraction. Even with this relatively mild treatment the protein is freed up as a suspension in the solvent.

As just one single example of this step of freeing protein, an already dry or wet ground germ may be slurried up in hexane and then agitated as desired. The hexane both washes out the protein particles contained within the cell and simultaneously extracts the desired oil constituent. The ground matrix or protein lean, fibrous material is then separated from the protein rich fraction suspended in the oil and hexane liquid mixture as further described. Thus, in essence, both extraction of germ oil and freeing of protein can be effected simultaneously.

The last step involves separation of the two germ fractions. The protein rich fraction of the germ is of a materially lesser average particle diameter, compared to the fibrous or protein lean fraction. Specifically, the particles forming the protein rich fraction have an average particle diameter of less than 40 microns. The greater proportion of the proteinaceous material has a particle diameter less than 20 microns, with the greatest percentage of the product falling within about 1 to 10 microns in particle size. The fibrous fraction, on the other hand, has a particle diameter substantially greater than 40 microns, and usually is of a diameter greater than 100 to 300 microns.

Thus, it is a simple matter to take advantage of this difference in physical size of protein rich and protein lean fractions to separate them by some classification means. If, for example, the mixed fractions are in dry form they may be screened or classified, such as via air classification.

If the protein has been freed by agitating it in slurry form such as by agitating hexane solvent, it is a simple matter then to separate the fractions by subjecting the entire mixture to centrifugation such as by means of liquid cyclones. The smaller protein rich fraction goes to the overflow with the hexane and oil and is thus separated out from the germ matrix or fibrous residue. The hexane may be then stripped off by evaporation, and the proteinaceous solids suspended in oil separated therefrom by filtration. The separation may also be effected through means of a centrifuge. To maximize the efficiency of this step the suspension before separation can be screened to remove the gross particles, and then the liquid suspension remaining may be subjected to centrifugal forces.

It was found that addition of water in small amounts, say 1 to 5% by weight of the total mixture, helps to effect a cleaner centrifugal separation. The water addition tends to swell the protein rich fraction, thereby increasing the separability of the proteinaceous solids from the mixture of germ oil and hexane or other suitable extraction solvent. Generally the water should be added to the suspension of protein fraction in organic solvent-germ oil mixture after the fibrous residue has been removed by centrifugation or other means.

Example 1: Corn germ was first softened by steeping for approximately 2 days in sulfur dioxide-containing warm water. The grain was then ground in a mill to liberate the germ. The germ was washed, dried, heated and subjected to a mechanical screw press to release the bulk of the oil contained therein. The residue was then flaked and extracted with hexane via a countercurrent flow method. 100 g of the above extracted spent corn germ flake was then wetted with sufficient water to obtain a slurry. In order to release the proteinaceous fraction the slurry was then subjected to abrasion by means of an attrition mill.

To classify the two fractions described above, the germ slurry was then screened on a bolting cloth having a 53 micron mesh size. The tailings retained on the cloth were again passed through the attrition mill and also screened. The fibrous protein lean fraction which was retained on the cloth was washed with tap water and freeze-dried. The proteinaceous fraction which passed through the screen was centrifuged in a Sharples Supercentrifuge. The solids of this fraction were also freeze dried.

Example 2: Both of the fractions obtained were a light tan color. The protein rich fraction had a pleasing nutlike flavor. About 36% of the material was recovered as a protein rich fraction and 50% as a protein lean or fibrous fraction. That only 24% of the starch in the original germ flake could be accounted for in these fractions can be explained. The relatively high temperature and pressure to which the germ is subjected during the oil expelling process gelatinizes and dextrinizes much of the starch and thereby causes it to become water soluble. This portion of the starch is recovered in the solubles derived from the spent flake during milling. The same sequence of steps is carried out here as outlined in Example 1 with the exception that the spent corn germ from the wet milling process is only expelled, that is pressed in a mechanical screw press, and is not extracted with solvent. Comparable results are also obtained here.

Example 3: In this series of runs three spent cereal germs were abraded to remove protein and then classified by means of a cloth bolting screen having 53 micron apertures. The first material was a wet milled sorghum grain that had been oil-expelled, that is pressed in a mechanical screw press, but had not been extracted with solvent. A second sample obtained by a wet milling process was a spent corn germ flake which had been both screw pressed and then extracted with hexane. Lastly, a spent corn germ was abraded and classified, which germ had been previously isolated by dry milling and then pressed to remove germ oil. The spent corn or sorghum germ flakes were again milled by attrition.

Example 4: In this run, dry corn germ was ground in a Bauer Double Disk Attrition Mill to produce particles of a size which passed through a 20 mesh screen. The ground material was slurried in hexane and the slurry processed through a series of 3" diameter hydrocyclones to effect a separation of the smaller protein particles which were thus liberated by the treatment from the larger germ matrix or fibrous residue. The hexane was removed from the oil-protein mixture by evaporation and steam stripping. A high protein (59.5%) substance was obtained by filtering the oil-protein slurry and washing the solids with additional hexane. The solids were then dried to remove residual hexane. The amino acid balance of this protein makes it very desirable for human nutrition. The protein particles obtained were generally of a size of about one to three microns, and the protein yield was in the neighborhood of 36% of the original protein present.

Fractions from Oats

A dry protein base is used for high-protein breakfast cookies which has the essential amino acids equivalent to that of whole egg. The process is described by *E.J. Turro; U.S. Patent 3,529,969; September 22, 1970; assigned to Central Soya Company, Inc.* To provide a blended base—from 30% to 50% of a ground cereal such as rolled, fine-milled oats is used. Other cereal flours such as high gluten wheat flour can be used in certain instances, but for the most part the oat product is superior in providing a dough which results in cookie-like textural properties.

From 20% to 45% of a soy protein concentrate containing not less than 70% protein (N x 6.25) on a moisture-free basis is incorporated. The soy protein concentrate is prepared from high-quality, sound, clean, dehulled soybeans by removing most of the oil and water-soluble, nonprotein constituents. Optimum results are obtained using 30% of the soy protein concentrate (Promosoy). Included also is 1½ of the weight of the soy protein concentrate of methionine which is not present in substantial quantities in the soy protein concentrate.

Example 1: In the preparation of high-protein cookies, the following protein base formulation was used.

Dry Blend	Percent	Weight, g
Rolled, fine-milled oats	39.5	900
Promosoy	28.9	660
Nonfat dry milk	13.2	300
Promine-D	7.9	180
Egg yolk solids	5.25	120
Egg white solids	5.25	120
Methionine (based on 1½% Promosoy wt)		9.9

The dry blend is mixed for 15 minutes in a Hobart A 200 (20 qt) mixing bowl, first speed, after which the following ingredients were added to the 160 g of the dry blend, the ingredients being mixed for 1 minute at the first speed of the blender; 4 g baking soda, 120 g sucrose, 4 g salt, 2 g protease (Red Star Yeast) and cinnamon. Thereafter, vegetable shortening in the amount of 28 g was added and mixed at 30 seconds at the first speed and finally 71 ml of water at 75°F were added and mixed two minutes at the first speed. The final dough temperature was 78°F and two hours enzyme lay time was used. The dough was made up into cookies 3" x 3" x ¼" using as little dusting flour as possible. The scaling weight was 58 g and the bakeout weight 45.5 g (approx 1.6 oz) resulting in a bakeout loss of 8.5 g or 15.75%.

Cookies made according to the formulation were found to be of equal or superior quality to those using milk protein in place of the soy protein concentrate. The same was true when a comparison was made with cookies utilizing whole egg solids in place of the soy protein concentrate. The amino acid content of the cookies in g/100 g are compared with those made with whole egg solids as can be seen in the table below.

Amino Acid Content	Whole Eggs	Example Cookie
Cystine*	0.30	0.36
Histidine	0.31	0.56
Isoleucine	0.85	1.04
Leucine	1.13	1.78
Lysine	0.82	1.02
Methionine	0.40	0.43
Phenylalanine	0.74	1.10
Threonine	0.64	0.91
Tryptophan	0.21	0.21
Valine	0.95	1.14
Other Analyses, Percentages		
Protein (Kjeldahl) 6.25	12.9	21.6
Moisture	73.7	8.6
Fat	11.5	9.0
Protein dry basis	49.2	23.6

*Nonessential but has methionine sparing action.

Example 2: In the preparation of a high-protein bread a base protein mix, similar to that of Example 1, was developed. To simplify the manufacture of a protein bread, the regular sponge can be utilized. This can be prepared either from clear flour or patent flour. An example of the sponge used is as follows: Flour—clear or patent, 560 lb; yeast, 24 lb; yeast food, 4 lb; water, 345 lb and enrichment tablets, (9).

The sponge is allowed to ferment in a fermentation room for 3 hours before it is made up into a dough. The protein mix is added at this stage. An example of the dough make up is as follows: Flour—patent, 240 lb; sugar (sucrose), 40 lb; yeast, 8 lb; mono-diglyceride, 1.8 lb; protein mix, 160 lb and water, 390 lb.

The dry protein mix is basically the same as that used in the high-protein cookie mix of Example 1. Nonfat dry milk and methionine have been eliminated, and egg protein has been reduced since its functional value in bread is not as critical as it is in a cookie; the protein level was then adjusted with soy protein. The composition of the dry protein mix used in this formulation is as follows:

Dry Blend	Percent	Weight, lb
Rolled oats	37.6	60
Promosoy	39.4	63
Promine-D	7.5	12
Egg white solids	2.5	4
Egg yolk solids	2.5	4
Salt	10.5	17

The normal procedure is used in the makeup of the dough. The fermented sponge is added to the mixer accompanied by the ingredients used in the dough-development. The dough is mixed for the prescribed amount of time to develop the gluten. The resultant dough is then rested, divided, molded, panned, proofed and baked. The above bread was air-dried and analyzed as follows: Moisture, 9.70%; fat, 0.10%; protein, 19.93% and fiber, 0.77%. A typical analysis of commercial white bread containing 3% nonfat dry milk using the same air-dry technique is as follows: Moisture, 8.15%; fat, 1.24%; protein, 12.25% and fibre, 1.00%.

The bread produced using the protein mix has over 50% more protein content than the standard commercial white bread currently sold using 3% nonfat dry milk, and substantially higher protein when compared to the commercial white bread produced with less than 3% nonfat dry milk.

The amino acid content of the bread crumbs was evaluated against a typical commercial white bread. The analysis shows that the bread produced with the protein mix has a much more favorable amino acid profile than the commercially produced white bread, especially for the essential amino acids necessary for growth. Lysine, which is normally deficient in wheat flour that constitutes approximately 70% of the weight of a loaf of bread, was found to be considerably lower in commercial white bread crumbs than in the bread crumbs containing the dry protein mix. The amino acid comparisons are given in the following table.

	- - Essential Amino Acid Content (g/100 g) - -	
	White Bread Crumbs	Protein Bread Crumbs
Lysine	0.27	0.62
Histidine	0.26	0.45
Threonine	0.35	0.63
Cystine	0.30	0.40
Valine	0.48	0.84
Methionine	0.21	0.28
Isoleucine	0.43	0.75
Leucine	0.80	1.34
Phenylalanine	0.56	0.96
Tryptophan	0.10	0.19

IRON FORTIFICATION

Sequestered Iron Method

Fruit cereal formulations of rice flour, apple puree and banana puree of a desired viscosity can be prepared as an attractive product through rigid control of the pH of the formulation during its preparation. The critical pH range for the combination of rice flour, apple puree and banana puree has been found to be 4.3 to 4.6.

F.W. Billerbeck, F.S. Hing and G.A. Purvis; U.S. Patent 3,429,711; February 25, 1969; assigned to Gerber Products Company state that it appears that by adjusting the pH so that it is more basic than 4.2, i.e., 4.3, 4.4, etc., acid hydrolysis, which seems to accelerate the breakdown of starch molecules thereby producing a formulation of undersirably thin consistency, is prevented. Further experimentation with variations in the pH of the formulation has also revealed that when the pH is adjusted to be more acidic than 4.2, the product develops an undesirable sour flavor. Additionally, when the formulation is more basic than 4.6, for example, 4.8 or 5.0, it is necessary to use longer cooking times and higher temperatures to attain the required levels of commercial sterility. As a result of these longer cooking times and higher temperatures, the formulation develops an overly sweet flavor and is subject to serious color degradation.

It has been found also that the critical pH adjustment can be most effectively accomplished through the use of a pH adjusting agent which can either be basic or acidic, depending on which pH adjustment is needed. Alkalizing agents which have been found to be effective include salts of a strong base and weak acid, for example, disodium phosphate, sodium carbonate, sodium bicarbonate, sodium ascorbate.

Disodium phosphate in its anhydrous form is especially effective although other forms of disodium phosphate, i.e., dihydrate and heptahydrate, are also advantageous. Where sodium ascorbate is used, the desired pH adjustment is accomplished while at the same time enhancing the nutritive value of the product by the incorporation of Vitamin C. Acidifying agents which can be used include food acids, such as citric, fumaric, malic, ascorbic, and adipic. Citric acid is especially effective. Ascorbic acid (Vitamin C) and its metal salts are especially effective in maintaining good color.

Certain diet supplements such as iron salts, in a sequestered form, can be incorporated with the formulations of rice cereal, apple puree and banana puree, provided that the critical pH range in maintained throughout the preparation and storage of the formulation. The capacity to incorporate a quantity of iron is especially significant as whole milk, the main component of infant diets, is ineffective as a carrier of iron. The iron, especially in a form easily attacked by infant digestive systems, is essential to a number of metabolic processes. For example, it is used in the formation of compounds such as hematin which constitute a large part of the chromo protein, hemoglobin.

When digestible iron salts which have not been sequestered are incorporated with the instant formulation, in the desired pH range as well as lower pH ranges, a reaction occurs which produces black flecks that impart an unattractive grey hue to the whole formulation. Although a large number of sequestering agents have been used as reactants with iron salts in other compositions, only certain sequestering agents have been found to be effective in the formulation. This limited group of effective sequestering agents includes ethylenediaminetetraacetic acid (EDTA) in the esterified form, such as disodium ethylenediaminetetraacetate dihydrate and disodium calcium ethylenediaminetetraacetate dihydrate. These specific sequestering agents react with the iron salts, thereby preventing catalytic oxidation activity of the iron and other undesirable reactions, such as the formation of iron tannate, when the sequestered iron salts are subsequently incorporated with the wet formulation.

As a further result of the effectiveness of the pH adjusting agents, it is no longer necessary to employ a blend of apples to produce a puree in a specific pH range. In this regard, the pH of the apple puree may vary within wide limits, for example, from 3.0 to 5.6. Attention can now be concentrated on attempting to incorporate a blend of apples having strong yet desirable flavor characteristics, rather than those which possess a certain pH.

The amount of pH adjusting agent required will depend on the initial pH range of the apple puree as well as upon the amount of sugar and other ingredients included in the formulation, and will vary in each individual application. The amount of apple puree utilized is not critical and may be used in whatever proportions are desirable. In fact, the amount of apple puree may be varied from 30 to 70% by weight of the formulation. Banana, in the form of a puree, is included for its flavor characteristics and may be added either in its frozen or aseptic puree form or dehydrated flake form. The amount used will depend, of course, on the intensity of the banana flavor desired and can also be varied considerably.

Although any cereal grain flour can normally be used, for example, oat, wheat, corn, barley, and mixtures thereof, it has been found that the advantages obtained are only germane to the employment of rice flour. Further advantage has been found in using rice flour to constitute about 10 to 15% by weight of the total formulation. Sugar, preferably granulated white sugar, may be added in amounts up to 25% to assist in bringing out the natural fruit flavor. Other ingredients such as starch, waxy maize or tapioca, either modified or unmodified; salt, especially up to 2% by weight; and miscellaneous vitamin mixes may be incorporated in minor amounts for their obvious advantages.

Iron salts which have been employed as diet supplements are those which readily disassociate and include ferrous carbonate, ferrous citrate, ferrous sulfate. They are normally utilized in minor amounts, such as up to 0.1% by weight of the formulation. Equivalent or equimolar amounts of ethylenediaminetetraacetic acid or a derivative, in either the disodium form or disodium calcium form, is used to complex with the iron salts prior to their incorporation with the formulation, thereby preventing the iron from reacting with the other components or catalyzing other reactions.

The formulation can be prepared by mixing the rice flour, apple puree, banana puree, granulated white sugar, waxy maize or tapioca, either modified or unmodified, starch, and the iron salts, sequestered with an ethylenediaminetetraacetic acid compound; adjusting the consistency with water; and then adding sufficient pH adjusting agent to adjust the pH to the critical range of 4.3 to 4.6. The formulation is heated at a temperature of 150° to 200°F to gelatinize the starch. Containers filled with the formulation are then processed according to the conventional processing methods used in the canning industry.

Example: A fruit-cereal formulation was prepared in the following proportions:

Ingredient	Quantity, wt %
Apple puree	58.00
Rice flour	10.00
Granulated white sugar	16.00
Banana puree, frozen and deseeded	5.00
Starch, modified food	5.00
Tapioca starch	5.00
Vitamins, seasoning, etc.	0.90
Disodium ethylenediaminetetraacetate dihydrate and iron sulfate ($FeSO_4 \cdot 7H_2O$)	0.10

Water was added to obtain a desired consistency and the pH was adjusted to 4.4 with anhydrous disodium phosphate. The slurry was heated to 195°F for 1 to 5 minutes. The formulation was pumped into small glass containers and processed in the manner conventional to the canning industry.

Reduced Iron in Sorbitol

An improved method to fortify cereal grains with iron is outlined by *L.P. Carroll, J.C. Novotny and A.W. Richards; U.S. Patent 3,806,613; April 23, 1974; assigned to The Quaker Oats Company*. Over the years many procedures have been developed for adding vitamins and minerals to grain products. Usually these procedures consist of mixing vitamins and minerals with the grain product or applying thin films of water, gum, zein, shellac, etc. by which the vitamins and minerals are bonded to the grain product.

The above procedures work quite well when the fortifying form of iron is ferrous sulfate, ferrous gluconate, ferrous fumarate, ferric orthophosphate, sodium iron pyrophosphate, etc. but are unsatisfactory for the addition of large amounts of reduced iron. Reduced iron in amounts in the order of 100 to 200 mg/lb when applied by conventional procedures noticeably tends to separate from the fortified grain product during shipment and accumulates in the bottom of the package. Reduced iron is a physiologically available form of iron which is less expensive per unit of iron than any of the above mentioned iron salts and is less likely to cause problems such as rancidity and off colors. In view of the above, there currently exists a need for a satisfactory method of applying reduced iron to grain products so that the reduced iron does not separate from the product upon shaking and in shipping.

The words "cereal grain product" include such products as white rice, corn grits, corn meal and barley. "An edible ester of glycerol or sorbitol and a fatty acid" includes any animal or vegetable oil, natural or hydrogenated of an edible type and having a melting point below 125°F, preferably below 80°F. Suitable edible esters include shortening, lard, butter, coconut oil, cotton seed oil, soybean oil, palm oil, corn oil, peanut oil, sunflower seed oil, safflower oil and olive oil. As indicated above, suitable esters may be based on fatty acids which include stearic acid, palmitic acid, myristic acid, and lauric acid for example.

Edible esters of glycerol and sorbitol having a melting point below 125°F are included such as Atmos 300, Tween 20, and Tween 60. Atmos 300 is a mixture of mono- and diglycerides which is formed by the reaction of glycerol mono-oleate with oleic acid, palmitic acid, and stearic acid, is a light amber liquid, and has a viscosity of 150 centipoises at 75°F. Tween 20 is a polyoxyethylene sorbitan monolaurate which is a yellow liquid at 75°F having a viscosity of 360 centipoises. Tween 60 is a polyoxyethylene sorbitan mono-stearate which is a yellow liquid at 75°F having a viscosity of 600 centipoises.

The slurry of reduced iron and an edible ester of glycerol or sorbitol includes molecular dispersions (solutions), colloidal dispersions, and particulate or mechanical dispersions. The amount of finely divided reduced iron applied in the process is not narrowly limited. Amounts of reduced iron much in excess of the adult minimum daily requirement for iron per unit of the cereal grain product reasonably suited for adult consumption probably should not be applied as a matter of good nutritional policy; however, in testing the process it was found that amounts much in excess of that amount can be applied. In general from 25 mg to 1,200 mg and preferably 80 mg to 250 mg of reduced iron per pound of cereal grain product may be applied in the process. It is essential that the slurry contain between 0.15 and 5.0% by weight of the fatty acid ester of glycerol or sorbitol based on the weight of the cereal grain product.

If much less than 0.15% of the edible ester is used, the reduced iron will not be effectively bound to the cereal grain product. If much more than 5.0% of the edible ester is used, the ester will interfere with the cook and the visual and organoleptic acceptance of the cereal grain product. That is, if more than 5.0% of the edible ester is used the grain product when coated with the slurry has an unnatural appearance best described as like wet sand. Above the 5.0% level the cereal grain product when cooked is oily and has an unpleasant mouthfeel.

The slurry is prepared by adding the reduced iron to the edible ester of glycerol or sorbitol, holding the mixture at a temperature sufficient to liquefy it, and agitating the mixture to produce the slurry. Other materials such as salt, vitamins, other minerals, antioxidants, colorants, flavorings, flavor enhancers, flavor potentiators, etc. which can be dissolved or suspended in the edible ester may be added. Temperatures above the temperature sufficient to liquefy the slurry can be used but are not necessary. When heat degradable vitamins such as vitamin A, vitamin B, vitamin B_{12}, vitamin C, and vitamin D or other heat unstable materials are added to the slurry, heating the slurry to temperatures much above that necessary to liquefy it or for prolonged periods of time should be avoided.

The slurry may be applied in liquid form to the cereal grain product by spraying, by tumble enrobing, or by any other conventional method. In each case the object is to evenly distribute the slurry around the product. After coating the cereal grain product with the slurry, the product is immediately ready for human or animal consumption or for packaging.

Example 1: A slurry was prepared from 0.10 lb of coconut oil having a melting point of 75°F, 2,500 mg of reduced iron, and 9.0 g of a vitamin and mineral premix. 10 lb of corn grits were tumbled with 56.8 g of the oil slurry prepared above. The corn grits and oil slurry were mixed until the slurry was uniformly distributed over the surface of the cereal product. The enriched cereal product was packaged and subjected to a 1,000 mile simulated rail trip. To determine the amount of separation of the reduced iron from the corn grits, samples were taken from the top, center, and bottom of the package and analyzed for reduced iron. The results are reported in the following table.

Example 2: The procedure of Example 1 was repeated in this example except that 10 lb of farina was substituted for the corn grits. The results are reported in the table.

Example 3: 10 lb of corn grits were tumbled with 0.10 lb of coconut oil having a melting point of 75°F. The corn grits and the oil were mixed until the oil was uniformly distributed over the surface of the cereal product, 2,500 mg of reduced iron and 9.0 g of a vitamin and mineral premix were added and mixing continued until the mixture was uniform. The enriched cereal product was then packaged, subjected to the 1,000 mile simulated rail trip, and samples were taken as in Example 1. The results are reported in the table.

Example 4: The procedure of Example 3 was repeated in this example except that 10 lb of farina was substituted for the corn grits. The results are given in the table.

Example 5: 10 lb of corn grits were tumbled with 2,500 mg of reduced iron and 9.0 g of a vitamin and mineral premix. The tumbling was continued until the vitamins and minerals were evenly mixed with the corn grits. 0.10 lb of coconut oil having a melting point of 75°F were then added and the tumbling continued until the oil was uniformly distributed over the surface of the cereal product. The results are given in the table on the following page.

Example 6: The procedure of Example 5 was repeated in this example except that 10 lb of farina was substituted for the corn grits. The results are reported in the following table.

Example No.	Top mg Fe/lb	Sample Middle mg Fe/lb	Bottom mg Fe/lb
1	264	254	278
2	217	254	281
3	195	213	315
4	185	184	303
5	250	279	423
6	363	449	694

Samples taken from Examples 1 and 2 show that substantially no separation of the reduced iron from the cereal grain product occurred during the simulated rail trip. Samples from Examples 3 to 6 inclusive show that samples suffered separation of the reduced iron from the cereal grain product. Examples 1 and 2 demonstrate the criticality of the particular sequence of steps in the process. It was found that it is absolutely essential to prepare first a slurry of the reduced iron in the above mentioned edible ester and second to apply the slurry to the cereal grain product to be enriched. Other sequences of steps as demonstrated by Example 3 to 6 inclusive result in an iron enriched cereal product from which the reduced iron separates during handling.

VITAMIN AND MINERAL COATING

H.D. Spangler; U.S. Patent 3,784,716; January 8, 1974; assigned to Merck & Co., Inc. prepares a coating composition of vitamins and minerals for enrichment of dry cereal products. The cereals to which the nutrient coatings are applied are represented, for example, by the ordinary corn flakes, puffed oat grains, and wheat shreds including shredded wheat biscuits. These flakes, puffed grains and shreds are found in a variety of shapes, sizes and grain substances on the grocers' shelves. These edible products may be consumed dry but ordinarily they are eaten after milk or other dairy product is combined with them.

Vitamin and mineral particles which are now being supplied with a coating to preserve them or mask their taste become tacky when they are heated. This tackiness causes them to adhere to the cereals on which they are dusted if the cereals are at the same temperature. The coating is a hard fatty acid such as commercial stearic acid; vitamin B_1, B_2, niacinamide and B_6.

At a temperature between 30° to 70°C, depending on the particular coating material it becomes tacky or sticky. These coated nutrient particles are mixed with ingredients which do not become tacky at this temperature and the coated particles should constitute from 20 to 25% of the total mixture weight as within this range the particles will not stick together but the entire mixture will adhere to the preheated cereal.

The other ingredients which do not possess this tacky coating may be vitamins such as A, C, D, and B_{12} and these uncoated vitamins may constitute from zero up to 25% of the total weight. The remainder of the mixture, which serves to prevent the coated particles from sticking together, is corn or other starch and it should amount to at least 50 to 60% of the total weight. In this mixture the particles will not adhere together when heated to a tacky temperature but they will adhere to preheated, dry cereal flakes, grains, shreds, etc.

To carry out the process the ingredients are first intimately combined using a blend-mill-blend procedure. The resulting composition is heated to its softening point and this generally is within the 30° to 70°C range, At this temperature the particles will be slightly tacky but they will remain discrete from each other and not combine and consequently the powdery consistency is maintained. The cereal also is heated to approximately this same temperature in a conventional jacketed rotating coating pan so that the temperature of its contents can be adjusted and maintained.

The preheated nutrient mixture is then dusted on the heated cereal to obtain a uniform distribution. This dusting operation is carried on in a manner similar to the sugar coating of pharmaceutical tablets in that the worker observes the rate of adherence to the cereals and makes sure the powder is added at the rate that it adheres to the cereal. Care must be observed to not add the powder faster than this rate because it may cause adherence.

After a sufficient amount of coating composition is added the heat supply to the jacket is stopped, and while rotation is continued, the cereal is allowed to cool to room temperature. The amount of nutrient coating to be added may be up to or slightly above that which will supply the daily intake of the nutrients to a person consuming an average serving of the cereal.

Example 1: A representative example of the vitamin composition is:

Ingredient	Grams
Vitamin A, 60/120 mesh	18.1
Vitamin D, 60/120 mesh	1.8
Vitamin B_1, Mercote (33⅓%)*	7.3
Vitamin B_2, Mercote (33⅓%)	11.4
Niacinamide, Mercote (33⅓%)	70.1
Vitamin B_6, Mercote (33⅓%)	14.3
Vitamin C	85.0
Vitamin B_{12}, Stabicote (1%)**	0.5
Cornstarch, USP powder	245.1
	453.6

*This percent describes the vitamin activity by weight in the commercial product.

**Spray-dried, gelatin coated product.

Example 2: One or more of the Mercote vitamins in Example 1 may be increased, reduced or omitted to the extent that the total weight of the remaining Mercote products remains within the range of 15 to 30% of the total weight of the mixture. Thus, if in Example 1 the vitamin B_6 is altogether omitted, the amount of starch could be within the range of from 160 to 250 g.

Example 3: The vitamin A, D, C and B_{12} may be increased, reduced or omitted, and be replaced by the amount of starch needed to maintain the Mercote products as 15 to 30% of the total weight of the mixture. Thus, if vitamin C is omitted in Example 1, it would be replaced with from 290 to 390 g of starch.

Example 4: Any one of the above coating compositions may be applied to commercial corn flakes as follows. The ingredients are combined, then blended, then milled and then reblended. The operation may be carried out in known ribbon blenders, Hobart mixers, Glenn mixers, Fitzpatrick comminutors, micropulverizers, ball mills, hammer mills or any other apparatus known to thoroughly blend and mix such powdered or granular ingredients.

After it is assured that an absolutely complete intermix is obtained, the mixture is heated to 30° to 70°C in a jacketed container, and preferably while in motion as in a tumbling barrel or drum or in a rotating coating pan such as is used in the sugar coating of pharmaceutical tablets. At the same time the corn flakes are heated to a temperature in the same range utilizing a jacketed rotating coating pan, for instance. Then, while the tumbling is continued, the coating mixture is dusted on the corn flakes, care being taken to not exceed the rate of addition at which the flakes tend to stick together.

As 0.2 g of the mixture of Example 1 provides the minimum daily requirement for these vitamins, this amount, if applied to one pound of the corn flakes would provide a cereal such that a five- to six-ounce serving would provide one-third of the minimum daily requirement. With this as a basis it is apparent that up to 0.6 g of the mixture can be applied to each pound of cereal. Because of the excellent adhering nature of the composition, it may be applied at any rate per minute below that which will cause the adherence of individual corn flakes. When the desired amount of coating has been added, the heating is terminated and while the tumbling operation is continued the coated flakes are allowed to cool to room temperature or, at least, to a temperature below that at which the flakes are tacky or sticky. Cooling air may of course by supplied to assist this. Then the product may be packaged.

VITAMIN COATING EMULSION

To overcome some of the difficulties associated with mixing the vitamins into the uncooked dough, vitamins have been sprayed in an aqueous solution or oil emulsion on the surface of a cooked dough usually having a moisture content between 25 and 35%. To reduce the moisture content of the cereal product, to provide the desired crispness, and to develop a desired flavor in the dough, the vitamin coated dough is subjected to an intensive heat transfer step. This heat transfer step in commonly referred to as drying, drying and toasting, drying and puffing, or drying, puffing, and toasting depending upon the precise manner in which it is carried out.

This process suffers from the same difficulties as the process wherein the vitamins are mixed in the uncooked dough. Namely, many of the more heat labile vitamins degrade during the drying step. Further shortcomings peculiar to this process are that there is a large material loss of vitamins during the spraying step due to spray over and that the distribution of vitamins on the cooked dough is customarily uneven. Furthermore, as sufficient vitamins are added to the cereal product to provide the adult MDR or the recommended daily allowance established by the U.S. Food and Drug Administration, the ready-to-eat cereal product develops a decidedly and wholly undesirable medicinal flavor.

The disadvantages of previous processes are eliminated in the method of *L.F. Duval and C.D. Stone; U.S. Patent 3,782,963; January 1, 1974; assigned to The Quaker Oats Company*. The improved process consists of three steps:

(a) Precoating a precooked cereal with an aqueous suspension or solution of sugar, the suspension or solution having a viscosity between 100 and 200,000 centipoises at 75°F.

(b) Drying the precoated cereal until the moisture content is between 0.5 and 3% by weight.

(c) Coating the dried cereal with vitamins while the cereal is hot and tacky, whereby the vitamins adhere to the tacky surface of the cereal.

The vitamins are coated with a fatty composition which is a saturated aliphatic fatty acid having between 12 and 20 carbon atoms, inclusive, or a glyceride having the formula:

$$\begin{array}{l} CH_2\text{—}O\text{—}R_1 \\ | \\ CH\text{—}O\text{—}R_2 \\ | \\ CH_2\text{—}O\text{—}R_3 \end{array}$$

wherein R_1 is an acyl radical derived from a saturated aliphatic fatty acid having between 12 and 20 carbon atoms, inclusive, per molecule and R_2 and R_3 are hydrogen or the same as R_1.

The precooked cereal is prepared by any of the known processes, such as extrusion, puffing, or flaking. The grain in the precooked cereal may be for example, rice, wheat, oats, corn, barley, or any combination of two or more such grains. The precooked cereal may be formed into a variety of shapes and sizes by well-known techniques in the art. The aqueous suspension or solution is composed substantially of sugar. The sugar may be any blend of mono-, di-, tri-, and other higher saccharides. For example, the sugar may be dextrose, lactose, sucrose, fructose, maltose, or combinations thereof; or any of the higher saccharides present in corn syrup and combinations thereof. When the sugar is sucrose, it is preferred that the aqueous suspension comprise 50 to 90% by weight of sucrose.

It is advantageous to heat the aqueous sugar mixture to just below 200°F to bring about the partial dissolution of the sugar. The partially solubilized sugar in the aqueous mixture does not clog the spray head when the mixture is applied to the precooked cereal by spraying. The aqueous sugar mixture may be heated to temperatures above 200°F to bring about complete dissolution of the sugar but this was not found advantageous since the resultant syrup cannot be sprayed on the precooked cereal but must be applied by tumbling the cereal with the syrup.

The aqueous suspension or solution may contain salt, fats, flavorings, stabilizers, flavor enhancers, flavor potentiators, etc. For example, certain flavoring ingredients such as honey, chocolate, peanut butter, and fruit juice concentrate can be added to the aqueous suspension to modify or enhance the characteristic flavor of the cereal product. The aqueous suspension or solution of sugar is coated onto the precooked cereal. The precoating can be accomplished by any known method, for example, by mixing and agitating the cereal with a quantity of the aqueous sugar suspension in an enrober or related device. The precoating can also be accomplished by spraying the aqueous suspension on tumbling precooked cereal in an enrober so that the aqueous suspension of sugar is distributed evenly around the product.

The precoated cereal is dried under temperatures and for times depending primarily on the nature of the cereal piece and the amount and viscosity of the aqueous sugar suspension or solution. It is believed that the vitamins adhere to the hot cereal product in the glassy, sticky sugar. While the temperature of the dried precoated cereal product during the vitamin coating procedure is not narrowly critical, it should be sufficient to prevent crystallization of the sugar. Much higher temperatures than necessary to prevent crystallization of the sugar should be avoided to prevent deactivation of the heat labile vitamins. The precoated dried cereal may be coated with vitamins while the cereal is between 120° and 300°F. After coating the precoated dried cereal product with vitamins, the cereal product is immediately cooled and is ready for human consumption.

As sufficient vitamins are added to the cereal product to provide the adult MDR or recommended daily allotments when consumed in an amount reasonably suited for adult consumption the cereal develops an unpleasant medicinal taste. By this process high levels of vitamins may be added by coating vitamins with a fatty composition which is a saturated aliphatic fatty acid or a glyceride of the above formula without the development of an undesirable flavor. The fatty composition also helps anneal the vitamins to the dried, hot cereal piece.

Suitable saturated aliphatic acids include for example, stearic acid, palmitic acid, myristic acid, and lauric acid, The saturated aliphatic fatty acid glyceride may contain between 0 and 2 hydroxy groups inclusive. The vitamins are coated by suspending the particular vitamin in a molten fatty material and chilling the resulting suspension to produce spheroidal particles of the vitamin having a coating of the fatty composition. The coated vitamins may be blown through a small orifice to provide beadlets of vitamins the cross section of which is determined by the diameter of the orifice. Higher temperatures than necessary to form the suspension are avoided to prevent deactivation of the heat labile vitamins.

The vitamins thiamine (vitamin B_1), riboflavin (vitamin B_2), niacinamide, ascorbic acid (vitamin C) and cyanocobalamin (vitamin B_{12}), for example, may be coated with the above described fatty composition. The coating of vitamin A, vitamin D, and vitamin E, with the fatty composition would not be advantageous since these vitamins are not sufficiently soluble in the fatty composition. Satisfactory and commercially available vitamins coated with the fatty composition are available as Mercote and Rocoat preparations.

While vitamin A and vitamin D are not coated with the fatty composition, they are often dispersed in gelatin. The gelatin is designed to act as a protective medium to prevent oxidation whereby the vitamins lose their vitamin activity. While the size of the particles of vitamins is not narrowly critical, they must be no larger than will adhere to the cereal piece. This size is dependent on the melting point of the fatty composition forming the coating if any, the geometry of the particle which need not be perfectly spherical, and the temperature of the dried cereal piece to which the vitamin particle is annealed. The maximum size can be easily determined.

Example 1: 50 lb of a precooked cereal product having a moisture content of 7% by weight was sprayed with an aqueous suspension comprised substantially of sugar which was prepared by mixing together 41 lb of sucrose, 7 lb of vegetable oil, 1 lb of salt, and 11 lb of water. The components of the aqueous suspension were heated to 190°F for 2 minutes and sprayed on the precooked cereal product while the cereal tumbled in an enrober. The precoated precooked cereal product was tumbled sufficiently to provide a uniform distribution of the aqueous sugar suspension around the precooked cereal pieces.

The precoated cereal product was then dried in an oven equipped with tray pans having a volume of approximately 3,000 in^3 and a bed depth of approximately 3 inches. The direction of air flow was upward through the product. A temperature of 280°F was maintained for 160 seconds. After the first 80 seconds elapsed, the tray was shaken vigorously to shatter clumps and redistribute the partially dried product in order to insure uniform drying. At the end of the drying period the moisture content of the dried cereal product was about 2.4% by weight.

While the dried cereal product was at a temperature of 160° to 180°F and the sugar coating still tacky, the cereal was tumbled with 2.7542 lb of a vitamin premix. The vitamin premix was composed of 1.9830 lb of confectioner's sugar and 0.7712 lb of a vitamin blend. Each pound of the vitamin blend was composed of vitamin A palmitate having an activity of 9.0720 MU (1 MU = 1,000 IU) in combination with vitamin D_2 in gelatin having an activity of 0.9072 MU, thiamine mononitrate having a glycerol monostearate waxy coat and an activity of 2.2680 g, riboflavin having a stearic acid waxy coat and an activity of 2.7216 g, niacinamide having a stearic and waxy coat and an activity of 22.6800 g, pyridoxine hydrochloride having a stearic acid waxy coat and an activity of 3.4020 g, sodium ascorbate having an activity of 68.0400 g, vitamin B_{12} in gelatin and having an activity of 4.9900 mg, and cornstarch to make up the balance to 1 lb.

The vitamin coated cereal product was tumbled for 10 seconds after the addition of the vitamin premix to insure even distribution of the vitamins on the surface of each cereal piece. The vitamin coated ready-to-eat cereal product prepared by the above process had the minimum daily adult requirements of vitamins per 1 ounce serving of the product. More specifically on analysis it was found that there is at least 16 mg/lb of thiamine; 480 milligrams per pound of vitamin C; 6,400 IU/lb of vitamin D; 64,000 IU/lb of vitamin A; 160 mg/lb of niacin; 1,912 mg/lb of riboflavin; 24 mg/lb of vitamin B_6; and 35.2μg/lb of vitamin B_{12}. The product was also subjected to taste evaluation. A comparison of taste of the product with the product obtained by conventional processes showed that the vitamin coated product produced by the process had a superior, nonmedicinal flavor.

Example 2: 97.2458 lb of dried cereal prepared as above by precoating a precooked cereal product with an aqueous suspension comprised substantially of sugar was coated with 2.7542 pounds of a vitamin premix while the temperature of the cereal was 250°F and the sugar was still tacky. Each pound of the vitamin premix contained vitamin A palmitate having a gelatin coat and an activity of 3.024 MU in combination with vitamin D_2 having an activity of 0.3024 MU, thiamine mononitrate having an activity of 0.7560 g, riboflavin having an activity of 0.9072 g, niacinamide having an activity of 7.5600 g, pyridoxine hydrochloride having an activity of 1.1340 g, sodium ascorbate having an activity of 22.6800 g, vitamin B_{12} having a gelatin coat and an activity of 1.6633 mg, and cornstarch to make up the balance of the pound. The cornstarch acts as a carrier to insure even distribution of the vitamins over the cereal piece.

BAKERY ENRICHMENT WAFER

It is a common practice to add enrichment wafers to bakery doughs, either to add them directly to the dough so that they will be intimately mixed throughout the dough or to add them to an amount of water in a separate container so that the wafer will be dispersed and to then add this dispersion to the dough so that it will be distributed more quickly. In either method, it is important that the wafer disintegrate as soon as possible to thereby hasten its distribution throughout the dough.

This rapid disintegration of the wafer could of course, be achieved if it is made as a fragile structure but such a weak wafer will not withstand ordinary shipping and handling. It has been found that if the wafer is compressed hard enough to make it sufficiently strong to sustain shocks it has a very slow disintegration rate. Past practices, therefore, have tried to arrive at a medial position between the conflicting aspects of a too hard wafer which disintegrates too slowly and a fragile wafer which disintegrates in the desired short time.

A hard, shock resistant wafer is prepared by *M.J. Boroshok; U.S. Patent 3,753,732; August 21, 1973; assigned to Merck & Co., Inc.* if the starch filler is added in two separate stages. Special advantages are found if the starches which are separately added are of different kinds, namely a pregelatinized starch such as that of tapioca starch and a nonpregelatinized starch such as ordinary cornstarch. The nutrients which are to be present in the wafer are the usual vitamins and minerals such as riboflavin, thiamine or one of its nutritionally acceptable salts such as the mononitrate, niacin or a derivative such as niacinamide, and iron such as ferrous sulfate. A tableting lubricant such as calcium stearate is also added.

In the process selected vitamins and minerals and the pregelatinized starch are first thoroughly blended together using a conventional mixing or blending machine. A lubricant such as calcium stearate is included in the mix. The product is then directly compressed into wafers and the hardness is immaterial as long as the wafers are fully formed and can be handled without immediate breakage. These wafers are then broken up and are comminuted or reduced through a sieve which can vary from a No. 6 to No. 30 mesh size. To this granular material is added the nonpregelatinized starch and additional lubricant such as calcium stearate. After thorough blending this preparation is tableted into wafers.

It is important to note that two different starches are added at successive stages. Best results are obtained if the pregelatinized starch, is added to the vitamins and the nonpregelatinized starch is subsequently added. The two starches are interchanged but best results are obtained if the pregelatinized starch is added first. The total amount of starch to be added is that which is conventionally used in the wafers, i.e., it should add up to 25 to 60%, preferably 40% of the weight of the wafer. The pregelatinized starch which is added to the vitamin-mineral premix should constitute from 15 to 35% by weight of that premix. The cornstarch added at the final stage should constitute from 10 to 25% of the final granular material to be compressed into the final wafer.

Example 1: (I) Formula –

Ingredients	Per Tablet mg	Quantity Per 100 Tablets g
1. Riboflavin	105	10.5
2. Thiamine mononitrate	155.8	15.58
3. Ferrous sulfate	2962.0	296.2
4. Niacin	1248.0	124.8
5. Tapioca starch (pregelatinized)	1500.0	150.0
6. Calcium stearate	0.3	0.30
Total	5971.1	597.11

(II) Process –

A. In a seven quart Hobart mixer bowl, blend (2), (3), and (4).
B. Add (1) to above and continue to blend. Add (6).

C. Pass through a No. 28 mesh screen.
D. Add (5) to above screened blend, and blend in Hobart mixer.
E. Compress on a Manesty single punch press equipped with 24/32 inch set of tooling.
F. Pass wafers through a No. 12 gauge sieve.
G. Combine: Product F blend, 542.0 g; cornstarch, 102.0 g; and calcium stearate, 2.0 g for a total of 646.0 g.

This was thoroughly blended and again compressed as above. There was no evidence of build-up on tablets. They disintegrated in 45 seconds in room temperature water. They had a hardness of 7 kg on a Strong Cobb hardness tester.

Example 2: This shows that the combined addition of the two starches results in a slowly disintegrating wafer.

(I) Formula –

Ingredients	Quantity Per 100 Wafers g
1. Ferrous sulfate	292.20
2. Riboflavin USP	10.50
3. Niacin	124.80
4. Thiamine mononitrate USP	15.58
5. Tapioca starch (pregelatinized)	150.00
6. Cornstarch	40.00
7. Calcium stearate	3.00
Total	640.00

(II) Process –

A. In the 7 quart Hobart bowl, add (1) and (2) and blend.
B. Add (3) and (4) to above and blend.
C. Add (5) and blend.
D. Pass above blend through a No. 28 mesh screen and reblend in the Hobart.
E. Add (6) and blend. Then add (7) and blend.
F. Compress on a Manesty single punch press equipped with a 24/32 inch set of tooling.

(III) Compression – Firm tablets were compressed. Filming was present on upper and lower punch. Disintegration in room temperature water extended beyond 3 minutes. Tablets had a mottled appearance.

ENRICHED AND DIETETIC BREADS

HIGH PROTEIN BREAD

Soy and Sesame Proteins

Conventional breads are generally quite low in protein, but quite high in carbohydrates. For example, ordinary white breads contain approximately 8½ to 9% protein, and rye breads generally no more than 9%. At the same time such breads have a carbohydrate level on the order of 48 to 52%. Conventional breads also have a very short shelf life, due to loss of moisture, loss of flavor and staling. Rye bread, for example, is noticeably drier after 2 to 3 days. With other breads, staling is noticeable in no more than 1 week.

More recently, the protein level of conventional breads has been increased by the addition of fortifiers such as more milk solids, extra gluten, soybean flour, cottonseed flour, that is, oil seed grains from which the oil has been removed. What is being done here is that some of the starches are being replaced by the fortifiers, thereby reducing the carbohydrates and increasing the protein level. Such techniques have increased the protein level to approximately 11 to 12% without, however, any improvement in processing time or procedure, and with no increase in shelf life.

With most grains such as wheat and rye, of the basic constituents, the greatest proportion is carbohydrates or starches. A smaller proportion is protein, and a very small portion is fat. In contrast, in oil grains seeds, the greatest proportion is fat; a smaller proportion is protein and the least amount is the carbohydrate. Therefore, when the fat is removed from the oil grain seed, one is left with a nonelastic, high protein content material.

Manufacture of leavened high protein bread with nonelastic protein materials is described by *C.S. Lynn; U.S. Patent 3,697,290; October 10, 1972.* This nonelastic type protein material is combined with an elastic type vegetable protein and various texture and flavoring agents including vegetables to produce a bread-like textured edible product. The taste, texture and structure of the product are created in a process in which the tendency of the elastic type vegetable protein, typically a wheat gluten, to expand or swell, is controlled or counteracted, within a body of nonelastic type protein material, typically sesame meal, alone or in combination with soybean meal, corn meal, etc.

The process involves the following procedure. In the first step, formation of the base material, a nonelastic type protein, oil, seasoning, flavoring and acid are combined and heated. The nonelastic type protein can be hulled, chopped, or comminuted sesame seeds, preferably in the range of 25 to 50% by weight. In one example extracted sesame meal was used in combination with sesame seeds. As noted above, the nonelastic material serves as a high protein source. Other nonelastic type proteins that can be used are soybean, cottonseed, etc.

A variety of oils in the range of 35 to 50% can be used, for example, vegetable oil made from sesame, cottonseed oil (preferred) and soybean oil. A quantity of salt is included for seasoning. The three elements are heated together with occasional stirring at a low simmering temperature, approximately 300° to 350°F for a short time. Then, flavoring agents (finely ground caraway seeds when making a rye-like bread) and a mild acid such as citric acid are added and the heating continued for a short time to bring out the flavor.

The caraway seeds are included mainly for their rye type flavoring characteristics although they are beneficial to the texture. Oil of caraway may be used instead of the caraway seeds. Similarly, oleoresin caraway (an oily liquid substance) may be used having the flavor character of caraway seeds. The resulting material is dark, brown, oily and sandy in appearance. The caraway need not be included in this step but could be added in the following step.

In the second step, formation of the intermediate product, vegetables, water, nonelastic protein meal, a thickening agent and a mold inhibitor are added to the base material formed in the first step. Initially, the vegetables are added to the base material along with water and boiled for a period of time depending on the vegetables used. A wide variety of vegetables of the leaf, root or grain type can be used. Vegetables are included for their flavoring characteristics and, of course, the taste characteristic will vary somewhat with the vegetables used.

After the boiling, additional water and nonelastic protein meal are added and the entire mass is brought back to the boiling temperature for another short period depending on the materials used. Examples of nonelastic protein meals are extracted sesame meal, cottonseed meal and soybean meal. At this point, a thickening agent, preferably farina made wheat or corn, is added and cooking continues. The thickening agent is required to create a desirable texture. Other thickening agents can be used such as wheat or cornstarch or wheat flour. To inhibit the propagation of mold, a suitable amount of inhibitor conforming with the Food and Drug Administration regulations can be added, for example, sorbic acid or potassium sorbate or a very small percentage of propyl paraben.

To the wet material there is next added an elastic vegetable protein, other protein materials and a chemical leavener. The material is blended, mixed and kneaded into a dough. Other materials can be added at this point to modify flavor and texture characteristics. Representative of these materials would be wheat flour, albumen, cheese, etc. A typical elastic vegetable protein is wheat gluten. A portion of gluten can be replaced with edible vegetable gums such as carboxymethylcellulose gum, seaweed gums, gum tragacanth, etc. Also added to the material are certain additional sources of protein, for example, whole eggs, a mixture of casein, lecithin, albumen and water and commercially available milk solids. A chemical leavener such as baking powder, bicarbonate of soda or a combination of both is blended in.

After thoroughly blending, the resulting mixture constitutes a dough which may be removed from the mixing bowl and cut into desired sizes and molded in the usual fashion into the desired shape of bread-like loaf or rolls, etc. where it is rested for preferably, but not necessarily, 10 to 20 minutes and then baked in an oven to form a baked leavened bread-like product. It should be observed that the wet base material already contains most of the water necessary to form the dough. Hence, if desired, the egg solids, nonfat milk solids, wheat gluten and chemical leavener can be combined in the dry form before adding to the wet base material. In this case a slight adjustment of the water content should be made to balance the moisture normally present in whole eggs. Merely adding the contents of the two combinations together both wet and dry, and thoroughly blending will produce the dough.

The cut and shaped material is then placed into any conventional oven at approximately 400° to 450°F. It requires a baking time very much like a regular bread or cracker, as the case may be. During this baking time, the dough rises and leavening occurs, as a result of which the loaf or roll or pastry has the appearance and taste of a conventional loaf of bread or pastry with the usual cellular structure.

Example:

Ingredients	Processing
Base material:	
Chopped sesame seeds, 34 grams Vegetable oil, 43 grams Salt (NaCl), 17 grams Citric acid, 1 gram	Combined and simmered at 300° to 350°F for 90 minutes
Caraway seeds, 5 grams	Heating continues for 30 minutes
Intermediate product:	
Parsnips, 120 grams Carrots, 30 grams Water, 450 grams Base material, 100 grams	Combined and boiled for 20 to 25 minutes at 215° to 230°F
Water, 450 grams Ground sesame meal, 150 grams	Cooking continues for 20 to 25 minutes
Farina, 60 grams	Cooking continues for 5 to 15 minutes
Final mix:	
Intermediate product, 1,000 grams Whole eggs, 100 to 200 grams Mold inhibitor, 2 grams	Blended
Milk solids, 105 grams Wheat gluten, 310 grams Baking powder, 15 grams	Blended to form dough, then rested for 15 to 20 minutes

The cut and shaped loaf is placed in a conventional oven at 400° to 450°F and baked. The dough rises, leavening occurs and the resulting loaf has the appearance and taste of a conventional rye bread with the usual cell-like openings in the loaf. The finished product has a protein content above 25% and a carbohydrate level of no more than 25%.

Stearoyl-2-Lactate Emulsifiers

Soy flour is a particularly valuable additive for bread not only because of its high protein content but also by virtue of the fact that it contains 3.2 to 3.8% lysine as compared with 0.375% in wheat flour. Thus, with the addition of 12 grams of soy flour to 100 grams of wheat flour, the lysine content of this mixture will be more than doubled to a value of from 0.76 to 0.83%. The resulting product provides an exceptional vehicle for nutritional improvement of human diet.

Although enrichment programs providing for the addition of vitamins and minerals have been successfully used for many years, provision of additional protein in the diet at a low cost has thus far evaded really successful implementation. This is true despite the fact that protein, both quantity and nutritional quality, is the most needed nutrient to improve man's diet. However, it has been found that the addition of any nonwheat protein or high protein wheat derived product to bread in an amount necessary to significantly improve its nutritive value caused a loss in load volume, poorer texture and a general loss of organoleptic qualities.

C.C. Tsen and W.J. Hoover; U.S. Patent 3,780,188; December 18, 1973; assigned to Kansas State University Research Foundation use sodium or calcium stearoyl-2-lactylate to improve the nutritive quality of the bread without a subsequent loss in loaf volume and organoleptic characteristics. Incorporation of this additive permits supplementation of the wheat flour based dough with any one or more of a number of protein additives which may, for example, be selected from the group consisting of soy flour, soy isolates, nonfat milk solids, whey products, fish protein concentrate, cottonseed flour, chick pea flour, sesame seed flour, corn-soy-milk blend flour, wheat protein concentrate, wheat gluten, defatted wheat germ, Torula yeast, wheat soy blend flour, edible single cell proteins, compatible with wheat flour for baking purposes, and mixtures.

It is common practice in the baking industry to regard "specific loaf volume" as an important parameter for gauging the marketability of bread. A marketable bread should have a specific volume over 6.00 cc/g, provided it also has acceptable appearance, crumb texture, grain and organoleptic properties.

Although incorporation of 12% soy flour (bakers' weight) into bread dough is adequate for most commercial purposes, it has been found that the level of soy flour can be increased to 16% if desired using only 0.5% sodium stearoyl-2-lactylate as an additive and levels as high as 20% soy flour can be readily adopted although in this instance it may be desirable to slightly increase the amount of lactylate used to assure retention of the necessary physical and taste properties of the bread. Similar results were obtained using 0.5% of polyoxyethylene (20) mono- and diglycerides of C_{14} to C_{18} fatty acids. Usually, the amount of calcium stearoyl-2-lactylate added should exceed the level of sodium stearoyl-2-lactylate in the same dough formulation to obtain generally equivalent results. For most purposes, it is preferred that the amount of protein rich additive added to the bread composition be at a level to provide an amount of protein in the finished product in the range of 4 to 8%.

The lower limit of protein supplementation which does not require addition of the sodium or calcium salts of an acyl lactylate of C_{14} to C_{22} fatty acids, or the ethoxylated glyceride additive is that amount which does not adversely affect the finished product in the absence of the lactylate additive. Although the protein supplementation which may be carried out without addition of sodium or calcium stearoyl lactylate, or the ethoxylated glyceride compostion to the formulation obviously varies with the supplement incorporated into the mixture, the upper limit of protein supplementation without inclusion of the lactylate or glyceride additive is usually about 2% by weight of protein added to the dough.

The upper limit of protein supplementation along with addition of the acyl lactylate or glyceride additive to the formula is usually dictated primarily by economic considerations. Normally, it is not essential to go above about 12% by weight of protein added to the final bread, baked or fried dough products. Thus, when the protein supplementation is at a level of about 2¼% or more (total protein added to the finished product), the additive should be employed at a concentration within the range of 0.1 to 3%.

For example, 12 parts by weight of soy flour may be added to a conventional bread dough composition for each 100 parts of standard wheat flour with the preferred amount of sodium stearoyl-2-lactylate, calcium stearoyl-2-lactylate, or the ethoxylated glyceride composition in each instance being 0.5% bakers' wheat (i.e., 0.5 part by weight for each 100 parts by weight of wheat flour in the dough). Best results are obtained when sodium stearoyl-2-lactylate is used as the additive.

However, calcium stearoyl-2-lactylate and the glyceride composition are equally usable although in certain instances and particularly with respect to the use of calcium stearoyl-2-lactylate as the additive, a somewhat larger relative amount as compared with the optimum quantity of sodium stearoyl-2-lactylate, may be necessary to obtain equivalent quality in the final baked or fried product.

In most instances utilization of 0.5% of an additive such as sodium stearoyl-2-lactylate, calcium stearoyl-2-lactylate or the ethoxylated glyceride composition based on the weight of wheat flour is adequate for permitting addition of a protein supplement in an amount to provide an absolute protein value of wheat flour. For example, when bread is prepared by the standard sponge dough method using a formula comprising in parts by weight, 100 parts of wheat flour, 12 parts of soy flour, 2 parts of yeast, 5 parts of sugar, 2 parts of salt, 0.5 part of a yeast food containing an oxidizing agent such as potassium bromate, (for example, Arkady yeast food), and 0.5 part of sodium stearoyl-2-lactylate, the resulting product compares very favorably with bread made from the same formula minus the soy flour and the lactylate additive.

Baking tests have shown that when the standard bread formula is supplemented with 12% soy flour (bakers' weight) the addition of 0.5% sodium stearoyl-2-lactylate (bakers' weight) causes the specific volume of a loaf of bread made from wheat flour supplemented with the soy flour to increase from 5.38 where no lactylate additive is used, to 6.25 cc/g when the lactylate is added to the formulation. The latter figure represents a volume that is even larger than bread made from the same amount of wheat flour without soy supplementation. In this case it can readily be appreciated that a significant amount of protein has been added to the bread composition since the soy contains 52.7% protein.

Added Wheat Fractions

D.P. Philips and J.S. Schlesinger; U.S. Patent 3,829,590; August 13, 1974 discuss alternate approaches to improve the nutritional quality of breads. The quality of breads and other baked wheat products is directly related to the protein content of the wheat flour used in the baking process. As contrasted with this desideratum, wheat cereals are generally made from a wheat stock of lower protein content. Efforts have been made to improve the baking quality of wheat through improving the varieties of wheat, horticultural practices, such as fertilization, and classification of particles of wheat flour by air currents and the like.

Attention has also been directed recently toward determining the correlation between wheat color and protein content. The importance of color differences is evidenced by the wheat classification system of the U.S. Department of Agriculture, which system is based largely upon color. Moreover, recent work at Kansas State University has confirmed that there is a relationship between the amount of so-called yellowberry kernels in a sample of wheat, and the protein content of the samples. The higher the percentage of the kernels which are yellowberry kernels (as contrasted with dark, hard vitreous kernels), the lower is the protein content of the sample.

Since, in general, relatively high protein wheat demands a high value in many markets, provided the grains are not damaged, or shrunken and broken, a colorimetric, or other method, for classifying bulk lots of wheat according to protein content of the kernels in each fraction would be a highly useful process. This is especially true since the high yellowberry content fractions left upon removal of the more valuable dark kernels generally encounter no price penalty or markdown in relation to the mixed parent lot. Moreover, very low protein wheats may, on occasion, command a premium when sold for certain end uses, such as for cereal production. No automatic colorimetric method for effecting such separation has been developed. Neither have other methods of isolating the valuable protein-rich kernels been proposed for commercial usage.

Numerous experiments and efforts extending over several years have been made to identify a characteristic or property of wheat kernels which could be made the basis of an effective, commercially useful method of classifying or dividing wheat kernels into fractions of varying protein content, and thereby upgrading bulk lots of the grain.

Broadly described, the process consists of classifying wheat in an unsegregated bulk lot to divide the kernels of wheat into at least two fractions of differing kernel size, which two fractions each differ in a repeatable manner from the average protein content of each other and of the parent unsegregated lot. It was found that there is a direct correlation between the size of wheat kernels and the protein content. It further appears that a fairly good correlation between the color characteristics of wheat kernels and kernel size exists. Thus, by separating the relatively larger kernels from the smaller size kernels, it has been possible with numerous bulk lots of wheat moving in regular channels of commerce to repeatedly isolate protein-rich fractions of the lot from fractions of lower protein content.

In a preferred process, the wheat is classified by screening, or other suitable kernel size classification method, into several fractions. The fraction of lowest protein content is one in which the kernels are retained on a screen having an opening between adjacent wires of from 0.10 to about 0.13 inch. A second fraction contains kernels passing through the first mentioned screen, but retained upon a second screen having an opening between adjacent wires of from 0.093 to about 0.094 inch. A third fraction is that which passes through the last mentioned screen. Further classification can be made using additional or intermediate screens, depending upon the selectivity sought, and the particular prevailing market demands.

Bulk lots of wheat obtained from storage elevators or railroad cars are subjected to a kernel size classification process which consists of passing the kernels through one or more screens to develop fractions in which the average particle size of one fraction differs from the average particle size of another fraction. The fractions developed differ from each other in average protein content, and thus differ in value, according to the market in which the grain is to be sold.

The classification according to kernel size is preferably carried out to obtain a fraction having relatively larger kernels and having a U.S. Department of Agriculture grade and classification which is at least the economic equivalent of the grade and classification of the original unsegregated wheat lot, and which is of an average protein content sufficiently high that this fraction is not subject to an economic penalty. A second fraction which is of smaller average kernel size than the first fraction has a substantially higher protein content than the average protein content of the first fraction, so that the second fraction can be marketed at a premium with respect to both the first fraction and the parent unsegregated lot.

The classification of the wheat kernels according to kernel size is accomplished by moving the kernels of the unsegregated lot into contact with a stationary, or preferably, a moving or vibrating screen, so that the kernels of relatively smaller size pass through the screen to develop the second fraction. This initial screening of the kernels is carried out with a screen constructed so that adjacent wires are separated by a distance of from about 0.10 to about 0.13 inch. Screens in which this size opening is utilized include U.S. Standard Testing Sieves (Tyler) ranging from a No. 7 to a No. 6 sieve.

In addition to the primary classification accomplished by the single screen pass described, it is desirable to subject the kernels to a second screening step in which those kernels which have passed through the first screen described above are further screened to develop an additional fraction of kernels sufficiently small to pass through the second screen. In the second screening stage, the second screen has an opening between adjacent wires of from about 0.093 to 0.094 inch.

Further classification, if desired, can be carried out by further screening stages or by the use of screens intermediate in their opening size between those described. In the most preferred method the screen of smallest openings between wires is defined as one in which the openings are from about 0.093 to 0.094 inch. By subjecting the kernels to screening using a screen of this type, the kernels which pass the openings in such screen have been found to be those kernels which tend to be shrunken and broken kernels which have a relatively high protein content, but are subject to downgrading in the usual standardized grading procedures because of their physical appearance.

Example 1: A 1,000-gram aliquot of No. 1 grade hard red winter wheat was subdivided into five 200-gram aliquots, and each of these was sifted for one minute on a Seedburo Sedimentation Sifter equipped with a No. 6 U.S. Standard Testing Sieve. This screen had a spacing between adjacent wires of 0.132 inch. The kernels of wheat which passed through the No. 6 sieve were collected, and the "thrus" from each aliquot were blended, and were sifted for one minute on a No. 7 U.S. Standard Testing Sieve in aliquots not exceeding 200 grams in weight. The kernels passing through the No. 7 sieve (0.111 inch openings) were then sifted on a No. 8 U.S. Standard Testing Sieve (0.094 inch between adjacent wires) in aliquots not exceeding 200 grams in weight. The fractions which were obtained were analyzed for protein content and the results of such analyses are set forth in the following table.

Fraction	Description	Wt. Percent of Total	Wt. Percent Protein
1	Control (Parent)	100	11.90
2	overs - No. 6	15.3	10.95
3	overs - No. 7	56.8	11.40
4	overs - No. 8	25.1	12.55
5	thrus - No. 8	2.8	13.85

The data reported in the table show that a substantial gain in protein content is achieved in that fraction of the wheat which is retained on the No. 8 sieve, and which constitutes about 25% of the parent sample. It was further observed that although the kernels passing through the No. 8 screen showed a significantly higher level of protein than those retained on this screen, the kernels in this "thru" fraction were for the most part shrunken and broken kernels subject to downgrading by existing grading standards due to this visual appearance.

It may be observed, however, that this fraction, constituting 2.8 weight percent of the total parent sample can be blended back into the second or third fraction to slightly increase the protein content in either case without resulting in the downgrading of either of these fractions due to the retained over-all visual appearance of the entire fraction.

Example 2: Twenty samples derived from various types of wheat moving in commerce through a grain elevator terminal point were obtained. The kernels passing through a No. 7 U.S. Standard Testing Sieve and retained on a No. 8 U.S. Standard Testing Sieve were separated from the parent fraction by screening aliquots of the parent samples not exceeding 200 grams for a period of one minute on a No. 7 screen mounted in a Seedburo Sedimentation Sifter. The thrus from the No. 7 sieve were sifted for one minute in aliquots not exceeding 200 grams upon a No. 8 U.S. Standard Testing Sieve.

Protein analyses show that in the case of each of the wheat classes, the fraction of the kernels passing through a No. 7 U.S. Standard Testing Sieve and retained on a No. 8 U.S. Standard Testing Sieve contains a markedly higher protein content than the parent fraction from which it was derived.

Example 3: Seven samples of soft white wheats from different sources were initially analyzed for protein. 100-gram aliquots from each of these samples were then sifted for one minute on a No. 6 U.S. Standard Sieve mounted on a Seedburo Sedimentation Sifter. The "overs" on the No. 6 sieve were weighed and analyzed for protein. The thrus were sifted for one minute on a No. 7 U.S. Standard Sieve. The overs and thrus from the No. 7 sieve were then weighed and analyzed for protein. From the data obtained in the test, it was seen that soft white wheat can be fractionated by kernel size classification to isolate fractions having a higher protein content than the parent lot.

Example 4: A 250-gram aliquot of yellow hard winter wheat was visually tested for the presence of shrunken and broken kernels, and these kernels were separated from this sample and weighed to determine the percentage of shrunken and broken kernels in the parent lot of the yellow hard winter wheat. This was found to be 3.1 weight percent. Six 200-gram aliquots of the parent sample were then each subjected to sifting for one minute on a No. 7 U.S. Standard Sieve mounted on a Seedburo Sedimentation Sifter.

The overs from the No. 7 sieve runs of the six aliquots were combined, blended, weighed and tested for shrunken and broken kernels in the same manner as the 200-gram aliquots taken from the parent sample. The thrus from the No. 7 screen of the six aliquots were each sifted for one minute on a No. 8 U.S. Standard Sieve. The overs of the No. 8 sieve runs were combined, blended, weighed and tested for shrunken and broken kernels. The thrus from the No. 8 sieve runs were combined, blended, weighed and tested for shrunken and broken kernels. The data obtained is reported in the following table.

Fraction	Weight, Grams	Weight Percent	Weight, gms Shrunken & Broken	Wt. Percent of S & B Kernels
Parent	1200	100	37.2	3.1
"Overs" on No. 7	841	70	0.8	0.1
"Overs" on No. 8	308	25.7	8.6	2.8
"Thrus" on No. 8	51	4.3	27.8	54.5

From the data reported in the table, it will be seen that most of the shrunken and broken kernels present in the parent wheat lot pass into the fraction which passes through the No. 8 U.S. Standard Sieve. The protein gain which has thus been found to occur in the case of the 7 to 8 fraction is not due to any concentration of the shrunken and broken kernels in this fraction, but rather is due to the concentration of small, sound kernels of high protein content. This example demonstrates that the kernel size classification procedure of the process can improve the United States Department of Agriculture grade of the wheat lot by removing shrunken and broken material from the fraction retained on the No. 7 screen, and also from the fraction which passes the No. 7 sieve and is retained on the No. 8 sieve.

Added Milk Protein

A major difficulty with the use of dried whole milk, dried skim milk or casein to produce bread having a high milk protein content arises from the fact that these products increase the acidity of the dough and require a greater quantity of water to make a workable dough having the proper plasticity. In the baking of the bread the water must be removed by evaporation in the hot oven, and the removal of the large amount of water so required lengthens the baking time so that a thick hard crust develops before the water is removed from the interior of the loaf, and as baking is prolonged the interior is drawn toward the crust without developing a normal texture or "crumb" characteristic of yeast-raised bread. Also the extended baking time along with the milk sugar content blackens the crust due to over-caramelization of the sugar.

Three essential ingredients are used by *F.F. Hansen; U.S. Patent 3,561,974; February 9, 1971* to overcome these problems when milk proteins are used to prepare high protein bread. The first is a commercially available milk protein concentrate obtainable on the market which contains the essential amino compounds necessary to supplement the protein in the flour and which is low in milk sugar or lactose. It requires less water than whole milk solids, skim milk or casein to form a dough.

Also sodium bicarbonate is incorporated sufficient to raise the pH to an alkaline level. This enables the dough to be made and worked with less water than is usually required in conventional or special doughs which are inherently desirably acid in character. To offset the taste disadvantage of the alkalinity in the finished loaf, sodium pyrophosphate is also included in the dough. It remains inert, however, until the dough is in the oven when at elevated temperature it reacts with the sodium bicarbonate to neutralize it and bring the dough to an acid pH, the reaction producing a harmless sodium phosphate so that the taste of the bread as well as its appearance and texture more nearly resembles the conventional loaf.

The usual ingredients for making bread are, of course, normal wheat flour, yeast, sugar (beet or cane or corn), shortening, such as lard or lard-like hydrogenated oil or other fat substances or mixtures, along with water, milk, salt, and optionally yeast foods. In a usual procedure a "sponge" comprising part of the total flour, usually about 60%, with all of the yeast is "set," and after a period of time it is mixed with the remainder of the flour and other ingredients to form the final dough. This finished dough is cut into pieces, and then formed into loaves which are placed in pans and baked. The normal wheat flour used has a flour protein content of from 10 to about 12%. The milk may be boiled fresh whole milk, dried skim milk, but the amount of protein introduced from the milk is quite a small percentage of the protein in the entire finished loaf.

There are commercially available flours in which the protein is artificially raised to a higher level, either by dissolving out some of the starch, or dissolving out from a normal flour the gluten, drying it and supplying it to other flour. One commercially available flour is thus enriched to about 40 to 42% protein. Another is available through selective milling having a protein content between 22 and 24%, and ordinary gluten bread utilizes such flour. However, in the solvent extraction process some other values are lost which are not replenished in the bread formula, and although bread made in this manner is termed high protein bread, it is actually low or entirely lacking in some essential proteins.

According to this process a more or less normal sponge is set. The sponge is prepared by combining ingredients as follows in the approximate proportions specified. The ingredients are mixed together and set for about 4 hours at 80°F.

	Pounds
Wheat flour running 22 to 24% protein	20
Wheat flour running 12 to 13% protein	20
Water	50
Yeast	3
Sugar (derived from cane, beet or corn)	0.4
Lard or other shortening (hydrogenated vegetable oils, for example, or mixtures)	3

In addition to the sponge, three other mixes, which are here identified as A, B and C, respectively, are prepared. In the following formula for these mixes, the weights required for the foregoing weight of the sponge are given.

There is available commercially a milk protein concentrate for use by bakers that is high in milk protein, practically devoid of the fat contained in whole milk, and which is low in milk sugar (lactose), whereas dried whole milk and dried skim milk contain about 38 to 45% lactose. Unlike casein, it is fine powder readily dispersible in water. It is derived from whey produced in the manufacture of cheese, or milk albumin to which its composition is comparable in many respects, and a principal commercial source is Sheftene B.P.S. (Sheffield Chemical) and is about 84% protein. A typical analysis as published by the manufacturer is:

	Percent
Nitrogen (as is basis)	13.0
Protein (N x 6.38, as is basis)	82.0
Ash	4.0
Moisture	5.5
Ether extractables (lipids)	4.5
Carbohydrates (lactose)	3.0
Sodium	1.0
Heavy metals (as Pb)	less than 10 ppm

Approximate amino acid content (calculated on as is basis) is as follows.

	Percent
Arginine	3.3
Cystine and methionine	4.4
Histidine	2.2
Isoleucine	6.1

(continued)

	Percent
Leucine	9.0
Lysine	7.8
Phenylalanine	4.5
Threonine	4.3
Tryptophane	1.6
Valine	5.8

It will be noted that amino compounds tryptophane, arginine and cystine are found in milk albumin and not in casein. For the amount of sponge as above specified, mix A comprises about 30 pounds concentrated milk protein as above described.

To this is added sodium bicarbonate and sodium acid pyrophosphate or sodium aluminum pyrophosphates which will neutralize the bicarbonate of soda only after the baking of the dough begins. These ingredients are combined in single acting baking powder which is more convenient to use than separately measuring and adding the two ingredients. Formula A therefore indicates 2½ pounds single acting baking commercial powder comprising sodium bicarbonate and sodium acid pyrophosphate. The total weight of mix A is therefore 32½ pounds. These ingredients of mix A should be intimately mixed to distribute the baking powder uniformly through the milk protein powder. Mix B comprises the following.

	Pounds
Water	50
Yeast (variable)	3
Sugar (preferably beet or cane sugar, amount may be varied to taste)	6
Baking molasses (dark) optional for flavoring	4
Salt (NaCl)	1

These ingredients should be mixed immediately before the final dough is prepared. Little time should elapse between the mixing and the subsequent use of the mix. Mix C comprises the following.

	Pounds
Flour running 22 to 24% protein	20
Concentrated milk protein as described	10

This can be prepared at any time.

After the sponge has been prepared and set as described above, it is charged into a regular bread mixer. Mixes A and B are combined in a second vertical cake mixer which is run at slow speed for 2 or 3 minutes. The combination of A and B is then added to the sponge in the first mixer and mixed at slow speed for about 5 minutes or until the two are thoroughly incorporated and no lumps remain. Then mix C is slowly added as mixing continues, and when this is added the entire batch is mixed at medium speed for about 10 minutes. Care must be taken to avoid overmixing and this condition can be determined by an experienced baker. During this time the dough should be kept between 70° and 80°F.

After mixing, the dough should be discharged into the "trough" and without any "floor time" immediately divided in the dough divider into separate pieces of the desired loaf weight. The pieces are then immediately charged into a proofer, and after proofing, molded and placed into pans. In sheeting the dough for molding, care should be taken that it is not sheeted as thin as dough made for ordinary white bread, because it tears easily. A baker will have no difficulty determining a thickness at which the sheeted dough will not be too thin.

The proofing temperature should be in the general area of, but not over 100°F, and the dough will proof very rapidly since the yeast content is purposely very high, two or three times as much as would ordinarily be used with this amount of flour in normal baking where the yeast is about 2 or 3% of the total weight of the flour. Rapid proofing is desirable because the final dough with the high protein content deteriorates rapidly and full proofing is necessary because the dough in baking evidences little or practically no "oven spring," which means that it has little inherent ability to expand in the initial stages of baking, as do normal bread doughs.

Baking should be effected in a temperature of about 400°F for 30 or more minutes, since this bread does not "bake out" as fast as ordinary white bread and could be doughy inside unless the higher temperature and longer baking time were provided. This, plus the high protein, results in a crust of very dark brown color and somewhat thicker than the crust of normal white bread, but nevertheless the crust is not hard and the interior of the loaf has the usual texture and crumb comparable to conventional white bread, although slightly more off-white.

In the example given here, the total weight of flour is 60 pounds to 40 pounds of milk protein concentrate, but the milk concentrate has about 84% protein as against a maximum of 22% in the flour, so that the milk protein is predominant.

In the sponge, 20 pounds of 22% protein flour provides 4.4 pounds of protein. 20 pounds of flour running 12% protein provides about 2.4 pounds of protein. There is some protein in yeast, but this need not be considered. In mix A there are 30 pounds of milk protein concentrate having not less than 84% protein, so $30 \times 0.84 = 25.2$ pounds of protein. In mix B there is no appreciable amount of protein. In mix C the protein in flour is about 4.4 pounds plus 8.4 pounds in the milk protein concentrate. Summing up, there are 6.6 pounds of protein in the sponge plus 25.2 pounds in mix A, plus 12.8 pounds in mix C, or a total of about 45 pounds of protein.

The sponge loses about 5% of its total weight during fermentation, so that the final weight of the sponge in the mix is about 91.6 pounds. The total ingredients on the dough side weigh about 127 pounds. Added together, these weights give a complete weight of the dough of 218.6 pounds. During proofing, machining and baking, the dough loses about 10% of its total weight, or about 22 pounds. Deducting this loss of weight from 218.6 pounds, the finished weight of the bread is 196.6 pounds. Dividing the total weight of protein, 45 by 196.6 gives 23½%. This 23½% is in turn made up of better than 50% protein derived from milk and is 17% of the total weight of the bread. It is more than the amount needed to fortify or nutritionally complement the flour protein. Meat, fish, fowl and cheese range between 18 and 20% protein, so that this bread with 23½% now exceeds the protein content of these foods.

A certain amount of flour is necessary to give the bread its volume and texture. In the sponge two commercially available flours are used, one with a higher protein than the other. No flour commercially available is known that would provide the average secured by mixing the two, which in the example is around 19 to 20%, but not as high as if all of the flour had 22 to 24%. No yeast food is used in the sponge because the bromate or oxidizer in yeast foods in the presence of high protein will toughen the gluten too much in the sponge stage.

The water content in the sponge is much higher than is customary because of the high gluten content, but considerably less than would be required if dried whole milk or dried skim milk or casein were used as a source of protein, and if the dough had not been made alkaline by the use of sodium bicarbonate. The yeast is in much higher percentage to develop active fermentation in the sponge and vigorous activity in proofing and fermenting the final dough.

The preparation of mix A of the dough procedure involves mixing the dry concentrated milk protein with the recommended amount of single action baking powder. Mix A is in subsequent procedure mixed with B which contains all of the water and water-soluble ingredients. The baking powder which is a fine powder mixed through the milk protein concentrate, enables A and B to be mixed in a short time without lumping of the concentrate. Stage B, as stated, is where the water and soluble ingredients of the dough stage are combined and mixed to assure that these ingredients are completely dissolved and other ingredients uniformly dispersed.

It should be prepared just before use to avoid fermentation at a stage where it would serve no purpose. Some of the sodium bicarbonate of the baking powder will react with the acid ions of the protein when the ingredients are combined, and reduce over-all absorption of water, which is greater under more acid conditions. As explained above, one of the most difficult problems is controlling water absorption which at best is high and difficult to take out without leaving the interior of the loaf doughy. On the other hand, the bread must have sufficient moisture to have good eating qualities and not appear dry or stale or mostly a hard crust.

Stage C of the dough procedure is made separate from mix A largely for mechanical reasons. When A and B are mixed together, they provide a reasonably wet mix that may then be combined with the sponge and thoroughly blended. The flour in stage C aids to keep the remainder of the milk protein concentrate from lumping in the final step of mixing into the dough, and it also provides the final measure of flour needed to complete the dough. It is not necessary that A, B and C mixes be prepared as separately compounded mixtures, but the ingredients may instead be weighed and introduced singly or in combination with the sponge when the sponge is returned to the mixer to complete the dough.

The milk protein concentrate normally tends to form a glutinous, rubbery cohesive mass when mixed with water and requires more water than could be baked out to give it a plasticity required for mixing into a dough. It was found that this concentrate, like other milk protein concentrates, is slightly acid, but is more readily dispersible in water, so that less water is required. By using a single acting baking powder, enough of the sodium bicarbonate will react with the protein at mixing temperature to render the concentrate slightly alkaline, whereupon it may be much more easily mixed with the other ingredients without requiring excessive water.

When the dough is baked, the sodium pyrophosphate will react with the remaining sodium bicarbonate to yield gas which will impart some oven spring and volume to the loaf, but this is incidental, its important function being to bring the dough back to the acid side of neutral, somewhere about pH 5.5. This is comparable to the pH of white bread and it is a condition important to imparting the desired flavor to the bread. If sodium bicarbonate alone were used, this desirable flavor would be lost and its flavor would be particularly unattractive to a person who is especially critical of the flavor.

It is also important that the baking powder be a single acting powder, as contrasted with the familiar double acting ones, since in the double acting powder one acid salt reacts with the sodium bicarbonate at room temperature and another at the baking temperature, and this would be detrimental to the process of converting the protein concentrate slightly to an alkaline pH and then, as oven heat penetrates the dough, bringing the dough back to a mildly acid side.

Other heat acting acid salts may be used in place of sodium pyrophosphate, but this one is preferred because of the better flavor it gives to the bread. Desirably there are 5 to 6 parts by weight $NaHCO_3$ to 4 to 5 parts $Na_2H_2P_2O_7$, and 100 pounds of the baking powder may contain about 3% cornstarch and 0.5% calcium silicate finely powdered to keep the powder free flowing. The active baking powder ingredients, as explained above, could be separately introduced so long as the $NaHCO_3$ was first available to contact the protein concentrate.

The final dough does not lend itself to twisting as does normal white bread and therefore is best molded and panned as one piece or is cross-panned where the one piece is cut into three or four pieces and each piece panned crosswise of the pan. The dough is somewhat more sticky than conventional bread dough, so that more dusting flour may be required in the dividing and forming operation, this stickiness resulting both from the more abundant use of water in the makeup of the dough, and because of the high protein content.

Some deviation from the formulas outlined here may be necessary where different conditions prevail, such as the mineral content of the water, character and age of the flour, and the humidity in the bakery. Also market requirements may result in flours of different gluten or protein content becoming available. The dry ingredients such as those used for the sponge and for mixes A and C may be mixed in advance and distributed to bakers who supply the remaining ingredients according to direction.

Important factors are the use of larger than normal amounts of yeast and the introduction of the milk protein concentrate to the dough only in the final stages of preparing the dough mix so as to minimize the time before proofing that the concentrate, water and yeast are together in the dough, and the use of the baking powder or its equivalent ingredients as explained is important to working the dough and securing the easier and more uniform diffusion of the milk protein concentrate through the dough. The protein content may be reduced somewhat below or slightly above to meet the baker's desire or the public taste, but should range between 15 and 25% and preferably be around 20 to 23.5%, but in any case the milk protein should predominate.

Also, where the quality of the wheat flour makes it desirable, and since most wheat flour is low or entirely lacking in one essential amino acid, lysine monohydrochloride (sometimes called lysine hydrochloride), it may be added to the sponge stage at about 0.1% based on total flour in the formula commercially available lysine. This amounts to about 1 ounce for the total of 60 pounds of flour in the formula described above, and thereby increases the body utilization of the protein in the bread.

Protein-Lysine Enrichment

Since nonfat dry milk solids are used in preparing bread and are rich in protein, it might be thought that the quantity could be increased sufficiently to provide a high protein bread. However, attempts to prepare a 20% protein bread in this way have not succeeded in producing a product which is acceptable in place of the unfortified bread. Nonfat dry milk is available which contains 35.6% protein. In order to attain 20% protein (dry weight) in the bread, using nonfat dry milk as the only supplementary source of protein, a minimum of 19 parts per 100 parts of nonfat milk solids must be added.

Experiments have shown that addition of more than 6 parts nonfat dry milk per 100 parts flour has caused progressive darkening, thickening and toughening of the crust, diminished loaf volume, coarser grain and rougher texture, and less soft crumb. Bread fortified with 20 parts of nonfat dry milk per 100 parts flour was sour and unpalatable. The crust had a very bitter taste. Furthermore, the dough was difficult to mix or manipulate properly in standard procedures for bread preparation. Similar difficulties were encountered with high protein milk products (70 to 90% protein), such as lactalbumin (73 to 76% protein) or calcium caseinate (88 to 90% protein). Even though a smaller proportion of either of the latter is required to obtain the desired high protein content, a commercially acceptable loaf could not be produced with either one.

Wheat gluten is available which contains 70 to 75% protein. Small additions of this material have a desirable effect on the bread structure, but as much as 4 parts per 100 parts of wheat flour caused an undesirable shiny open grain in the bread. In addition, wheat gluten is deficient in lysine in the same way as wheat flour. The above proportions would require the addition of 0.315 part lysine monohydrochloride to correct this deficiency, i.e., 0.021 times the total weight of wheat protein. Gluten contributes to the taste of breads. Bread fortified to a 20% protein content with this combination was found unsatisfactory as to taste, cost and appearance in comparison with the usual commercial product.

S.R. Ehle; U.S. Patent 2,989,399; June 20, 1961; assigned to E.I. du Pont de Nemours and Company uses a special mixture of wheat gluten, nonfat milk solids and other milk proteins to fortify a typical American loaf to 20% protein or higher (dry basis). These materials are mixed in proportions with the other bread ingredients in an amount providing a total of at least 20 and preferably not more than 28 parts protein per 100 parts of dry ingredients.

For example, the addition of 9 to 11 parts nonfat dry milk, 2 to 4 parts lactalbumin or calcium caseinate and 4 to 6 parts wheat gluten per 100 parts of flour containing 12.7% protein will provide a high protein bread containing at least 20% protein on a dry basis. The lysine deficiency of the wheat protein should also be corrected by adding lysine, as such or in the form of lysine monohydrochloride, lysine propionate, or other compound. An amount should be added equivalent to 0.2 to 0.7 part of L-lysine per 12 parts of wheat protein.

The above fortifying ingredients may be added separately to the usual bread ingredients or can be added as a premixed composition. The composition can include other ingredients in addition to the nonfat dry milk, high protein milk product and wheat gluten. The composition may include lysine, flour or other dry materials used in bakery products. The fortifying ingredients can be incorporated in a ready mix of the type sold for use in homes, which uses baking powder instead of yeast and requires only the addition of water.

Protein bread may be prepared commercially by the sponge dough, straight dough, or continuous fermentation processes. In the sponge dough process the supplemental proteins may be added to the dough or it may be preferable to add the supplemental wheat gluten and high protein milk product separately to the sponge. A convenient premixed composition for fortifying commercial bread combines the fortifying ingredients in such proportions that the mixture contains 50 to 60% protein and 4.5 to 8.0% L-lysine. For example, such a composition containing 55% protein and 5.0% L-lysine can be added to the conventional mix in the proportions of 30 parts per 100 parts of wheat flour (12% protein) to give a bread containing more than 20% protein (dry basis) and a proper lysine balance.

The wheat gluten should be of the type customarily used in bread, i.e., the gluten should not be denatured or it will not have the gas retention properties which confer the desired volume and cell structure to the loaf. Further, the nonfat dry milk should be the spray-dried bakers grade, although the roller dried material may be used if it can be shown by analysis that the amino acids are in the expected proportions in the protein. It is sometimes advantageous to add 0.5 lb/100 lb flour of active dry yeast to the dough to increase the acidity of the bread and thereby decrease crust carmelization. Further, increasing the time of fermentation can also increase the acidity of the bread.

Example 1: In the production of high protein bread by the sponge-dough process, a sponge is prepared, containing 70 parts of wheat flour (12% protein), 42 parts water, 2.5 parts yeast, ½ part yeast food (starch and materials) and a small amount of fungal enzyme, and is set at 78°F to ferment for 4¼ hours. It is then made into a dough with 30 parts wheat flour, 28 parts water, 6 parts sugar, 2 parts shortening and 2 parts salt.

A fortifying mixture of 5 parts wheat gluten, 10 parts nonfat dry milk, 0.33 part L-lysine monohydrochloride and 3 parts calcium caseinate is added to the dough. The dough is mixed for 8 to 10 minutes at 80°F and allowed to rise for a floor time of 25 minutes. The dough is baked in the usual manner to yield a bread having good volume, close grain, soft texture and good taste. The bread analyzes 3.52% nitrogen and 1.18% lysine on a dry weight basis, corresponding to better than 20% protein of balanced lysine content.

A feeling study was conducted in which 21 day old male albino rats of Carworth Farms were fed either a diet of standard white bread or one containing the above-mentioned protein bread. The diets were made up as shown on the following page.

	Diet I—Standard Bread	Diet II—Protein Bread
	Percent	*Percent*
Bread	67.0	45.0.
Corn Oil	7.2	9.4.
Vitamin Diet Fortification Mixture (mfd. by Nutritional Biochemicals Co.).	2.2	2.2.
Salt Mixture USP No. 14	4.0	4.0.
α Cellulose	5.0	5.0.
Corn Starch	Make up to 100.	Make up to 100.

On analysis the diets showed:

	Diet I—Standard Bread	Diet II—Protein Bread
Percent N (dry)	1.66	1.70
Percent Protein (total wt. basis)	8.8(N × 5.70)	9.2(N × 5.85)
Percent Fat (dry)	10.0	8.4
Percent H_2O	7.21	7.52

Ten rats were picked at random for each group, so that the mean weights were the same for both groups, and the growth study was conducted for 4 weeks. The weight gain on each rat was measured at weekly intervals and the food consumption was checked twice weekly. The rats were fed ad libitum. The quality of the protein in the bread is apparent from a comparison of the protein efficiency ratios obtained on the rats. The protein efficiency ratio is defined as follows. PER = gram weight gained divided by total grams of protein eaten. For the standard white bread the PER was found to be 1.1. For rats fed the protein bread the PER was 2.7. This value compares favorably with values in the literature for meat proteins, e.g.:

Meat	Ave. PER	Reference
Bacon—uncooked	2.8	McBride, Guthneck, Hoffert, Knickel, and Schweigert, J. Nutrition 45, 393 (1951).
Bacon—cooked	2.5	
Cured—Ready to eat ham	2.7	Dunker, Berman, Snyder and Tubiash, Food Technol. 7, 288 (1953).
Beef liver	2.7	(M. Sahyun, "Proteins and Amino Acids in Nutrition," p. 66, Reinhold Pub. Corp. N.Y. (1948).
Beef heart	3.1	
Beef muscle	3.2	

A market survey was conducted in which residents of Wilmington, Delaware, were asked how this bread compared with the bread they are now using as to appearance, texture, aroma, taste and as toast. Preferences for the protein bread were as follows.

Quality	Better	Same	Less	Undecided
	Percent	*Percent*	*Percent*	*Percent*
Appearance	24	61	13	2
Texture	40	49	11	nil.
Aroma	13	72	7	8
Taste	34	56	10	nil.
Toast	40	45	9	6

Example 2: A mixture of 5 parts wheat gluten, 8 parts nonfat dry milk, 0.33 part L-lysine monohydrochloride and 8 parts calcium caseinate is added to the dough, instead of the fortifying mixture of Example 1, to give a protein bread containing at least 20% protein, and which is comparable in quality to the bread of Example 1. For comparison with the results obtained by this process, a fortifying mixture of 20 parts nonfat dry milk and 0.25 part L-lysine monohydrochloride when added to 100 parts of flour (containing 12% protein), 4 parts shortening, 60 parts water, 2 parts salt, 6 parts sugar, 2 parts yeast,

and 0.37 part of yeast food gave a sour bread which was inferior in volume, which had a thick black crust, and whose grain and texture were poor. When wheat gluten was used instead of nonfat dry milk to give 20% protein bread, undesirable large loaves with a leathery texture and a large open grain resulted. Also, the amount of L-lysine monohydrochloride had to be increased to 0.46 part per 100 parts of flour because of the lysine deficiency of the wheat products.

Added Nonwheat Proteins

The largest deficiency in protein quality of bread is in respect to L-lysine and, by adding L-lysine to the bread formulation, the utilizable protein of the bread is increased and the ratio of utilizable protein calories to total calories is increased by more than the amount added by the L-lysine. After the deficiency of L-lysine is eliminated, another essential amino acid deficiency becomes controlling and a further increase in protein quality is obtained by adding that particular acid. Addition of the acids themselves is inherently expensive, and furthermore, since wheat protein is relatively deficient in several essential amino acids, it would be preferable to add a raw protein source and, in particular, one which is relatively rich in those essential amino acids which are relatively deficient in wheat protein.

The most efficient raw proteins, from a technical point of view, would be those which were rich in those essential amino acids in an amount proportional to their deficiency in wheat protein. However, cost factors are of much more importance. Accordingly, the nonwheat protein sources used in the bread dough formulation are those which are relatively rich, upon hydrolysis, in essential amino acids in which wheat protein is relatively deficient. By relatively rich and relatively deficient are meant the amount of the acids in question relative to that which can be utilized by a human consuming the protein in question. In other words, a protein rich in L-lysine includes an amount of L-lysine above the amount in a given protein which can be utilized by a human consuming that protein. Preferably, the nonwheat protein is relatively rich primarily in L-lysine such as soy and fish.

S.K. Wolf and R.M. Cavanaugh; U.S. Patent 3,756,832; September 4, 1973; assigned to International Food Technology Inc. describes the use of soy protein as a nonwheat protein source to improve the protein quality of bread. Corn flour or corn meal is used to mask the soybean taste. Any of the conventional edible grades of soy flour may be used ranging in fat content from 18 or 22% by weight fat, in the full fat grade, to 1% or less in the defatted flours. It is best to use defatted soy flour, although full fat soy flour can be used since it is available as a stable, finely milled flour.

It is preferable that the soy flour be toasted, as toasting improves its flavor and renders it more digestible by inactivating antitrypsin enzymes usually present in soy flour. The term soy flour includes defatted and full fat soy flour, soy protein concentrates and isolated soy protein, milled to a particle size which will pass through an 80 mesh screen (U.S. Standard). The corn can be any edible grade, including corn flour and corn meal. Corn flour is preferred and, if corn meal is used, it is used in an amount of from 1 to 4% by weight.

It has been known for some time that the addition of a nonwheat protein, such as soy flour, into a broad formulation, "burdens" the bread which results in poor physical and eating qualities. Large quantities of soy flour, for example, cannot be readily incorporated into a bread dough formulation. However, it will be remembered that the object of adding the nonwheat protein is to add a source of essential amino acid in which the additive is rich and in which the wheat protein is deficient. This means, in practice, that relatively large amounts of nonwheat protein, generally at least 6 parts by weight per 100 parts by weight of wheat flour, must be added to obtain a substantial increase in protein quantity.

The addition of this large amount of nonwheat protein generally results in poor bread quality and various additives, generally classed as conditioners, emulsifiers, or softeners in the baking industry, have been proposed to remedy this defect. Among the proposed additives are sodium stearoyl-2-lactylate and calcium stearoyl-2-lactylate. However, these are expensive materials and they are used in relatively large amounts. It is preferred to use materials which can be used in lesser amounts and/or at lesser expense. It was found that several other materials can be so used including conventional mono- and diglyceride bread additives such as Atmul 500 and cellulose ethers such as sodium carboxymethylcellulose in amounts up to 0.5% based on the wheat flour weight. The bread formulation also includes essential vitamins and minerals to provide an enriched bread.

Examples 1 and 2: Bread is made by the sponge-dough technique from the following formulation. Percent is by weight based on wheat flour weight.

	Percent
Sponge (ingredients):	
Wheat flour	62.5
Water	46.0
Yeast food (flour, salt, NH_4Cl, $CaSO_4$, $KBrO_3$)	0.5
Yeast	2.5
Dough (ingredients):	
Wheat flour	37.5
Water	36.0
Sugar (sucrose)	7.0
Salt (regular)	2.0
Shortening (vegetable)	3.0
Ca propionate	0.1
Hofmann-La Roche vitamin mix	1.25
L-lysine monohydrochloride (Ajinomoto)	0.25
Fish protein concentrate (U.S. Department of Commerce 90% protein)	2.0
Nonfat dry milk	2.0
Soy flour (Central Soya Textrol)	6.0
Corn flour (Ceradex 265, Illinois Cereal Mills)	5.0
Emulsifier (Atmul 500, Atlas Chemical Products)	0.2
Sodium CMC (Dupont P-95-SM carboxymethylcellulose)	0.1

The process conditions are as follows.

Sponge formulation time	4 hours
Sponge temperature (into fermenter)	78°F
Sponge temperature (out of fermenter)	84°F
Dough mixing time	7 minutes
Dough temperature	85°F
Dough proofing time	55 minutes
Dough scaling weight	430 grams

The dough is baked at 400°F for 18 minutes. The baked bread weighs 487 grams with a volume of 1,900 cc, or a volume/weight ratio of 4.90, crumb color is yellow, grain is rather open with round cells. The flavor is level without noticeable soy taste, fish taste, or corn taste. A second sample is made in the same manner except that the corn flour is omitted. Process conditions are substantially identical. The baked bread has a noticeably nonwheat flour taste imparted by the soy and fish flour.

Examples 3 and 4: Bread is made according to the straight dough method from the following formulation. Percent is by weight based on wheat flour weight.

	Percent
Brew (ingredients):	
Water	68
Salt	0.75
Sugar	1.0
Yeast	2.5
Yeast food	0.66

The brew is left for 1 hour and 45 minutes during which time the temperature rises from 80° to 87°F with about 2.0% weight loss. The pH is monitored by litmus and is about 4.9. The brew is then placed in the mixer with the remainder of the formulation as follows.

Brew:	
Water, pounds	20
Flour, pounds	100
Sugar	7.5
Salt	1.50
Shortening	3.0
Ca propionate	0.25
La Roche vitamin mix	1.25
L-lysine monohydrochloride	0.25
Soy flour (Textrol)	10.0
Corn flour (Ceradex 265)	5.0
CMC (P-75-SM)	0.2
Emulsifier (Xpando, a polyoxyethylene glyceride type)	0.25
$KBrO_3$ wafer	*

*60 parts by weight per million parts by weight of wheat flour

The formulation is mixed for one minute at low speed and to clean up (10 minutes) at high speed. Dough temperature is 80° to 82°F. The dough is proofed for 60 minutes at 105°F, 90% relative humidity and baked for 10 minutes at 425°F. The baked bread is free of any penetration of flavor of soy or corn. A second sample is made as in Example 3 except that the corn flour is omitted. The baked bread has the characteristic color and beany taste associated with breads having 10% by weight of soy flour.

All of the bread produced meets the objective of 12% minimum for utilizable protein calories based on total calories. Some bread samples are as high as 18%. A principal object of the process is to provide a bread having enhanced protein quality, (i.e., the ratio of utilizable protein calories to total calories in the bread). In general, it is desirable to at least double this ratio and, in order to do so, the amount of nonwheat proteinaceous source will be between 6 and 15% by weight based on the wheat flour weight.

The amount of corn flour used will be from 1 to 8% by weight, same basis. Where corn meal is used, the maximum amount is 4%. The nonwheat protein includes soy and fish of between 6 and 12% by weight, in which case the amount of corn flour is from 2 to 6% by weight. In general, L-lysine may be added in an amount of up to 1% by weight based on the wheat flour weight, and preferably in an amount of up to ½% by weight.

Compared with standard breads, bread produced by this process is lower in calories and higher in vital protein building nitrogen. It has a much higher (12%) ratio of usable protein to total calories. It also may be enriched with all of the 19 essential vitamins and minerals listed in the *Federal Register*, March 30, 1972 by the Food and Drug Administration from the National Academy of Sciences "Recommended Dietary Allowances." Analysis of a typical bread prepared by this process is shown in Table 1 on the following page. The protein quality is indicated in Table 2 on the following page.

Vitamin C (ascorbic acid) may be sprayed onto the bread as an aqueous solution after slicing and before packaging. The package in this case is a gastight plastic film such as polyolefin film. In order to preserve the ascorbic acid from oxidation, the package is provided with an atmosphere of inert gas such as nitrogen. In normal practice, the empty packages are opened with air prior to insertion of the bread loaf. Nitrogen is preferably provided in the package by using nitrogen, in lieu of air, to inflate the empty package.

TABLE 1

Nutrient	Fraction of recommended dietary allowance, percent: One-meal serving (two-slices)	Daily intake eight slices)
Vitamin A	10	40
Vitamin C	15	50
Thiamine (Vitamin B_1)	15	60
Riboflavin (Vitamin B_2)	10	50
Niacin	10	40
Calcium	10	30
Iron	15	50
Vitamin D	10	40
Vitamin E	10	30
Vitamin B_6	10	40
Folacin (Folic Acid)	10	30
Vitamin B_{12}	10	30
Biotin	10	30
Pantothenic acid	10	30
Phosphorus	10	30
Iodine	10	40
Zinc	10	30
Magnesium	10	40
Copper	15	50

TABLE 2

Contents	Serving size: One meal, 2 slices (approx. 2 ounces, or 57 grams)	Daily intake, 8 slices (approx. 8 ounces, or 227 grams)
Calories	130	510
Grams total protein	6	23
Grams casein-equivalent protein	5	18
Grams fat	2	6
Grams available carbohydrate	23	91

Glycolipids Replacement for Shortening

In spite of efforts to overcome the detrimental effects, the amounts of protein supplements in baked products remains relatively low. Only a small percentage of total edible soy flour and meal produced in the United States is used in yeast leavened baked products, and the level of supplement added to dough formulations is not usually over about 3 to 4%.

Y. Pomeranz and K.F. Finney; U.S. Patent 3,679,433; July 25, 1972; assigned to the U.S. Secretary of Agriculture show how the nutritional quality of bread can be improved by addition of protein concentrates to the dough formulation in combination with glycolipids. In a straight dough process protein concentrates replace nonfat dry milk and natural or synthetic glycolipids replace shortening in the standard formulation to make breads that have greatly increased nutritional value without decreasing any of the over-all quality of baked products as evaluated by loaf volume, crumb grain, and texture. The protein concentrates and glycolipids are added to the formulations before mixing, but they may also be added to the flour at the processing plant before shipment to the bakeries.

Baked products of good quality were produced from dough formulations that contained as little as 0.25 part glycolipids and 4 parts protein concentrate per 100 parts wheat flour. In similar products containing 6 parts glycolipids and 16 parts protein concentrate per 100 parts wheat flour, the quality was exceptional. Flours from several wheat varieties, such as soft winter and spring wheat, were found to give results comparable to the untreated Regional Baking Standard (RBS) wheat flour used in the examples. RBS flour is a representative composite of hard winter wheat from the Great Plains of the United States.

Protein concentrates used in the examples contained from 27 to 90% protein and included soy flour, isolated soy protein, sodium soy proteinate, defatted cottonseed flour, nonfat dry milk, air-fractionated wheat flour, defatted wheat germ, torula yeast, expeller-extracted sesame seed flour, fish protein concentrate, and wheat gluten. Each of these gave breads having acceptable loaf volume, crumb grain, and texture. Soy flour containing 50% protein is preferred because it is most available and inexpensive. Unlike previous studies with soy flours, finely ground protein concentrates are preferred for use in this process.

The natural glycolipids used in the examples were from two separate sources. Free polar lipids rich in glycolipids were obtained by fractionating free lipids (petroleum ether extracted) from RBS flour by silicic acid chromatography and by exhaustive extraction of *Briza spicata* seeds with petroleum ether. Synthetic glycolipids (sucroesters) were obtained commercially and included sucrose monolaurate, sucrose dilaurate, sucrose monopalmitate, sucrose dipalmitate, sucrose monostearate, sucrose distearate, sucrose tristearate, sucrose sesquilaurate, sucrose sesquistearate, and sucrose tallowate.

All of the glycolipids listed above gave improved baked products when compared to similar products containing no lipids. However, free polar wheat flour lipids and the synthetic glycolipids with fewer and shorter chain lengths of fatty acids attached to the sucrose molecule are preferred because they show the greatest improving effect.

Sucroesters generally increase loaf volume of baked products without protein concentrates to the same degree as vegetable shortening, but in protein-enriched baked products the improving effect of sucroesters with short-chain fatty acids and of free polar wheat lipids was generally greater than that of vegetable shortening. This seems to indicate that with nonenriched baked products the effect is related to physical rather than chemical properties of lipids, but when protein concentrates are present, it seems that the chemical properties of the glycolipids interact with the protein to make it functional and compatible with wheat proteins in the production of bread and related products.

With the method described above, baked products may be enriched to high levels with protein concentrates from many sources to give products of equal or superior quality to those with no enrichment. Acceptable high protein products may be made with lean formulations by the method in areas where shortening, sugar, and malt are in short supply. The method also provides a way of controlling protein enrichment in yeast-leavened baked products.

Example 1: Bread was made essentially by the AACC standard straight dough method using untreated Regional Baking Standard wheat flour. The dough formulation consisted of the following.

	Grams
Wheat flour (12.9% protein)	100
Salt	1.5
Yeast	2
Nonfat dry milk (NFDM)	4
Sucrose	6
60° L malt syrup	0.5
Shortening (vegetable)	3
Water	as needed
Potassium bromate	optimum*

*see Finney, K.F., *Cereal Chem.* 23, 96 to 104 (1946) and Finney et al, *Cereal Sci. Today* 8, 166 to 168 (1963).

The above formula was mixed and allowed to ferment for 3 hours at 30°C. Kneading and shaping were performed mechanically. The bread was then baked at 218°C for 24 minutes. Soy flour enriched loaves of bread containing synthetic glycolipid, sucrose tallowate, had as good or better loaf volume, crumb grain, and texture as loaves made with the standard formulation and were considerably better than loaves with no lipids. Crumb softness in bread loaves enriched with soy flour and containing 0.5% synthetic glycolipids was equal to or better than bread baked with 3% commercial vegetable shortening.

Example 2: Vegetable protein enriched breads were baked by the standard method described in Example 1. Eight grams of protein concentrates ranging from 27 to 90%

protein were added to the dough formulation, which also contained either no lipids, 3 grams of shortening, 0.5 gram of natural glycolipids, or 0.5 gram of synthetic glycolipids. Sucrose monolaurate was added to NFDM, air-fractionated flour, and defatted wheat germ enriched formulation in the amount of 1 gram instead of 0.5 gram. It was found that essentially all protein-enriched loaves containing synthetic and natural glycolipids are improved over the control loaves containing no lipids and those containing 3 grams of shortening.

Example 3: Samples of bread were prepared using the standard procedure described in Example 1. The dough formulations were also the same except that in most samples the shortening was replaced by a natural or a synthetic glycolipid. Some of the formulations were made without soy flour for comparative purposes.

Breads enriched with 8 grams of soy flour and containing 0.5 gram of either free polar wheat flour lipids, sucrose monolaurate, sucrose monopalmitate, or sucrose monostearate were subjected to compressibility tests, and the results were compared to those from similarly enriched breads containing either shortening or no lipids. Softness was improved most by free polar lipids. Crumbs of bread baked with 0.5 gram synthetic glycolipids were equal to or softer than those of bread baked with 3 grams of shortening. Bread baked with no lipids was the least soft.

LOW PROTEIN BREAD

The product of *D.A. Wernecke; U.S. Patent 3,567,461; March 2, 1971; assigned to General Mills, Inc.* is a dietetic low protein bread mix containing wheat starch, sugar, shortening, and a structure preservative and is designed to be similar to conventional gluten-containing bread in taste, structure and appearance.

The product is designed to enable people on diets restricted in protein or specific amino acids to benefit from and enjoy a highly palatable, low protein bread. Examples of people who must restrict or modify their protein intake are: (1) those suffering from celiac disease and nontropical sprue, i.e., those who must restrict their intake of certain specific grain proteins; (2) individuals suffering from PKU or phenylketonuriac disorders, i.e., those who must reduce their intake of the amino acid phenylalanine; and (3) those suffering from uremia or severe kidney disorders, the latter class being unable to dispose of urea and other nitrogenous waste products due to kidney malfunction.

The problem of producing a palatable baking mix for bread acceptable to those suffering from uremia is the most difficult. Those suffering from celiac disease may have milk protein and milk protein aids significantly in bread production. This is not so with uremia sufferers because these patients must severely limit their intake of all protein.

Uremia sufferers and others such as heart patients must restrict their intake of sodium and potassium. Because the mix uses yeast there is no sodium or potassium residues from chemical leavening agents. No salt needs to be added to the mix. Salt is normally added to control fermentation. Uncontrolled fermentation in a gluten-containing bread is not desirable because the gluten must be aged to fully attain its elastic properties and if it is not, it will not function properly as a cell structure elasticizer.

Fermentation must therefore be slowed to allow the gluten to attain its desired properties and salt is used to slow this process. Since no gluten is used here, yeast growth may be allowed to proceed normally and salt is not needed. As a matter of fact, the mix has a sodium content of about 38 milligrams Na per 100 grams of mix. (The mix contains 8.8% moisture.) This is compared to a conventional gluten bread which contains 507 milligrams Na per 100 grams of bread (at 35.6% moisture). Also the mix has about 8 milligrams K per 100 grams as opposed to 105 milligrams K per 100 grams using the moisture figures above.

Although attempts have been made in the past to produce low protein bread by eliminating the gluten from wheat flour and by eliminating milk or milk protein from the bread mix, loaves made in this manner have proven to be unacceptable. These low protein loaves have a very close grained structure, lack the typical resiliency found in bread, are crumbly with a cake-like texture and are very dry and highly compact.

The ingredients used for making this mix are sugar, fat, wheat starch, and a structure preservative selected from a class consisting of pectin, pregelatinized waxy maize starch, pregelatinized tapioca starch, pregelatinized wheat starch, methylcellulose and carboxymethylcellulose. The wheat starch used is commercially available and is usually made by subjecting wheat flour to a series of water washing steps to separate the starch from the gluten. An example of this type of wheat starch is the product Paygel (General Mills).

The structure enhancer or structure preservative is designed to perform a function similar to the gluten and other proteins present in conventional bread. Bread normally contains a specific type of wheat gluten noted for its elasticity. It often contains milk protein. The gluten and added milk protein provide elasticity of structure to the cell walls of the foam created when the dough is subjected to a leavening agent.

By providing this strength and resiliency, gas produced during the leavening process is able to expand the cell walls to a point where the bread rises and a soft, springy, relatively open even cell structure characteristic of bread is present on the inside of the loaf. The outside of a normal bread loaf is characterized by a smooth tan or brown color with no apparent fissures on the surface due to uneven cell wall expansion. Therefore, this smoothness is also enhanced by the presence of protein.

It was found that by mixing a low protein wheat starch, sugar and shortening with one of the structural enhancing agents, it is possible to produce a low protein bread which has an even cell structure, a smooth brownish crust, a fair degree of resiliency and elasticity, an open crumb structure, and a moist white inner surface thereby producing a loaf highly similar in every respect to normally available gluten-containing wheat bread. Other related, binding agents such as pregelatinized corn, potato, and arrowroot starch and vegetable gums such as carrageenan do not provide the desired results. Some of these agents settle to the bottom of the loaf, others do not retain sufficient moisture or else they produce a loaf which is unappetizing in color, while others do not provide sufficient strength for proper rising of the bread or result in uneven expansion leaving fissures on the crust and/or throughout the interior of the loaf.

The importance of this product cannot be overemphasized. Most people who suffer from disorders requiring reduction or modification of protein intake are on extremely limited diets. For example, those suffering from uremia are restricted to as low as 20 grams of protein per day. Bread is a significant part of the American adaptation of this diet and low protein bread based on a wheat starch-containing mixture is strongly recommended.

However, the disadvantages of this bread from a palatability standpoint have been previously pointed out. The bread produced from the mix has approximately less than 10%, and preferably less than 6%, of the protein in the conventional enriched white bread (about 3.8 grams for a 1.75 pound loaf of this bread) and is still tasty and bread-like. A conventional 1.75 pound loaf of white bread contains approximately 8 to 9% protein.

Examples of the product follow. The procedure for preparing the mix, the proofing and the baking conditions were the same for all. First, a premix was made consisting of 86.69% wheat starch (0.3% protein), 7.08% sugar (granulated), and 6.23% shortening. To this premix was added the structure enhancer (the amounts of the structure enhancer will be given as percent of the total dry mix excluding yeast). Preferably, the resulting mix has a maximum wheat gluten content of about 0.3% by weight of the mix. The mix was added to an aqueous yeast solution, the yeast being present at a level of 4% by weight of the starch and the amount of water sufficient to make a dough, here an amount about equal in weight to the mix. The mix and the yeast solution were then blended until homogeneous at low speeds in a conventional mixer. The bowl was scraped with a rubber blade and blending was then continued for one-half minute at medium speed. The dough was then poured into greased pans and allowed to proof at a relative humidity of 80 to 85% until the pans, which were initially half-full, were full to the top with dough. This proofing generally takes about 30 minutes. The dough was then baked in a conventional oven at 400°F for 30 minutes. Of the total protein content, less than one-half was due to the gluten in the wheat starch.

Example: A number of loaves were prepared by the procedure outlined above. Each loaf contained one of the structure enhancers at the levels indicated below.

Structure Enhancer	Percent
Pectin (highly purified, unmodified)	0.5, 1.0
Waxy maize (pregelatinized)	*2 - 10
Tapioca starch (pregelatinized)	*2 - 10
Wheat starch (pregelatinized)	5
Methylcellulose	0.5, 2
Carboxymethylcellulose	0.25, 2

*In 1% gradations

When waxy maize or tapioca was used, it provided the best loaf at a 5% level and outside of the shiny appearance of the inside crumb structure was adjudged identical with the 5% wheat starch sample. Moisture levels were taken after 20 hours of storage at ambient temperature on the 5% tapioca starch loaf and the loaf was found to contain 43.5% moisture. At a level of 10% the loaf developed a more open crumb structure and was noticeably more moist than the loaf containing the 2% level. This was true with all of the loaves as the level of structure enhancer increased.

The major difference between the pectin, carboxymethylcellulose and methylcellulose loaves in this example were crumb structure and moisture following the pattern of the caveat above. Both cellulose derivatives and pectin prove to be useful at lower levels. Within the ranges given for the particular mix formulation the 10% waxy maize and tapioca showed more of a wide open crumb structure than loaves containing either 2% of the cellulose derivatives or 1% pectin. Therefore an increase of up to about 50% of the amount in the above table of either the cellulose derivatives or pectin is possible if these structural enhancers are used and an open crumb structure is desired.

All of the samples showed excellent moisture retention, i.e., 43.5% by weight of the loaf for 5% tapioca after 20 hours at room temperature, good leavening and crumb structure, desirable color both inside and outside the loaf. The pregelatinized starch containing samples produced loaves with the best crust color. The loaf containing pectin produced a crust color less brown than the loaves containing pregelatinized starch but better than the loaves containing cellulose derivatives, although the crust color on the latter loaves was still deemed acceptable.

The pectin-containing mix also produced a loaf which rose higher in the middle than any of the other loaves thereby producing a loaf more nearly approximating the exact configuration of a conventional gluten-containing wheat bread loaf. Pectin at a level of 0.5% and the methylcellulose present at levels as low as 0.5% produced binding characteristics superior to a 2% level of tapioca or waxy maize. For a relatively dense loaf which has a uniform crumb structure but is still not crumbly, a reduction up to about 50% in the amounts of these enhancers is possible.

LOW CARBOHYDRATE BREAD

Gluten Replacement for Starch

Some dietetic requirements for nutritious breads and buns are that little or no starch be used for diabetics and reducing diets, or high protein, low starch, and normal or high fat content for certain slimming diets. Other requirements may be a high content of unsaturated fats or essential fatty acids for atherosclarosis and the provision of vitamins such as the vitamin B group, vitamin E, niacin, with the minerals iron and calcium. Again the provision of bulk in the diet is often required and best met with cellulose compounds either synthetic or natural such as bran which has laxative properties and is a good source of iron and niacin. Most of these requirements can be supplied by the bran and germ of cereals and the protein of milk and such oil seeds as soy and groundnut and it has been found that it is possible to combine these materials, suitably treated, with wheat protein so that they replace starch in the manner described.

Replacement of starch with combinations of gluten with wheat, maize germ and protein materials is described by *C.H. Evans; U.S. Patent 3,348,951; October 24, 1967; assigned to Spillers Limited, England.* The protein of wheat may be used either as wet gluten or dried gluten powder provided this is "vital" or retains its properties of elasticity. Wet gluten, which is a by-product of the manufacture of wheat starch, normally contains dry matter from 50 to 70% or more wheat protein and about 5% fatty substances and a certain amount of unremoved starch. The fatty substances are present in the dry powder and are an essential part of the protein matrix. Wet gluten may be used if this is available or dried vital gluten may be used. The latter differs only from the former in the removal of water and wet gluten normally contains two-thirds of its weight of water.

If dry gluten is used, time should be allowed for its hydration with water whether the dry gluten is reconstituted by itself or as part of a dry premix which may be convenient. In considering the amount of dry gluten to use, account must be taken of the quality and for the purpose of the examples the protein content should not be less than 70%. The ratio of protein to other substance is best described with reference to the latter product as wet gluten is somewhat variable but normally contains about one-third of its weight of dry solids, mainly protein, but with some lipids and other substances. The quality of the gluten, i.e., strong and elastic, or soft and extensible, depends on the wheat variety from which the flour was milled and this quality must also be taken into account as, in fact, is the case with ordinary flour.

In wheat flour this ratio of starch to protein varies very considerably and may be in the range of from 10:1 to 5:1. It was found that for satisfactory results with starch substitute the lower ratio is more appropriate but must be varied both with the percentage of protein in wheat gluten and starch substitute substances.

Example 1:

	Parts by Weight
Wet gluten (or equivalent weight of dry gluten)	200 - 250
Finely ground bran containing 5% by weight of lecithin	400
Salt	2
Yeast	20
Water (varied to suit dough and increased by about twice the weight of dry gluten if dry gluten is used)	300

Wet gluten can be regarded as containing two-thirds volume water so that equivalent weight of dry gluten would be 66 to 83 parts by weight. The total weight of dry ingredients would be 494 to 511 parts and the percentage of dry gluten from 13.36 to 16.24%.

The wet gluten is placed in a bakery mixer of a high speed type and the bran added slowly until the gluten begins to mix, the yeast is then added in a part of the water. As the mixing proceeds, more bran and water are added until the dough clears. The salt and lecithin are conveniently added to the bran beforehand. The lecithin may be commercial lecithin as normally supplied to the baking industry and as such is usually soybean lecithin. The basic mixing is improved by the addition of the lecithin and may be still further improved by the addition of 50 parts by weight of an oil, such as cottonseed, soy or maize. The amount of wet gluten can be varied beyond the above limits, which cover natural variations in quality, if a different type of product is required. Less gluten will quickly lead to a brittle dough of little cohesion which cannot be handled but more gluten will give increased volume until the toughness makes the product approach a gluten roll nature. If the rather bitter taste of the bran is undesirable, the bran may be given any extractive treatment required before grinding.

There is sufficient starch adhering to the bran to allow a limited fermentation with yeast. If, therefore, wet gluten is used or gluten is hydrated with water, the dough should be mixed until it resembles a normal yeast and flour dough and then, after a brief recovery period, the dough is sealed off, tinned and after proving at about 100°F for a suitable period, normally about 45 minutes, the bread is baked in the normal way. If, for convenience, the dry gluten is incorporated with the other ingredients as a premix, then, on adding the water and making the dough, time must be allowed for the gluten to hydrate. Normally 30 minutes is adequate after which, if the mixing has been correct, the dough also resembles normal dough and can be handled in exactly the same way.

The product resembles brown bread and has quite good eating qualities and keeps very well if baked in cellophane. The dietetic value lies in the low calorie content due to the indigestible portion of the bran and also the bulk and laxative properties of the bran and also to the very low starch content.

Example 2: For dietetic regimens where a baked product with a minimum starch content is not quite so essential, for example, for diabetics for whom a small percentage of starch may be permissible and for reducing diets, a combination of soy flour and wheat flour is used to produce a baked product which still has a greatly reduced starch content as compared with conventional white or whole meal bread.

	Parts by Weight
A	
Full fat soy flour	300
Wheat flour	150
Lecithin	15
Calcium or sodium caseinate	30
Salt	12
B	
Dry gluten	120
Water	480 (approx)
C	
Yeast	40

The total weight of dry ingredients is 667 parts so that the proportion of dry gluten is $^{120}/_{667}$ or 17.99%. The ingredients listed under A are mixed together in dry state to form a premix. The gluten is hydrated for at least one hour and placed in a bakery mixer of the high speed type. The premix together with the yeast is added and the dough formed. There is a certain amount of sugar in soy flour to allow a limited fermentation with use. Alternatively, if the dried gluten is added to the premix, an appropriate quantity of extra water will be needed and then, on adding the water and making the dough, time must be allowed for the gluten to hydrate. Normally a period of about 30 minutes is adequate. The dough is divided or scaled off and yeast fermentation of about 45 minutes is then allowed prior to baking in the usual manner. Pastry and biscuits can be made from this formula with the omission of the yeast and the additon of fat in conventional manner.

Jerusalem Artichoke Flour

Many persons are afflicted with various degrees of difficulty in carbohydrate metabolism. Such difficulties are often accompanied by hyperglycemia (commonly associated with the name "diabetes" or "sugar diabetes" and characterized by a high sugar content of the blood of the victim) or hypoglycemia (low blood sugar). However, with early medical diagnosis and treatment, the difficulties attendant upon these disorders can generally be kept under control. The measures for control include continuing periodic medication for those who are seriously afflicted, and careful control of diet.

For either the diabetic or the hypoglycemic patient, the most important dietary requirement is to avoid foods which cause rapid changes in blood sugar. This involves very careful selection of foods because many commonly used foods either contain simple sugars which are directly and immediately absorbed by the body, or they contain other carbohydrates which are very rapidly changed by the body into simple sugars and immediately absorbed. Once absorbed, these sugars are immediately available to the blood. Normal individuals can stand the rapid change in the available sugar, but in diabetic or hypoglycemic patients, very serious trauma can result.

Unfortunately, wheat starch, which is one of the main constituents of ordinary bread, is one of the carbohydrates which is rapidly converted to simple sugars by the body. The conversion of wheat starch to simple sugars in the body is so rapid that ordinary breads, virtually all of which contain a high concentration of wheat starch, cannot be tolerated by those having carbohydrate metabolism difficulties. However, bread is one of the most popular foods, and it is a serious problem with any special diet to provide enough foods which are interesting and palatable and which have adequate nutritional values.

One of the important features which makes bread palatable and desirable as a food is the leavening which is usually provided by yeast. While other leavening agents can be used, yeast is preferred, particularly because it accomplishes various desirable purposes in addition to the leavening or increase in porosity of the bread dough. For instance, the fermentation of the yeast renders the gluten of the flour more elastic and changes some of the insoluble proteins of the gluten into soluble forms. The flavor, color, and texture of the bread are all controlled, in large measure, by the action of the yeast.

The growth of the yeast plant in the fermentation accompanying the leavening of the bread dough requires the presence of a number of different substances, most of which are usually found in wheat starch. The most important basic substances which are consumed in the yeast fermentation are the simple or monosaccharide sugars. In order to provide this constituent in the bread dough for the growth of the yeast, sugar must be added to the dough mix, or reliance must be placed upon conversion of wheat starch to sugar by a constituent of the starch called diastase. This conversion of the starch to sugar can be enhanced by the addition of malt to the dough mix, since the malt contains a high concentration of diastase. It is common to add sugar, or malt, or both, to bread mixes to satisfy this requirement.

Unfortunately, the monosaccharide sugars which must be present, or produced, in the bread dough for consumption by the yeast in the leavening process are the very sugars which are rapidly absorbed by the body of the consumer and which therefore provide rapid changes in blood sugar. The fermentation action of the yeast in the leavening process is thus similar in many ways to the consumption of sugar by the body of the consumer. Unfortunately, when the sugars and sugar producing materials are available in the bread mix for the consumption by the yeast, they are usually available in a high concentration in the final bread product.

A bread mix with a minimum of available carbohydrates but sufficient for the leavening action of the yeast is provided by *H.J. Schaefer and J.W. Tintera; U.S. Patent 3,497,360; February 24, 1970.* The main constituents of white wheat flour ("patent" flour) are gluten and wheat starch. The gluten is primarily a protein material which has coherent and adherent properties which tend to hold together the bread dough and the finished loaf. The wheat starch is primarily carbohydrate material which provides the carbohydrate nutrients for the action of the yeast and for the body of the consumer. These wheat flour constituents are both important in providing leavened bread which has the appearance and flavor which is appealing to most consumers and which is expected by them.

It is for this reason that most "fancy" breads which contain other flours and constituents, such as those made from rye, corn, soybeans, oats, and others, nevertheless contain a very high proportion of wheat flour components. Thus, most of these fancy breads are basically wheat flour breads with very small fractions of other constituents added for coloring and flavoring properties. These other constituents are present only in token amounts which usually do not substantially change the nutritional values or effect of the bread.

The process provides a method for producing a yeast bread including the steps of mixing a bread dough having a minimum of wheat starch in the order of 8% of the dry ingredients, and having a high concentration of yeast in the order of 5% of the dry ingredients. The dough is proofed only once for a limited period sufficient to substantially double the dough volume and then immediately baked. The composition of the bread dough exclusive of water may include from 8 to 14% wheat starch, from 5.7 to 7.5% Jerusalem artichoke flour, from 9 to 16% wheat protein, sufficient yeast to provide a yeast to wheat starch ratio of at least 1:3 and the remainder of the dry constituents including nonwheat flours such as oat flour and soy flour.

When the wheat starch is present in this low proportion, it is very substantially consumed by the yeast in the leavening process so that there is very little wheat starch remaining in the leavened loaf of bread after baking. The wheat starch may be present initially in a proportion up to approximately 14% by weight. However, it is kept to an absolute minimum which appears to be approximately 8%. In order to make the bread dough rise sufficiently with so little wheat starch available to "feed" the yeast, a high ratio of yeast to wheat starch is used. This ratio is preferably from one-third to two-thirds. This refers to commercially available compressed yeast.

In regular bread making processes, it has been common, not only to add sugar as a yeast food, but also to add various other materials, such as malts, which are effective in converting wheat starch to sugar so that it may be directly effective as a food for the yeast. In this process it is particularly desirable to avoid any conditions or any constituents which may cause the presence of residual sugar in the finished bread product. Malts, and other so-called yeast foods which promote the conversion of wheat starch to sugar are likely to do just this. These materials enhance the action of the yeast simply by accelerating the conversion of wheat starch to sugar so that there is plenty of sugar available as a direct food for the yeast.

Therefore the use of malts and other yeast food materials which enhance the conversion of wheat starch to sugar is avoided, and instead, a high proportion of yeast is employed in order to obtain the desired leavening action by consuming virtually all of the sugar which is available from the normal conversion of wheat starch to sugar.

Stated another way, when malts, and other materials of a similar nature are added to the composition, the objective is generally to obtain sufficient leavening of the bread with a smaller amount of initial yeast added to the mix. In those instances, with the malt present, the conversion of wheat starch to sugar is likely to outrun the consumption of that sugar by the yeast in the leavening process so as to result in a higher proportion of residual sugar in the finished product.

It has been found that in order to obtain a satisfactorily light loaf with so little carbohydrate available to sustain the leavening action of the yeast, the leavening must be carried out quickly, and the bread promptly baked after the fast leavening process before the gases have a chance to escape and the bread has a chance to fall. Generally, the proofing time is only about 30 minutes or until the bread dough mix has approximately doubled in volume.

Example: All of the ingredients listed on the following page except the water are mixed together in a dry blend. Immediately after dry blending, the water is added at lukewarm temperature and mixed in for about 7 or 8 minutes. A standard 4-speed bread mixing machine is used. It is run on low speed for about 6 minutes and then at second speed for about 2 minutes. After mixing, the dough is scaled into 1-pound, 2-ounce loaf sizes and then proofed for about 30 minutes in steam so that the volume of the scaled dough approximately doubles to fill the pan. The bread is then baked at 360°F for a maximum of about 55 minutes. The baking is carried out without steam. With the quantities given below, 36 loaves of bread are obtained which are very palatable and appetizing in appearance and which can be tolerated by individuals having serious carbohydrate metabolism difficulties.

	Percent of Dry Ingredients
Wheat flour (gluten), 4 lb	17.1
Wheat flour (vital gluten), 2 lb	8.5
Oat flour, 12 lb	51.5
Soy flour, 2 lb	8.5
Jerusalem artichoke flour, 1½ lb	6.4
Salt, 9 oz	2.4
Sodium propionate, ¾ oz	0.3
Yeast, 1 lb, 4 oz	5.3
Water, 9 qt	

According to the United States Department of Agriculture Handbook No. 8 entitled *Composition of Foods* and revised December 1963, the percentages by weight of the protein, fat and carbohydrate contents of the various wheat flours and the other major constituents of the above example, which are available to the body, are as follows.

	Percent protein	Percent fat	Percent carbohydrate
Wheat flour (standard patent)	12	1	75
Wheat flour (gluten)	41	2	47
Wheat flour (vital gluten) (from another source)	96	1	1
Oat flour	14	7	68
Soy flour	37	20	30
Jerusalem artichoke flour	2	0	0*

* Handbook states 16.7% carbohydrate—of doubtful availability.

Applying the food value analysis percentages given above to the composition of the example, the percentages of total dry constituents representing proteins, fats, and carbohydrates derived from the various major sources in the example can be computed as follows.

	Percentages of total dry constituents		
	Protein	Fat	Carbohydrate
Wheat flour (gluten)	7	3.4	8.0
Wheat flour (vital gluten)	8.2		
Total wheat flour food constituents	15.2	3.4	8.0
Oat flour	7.2	3.6	35.0
Soy flour	3.2	1.7	2.6
Jerusalem artichoke flour	0.1	0	0
Total	25.7	8.7	45.6

The Jerusalem artichoke flour is a dehydrated product which consists of pulverized tuberous roots of the Jerusalem artichoke plant. This material is commercially available under the name American Jerusalem Artichoke Flour. It is well known for its high content of a nonnutritive sweetener called inulin.

The soy flour referred to in the above example is a flour produced from soybeans. Soybean flours are regularly produced with various amounts of fat content. Since fats are quite acceptable to diabetics and hypoglycemics, the full fat flour is preferred and is intended to be used in the above example. Furthermore, shortening may be added in addition to the high fat content of the soy flour. The high fat content improves the palatability and the nutritional value, and accordingly is preferred. However, it is possible to produce an acceptable product with soy flours having reduced fat content and without shortening.

Added Rice and Soybean Hulls

Among the numerous nonassimilated carbohydrate materials proposed in food formulations such as low calorie bread is purified cellulose that has been subjected to acid hydrolysis. The material is believed to be identical with the product now being sold under the trademark Avicel (FMC Corporation). Avicel, an alpha-cellulose material, has been tried experimentally in making bread but is quite expensive. In addition, Avicel has a tremendous absorbing capacity for water. If enough is used to produce a 25% calorie reduction, a commercially satisfactory finished product cannot be produced using standard dough formulations or any dough formula. Groundnut protein or wheat bran can be used to partially or completely replace starch. Groundnut protein is expensive and wheat bran contains a good deal of nutritive material. Accordingly, its calorie lowering capabilities are limited.

Other nonnutritive food substances such as cellulose, rice hulls, oat hulls, corn cobs, bean hulls, etc. have been proposed for use separately but not in conjunction with one another in making biscuits or crackers. Biscuits and crackers are, however, hard and usually friable. Accordingly, performance requirements are relatively easy to meet. Moreover, gas development and expansion of the dough is a relatively minor factor. In the case of bread, on the other hand, a smooth, light, and resilient texture is required. The volume must be high and the grain structure even. In addition, each loaf must retain just the proper resiliency and gas holding ability while expanding during proofing and baking.

K. Tsantir and R.A. Gorman; U.S. Patent 3,767,423; October 23, 1973; assigned to The Pillsbury Company use finely ground rice hull and soybean hulls as replacements for regular flour to produce low calorie bread. The major ingredients present in the compositions are flour, finely ground soybean and rice hulls, sugar, water, and leavening. A great many minor ingredients can be used to provide optimum performance and to impart special characteristics. These include flavors, egg yolk for tenderizing the dough, emulsifiers which produce tenderness, gluten for strengthening the dough and making it more resilient, yeast food, an antimycotic agent such as sodium diacetate and color among others.

Color, flavor and egg yolk are used in conventional amounts, emulsifiers such as a monoglyceride bread emulsifier is used in the amount of about 4 parts for each 100 parts of flour and gluten is used in amounts from about 2.5 to 3.5 parts for each 10 parts of flour. In a typical formulation including 100 parts of flour, the minor ingredients might consist of 8 parts yeast, 1 part yeast food, 5 parts salt, 4 parts of monoglyceride bread emulsifier, 1.5 parts sodium diacetate and 0.4 part color. In general, the use of increasing amounts of emulsifier reduces the amount of gluten required.

It has been found that from about 0.9 to 3.6 parts sugar can be used for each 10 parts of flour but preferably from about 1.6 to 2.0 parts is used. If too little sugar is used, the yeast action may be inhibited and the bread will suffer from poor volume. If too much is used, the taste may be too sweet and the caloric content excessive. The function of the sugar is primarily for yeast action and flavoring and thus allows more hulls to be added.

There is, of course, no lower limit in the amount of hulls that can be used since any amount of added hulls has a small but measurable effect in reducing the caloric content of the bread. However, based upon dietary requirements as well as nutritional requirements that have been carefully gathered from nutritional experts consulted concerning the formulation, caloric content reductions of over 10% and preferably above 25% are preferred. This process uses from about 1.2 to 4.5 parts hulls for a bread having a 25% calorie reduction and up to about 20 parts or more but preferably up to about 12 parts hulls for each 10 parts of flour. The greatest effect noticed by the addition of hulls is the reduction in calorie content. The addition of hulls does, however, reduce favorable volume, grain, texture and hedonic scores and accordingly, no greater amount is used than is necessary to provide the requisite calorie reduction.

It is preferred that both types of hulls be present. If 100% rice hulls are used, performance with regard to bread volume and grain is usually very good but the bread sometimes exhibits a gritty character in the mouth. If 100% soybean hulls are used, the bread may have a beany flavor. Additionally, the dough absorbs and retains a higher percentage of water and requires more gluten for a given quantity of flour. Furthermore, the use of soybean hulls has a tendency to depress the volume of the bread. When a higher calorie reduction is desired, e.g., 20% or more, the ratio of rice hulls to soy hulls should be from about 20%:80% to 80%:20%, respectively. In general, the best results are usually obtained with about equal parts of rice hulls and soybean hulls. The optimum ratio of rice hulls to rice hulls plus soy hulls is between about 0.1 and 0.85.

The rice and soybean hulls are dried, cleaned and ground into a size similar to that of flour, for example, from about 5 to 160 microns. They can be ground to an average particle size below about 20 microns and are typically ground to an average particle size of about 5 to 25 microns in diameter. The particles preferably have an average diameter below 10 microns.

An important aspect of the process is the fact that it can be compounded using conventional methods, for example, the straight dough or sponge process. For every 100 parts of solids in the dough, there is usually about 65 to 85 parts and preferably from 70 to 80 parts of water in the dough before baking. The final moisture content of the bread is about the same as a conventional bread. The amount of water used in the composition before baking has a great effect on achieving a favorable taste panel test. If too little is used, the finished product tends to be hard, dry or have poor volume. If too much is used, the finished product may be too moist or have a poor structure, volume or hedonic score.

The variables are defined as follows in parts by weight. H = [Hulls/(Hulls + Gluten + Flour)] G = [Gluten/(Hulls + Gluten + Flour)] While diet bread currently being sold commercially offers a calorie reduction compared with conventional white bread of from about 6 to 8% at the most, this process provides a calorie reduction well in excess of this amount and is especially useful at levels of calorie reduction of over 10% and in the case of many typical formulations of over 25%.

With regard to the composition of the nonnutritive hulls fraction, it was found that it is possible to produce a bread which is satisfactory from the standpoint of performance criteria without rice hulls being present provided the H ratio is below 0.16 but under these circumstances little calorie reduction would be achieved. As the proportion of rice to soy is increased, it can be seen that good performance can be obtained with higher levels of hulls being present and accordingly greater calorie reduction can be accomplished. When the ratio of rice to rice and soy is increased above about 0.5 or 0.6, the amount of hulls that can be used and yet obtain a satisfactory bread decreases.

Also, relatively high levels of calorie reduction, for example, over 25%, can be obtained with gluten (G) present at its best ratio which is about 0.14 to 0.18 with water in its optimum range of from about 72 to 79% of the dry mix. Of course, it is also possible to achieve relatively low levels of calorie reduction, for example, 8%, in which case the rice to soy and rice ratio can be varied much more widely, for example, from 0.1 to 0.9.

With reference again to the effect of changing the ratio of the several major ingredients, it was found that increasing the H value, i.e., [H/(H + G + F)], decreased the over-all appeal of the bread as indicated by the hedonic scores achieved by a standard taste panel. It is also detrimental to aroma unless only soy hulls are present in which case increasing hull content does not reduce the aroma preference. Increasing the hull content also reduces flavor and decreases volume, reduces symmetry, harms the grain especially when the ratio of [H/(H + G + F)] is between 0.3 to 0.5 but a softer texture is obtained with larger [H/(H + G + F)] ratios.

From the standpoint of over-all appeal the middle level of gluten is better liked than either extreme. The best ratio of gluten to hulls plus gluten plus flour was from about 0.14 to 0.28 with a peak at 0.17. It was also found that when gluten is present, the deterioration of break and shred is entirely a function of the (H) ratio and is not dependent on the rice to soy ratio. It was found to help the grain of the bread at a water level of about 75% of the dry mix and to produce a more uniform grain. As mentioned above, gluten is not essential at the lower calorie reduction level.

The over-all acceptance of the bread is less when using the formula with all soy hulls or all rice hulls and particularly when using all rice hulls, especially when the formula has a low level of sugar and a high level of gluten. It was found that high levels of sugar allow more rice hulls to be used without losing preference. While rice hulls are perceived to be more gritty than soy hulls, bread will have a much larger volume with rice hulls than with soy hulls, especially at low water levels. On the other hand, soy hulls have a better flavor quality than rice hulls but rice hulls tend to harm the grain less than the soy hulls.

At the middle level of water, about 75% of the dry mix, high and low sugar levels are better than the middle levels of sugar. The sugar is especially valuable in allowing more rice hulls to be used without losing preference but sugar if present in the amount of about 8% of the dry mix decreases volume and in increasing amounts reduces the symmetry. Relatively high amounts of sugar also are detrimental to the grain description and the grain uniformity as well as giving the bread too soft a texture if large amounts are used but will make the bread more elastic if water is present in the amount of from about 75 to 80%.

As mentioned above, the best water level is about 70 to 80% of the dry mix. Using 85% water was found detrimental to the over-all acceptance of the bread and a poorer aroma is obtained with a high water level but at low water levels, for example, 65%, a better symmetry is obtained with rice hulls than with soy hulls. It was also found that when water is between 75 to 85% on the basis of the dry mix, a high sugar level, e.g., 12% is detrimental to the break and shred characteristics. A high water level also tends to harm the grain but makes the finished product more elastic.

Iron Enriched Flour

Ferrous sulfate has been used to a limited extent for iron fortification of many food products including bread and other bakery goods; but to fortify flour itself, the practice has been to use reduced iron, sodium iron pyrophosphate or ferric orthophosphate, all of which are insoluble. The bioavailability of the iron in these insoluble agents is questionable and, furthermore, even higher levels of enrichment are now considered desirable.

The bioavailability of reduced iron, the fortifying agent now most commonly used for flour, is not only questionable, but this agent causes other problems as well. For instance, because of its density, it tends to separate from mixtures with the much lighter flour and it also activates the tramp metal detectors used to inspect packaged flour.

While the bioavailability of iron in soluble ferrous sulfate is known to be much greater, the simple hydrates of this salt have been found unsuitable for use in flour at the desired concentrations. The heptahydrate of ferrous sulfate is customarily produced in the form of coarse crystals or granules, and attempts to reduce them to a fineness suitable for mixing with flour leads to severe caking problems. The monohydrate is physically much better suited to the purpose but it accelerates deterioration of the flour as evidenced by discoloration and a rancid odor. For these and other reasons there is a need for better iron fortifying agents, especially for use in flour.

A.D. Bell; U.S. Patent 3,803,292; April 9, 1974; assigned to Mallinckrodt Chemical Works describes improved compositions to fortify flour and other food products with iron. The ferrous sulfate compositions have a density closely approximating that of flour, and stable, intimate mixtures of flour with such compositions are readily prepared. The resulting enriched flour is characterized by the increased bioavailability of the contained iron compared with iron enriched flours previously known. Moreover, the enriched flour has a storage life comparable to that of the pure flour itself. Therefore, the further addition of stabilizers or other foreign additives which are of no nutritional value and which may be objectionable on other grounds is unnecessary.

Fortified flour contains about 0.003% iron (Fe), of which about 0.001% is naturally present in flour. It has been proposed to increase the level to about 0.01%. A nutritionally acceptable level of added iron in fortified flour would therefore fall within the range of approximately 0.002 up to 0.025% Fe, the preferred level based upon recommendations being approximately 0.01% Fe.

The improved iron-enriching agents may be prepared using conventional spray drying apparatus and methods. The particle size of the ferrous sulfate composition also appears to have some effect on the stability of the enriched flour and it may be varied in the usual manner. Optimum stability has been observed when the ferrous sulfate has a particle size within the range of 4 to about 25 microns as measured using a Fischer Sub-Sieve Sizer. Generally speaking, the larger the particle size the greater will be the stability of the fortified flour. It is also desirable to adjust the size of the spheres so that their bulk density approximates that of flour as nearly as possible, i.e., approximately 5.3 pounds per gallon for loose flour.

In the spray drying process each droplet formed at the edge of the atomizer wheel goes through the following heating and cooling cycle. As the temperature of the droplet increases, water is removed and the concentration increases to the point at which ferrous sulfate monohydrate starts to crystallize. Crystallization of the monohydrate continues until the temperature of the particle drops below the melting point of ferrous sulfate heptahydrate (64°C). At this point, any remaining water in the droplet crystallizes as the ferrous sulfate heptahydrate on the surface of the dry particle. This heating and cooling cycle, described above, for the case where the feed stock is a pure solution, is also true for the case where a ferrous sulfate monohydrate slurry is used as a feed stock to the dryer.

Alternatively, a hot slurry of ferrous sulfate monohydrate, which may be prepared by conventional procedures, is first centrifuged. The wet cake so obtained is then immediately broken up to less than 100 mesh and cooled to room temperature. The resulting free-flowing product consists of ferrous sulfate monohydrate coated with ferrous sulfate heptahydrate and has a light color similar to that of the spray dried product.

Example 1: Spray dried ferrous sulfate is prepared using conventional spray drying equipment and techniques. For example, using a Nichols-Niro Utility Spray Drier No. 6903 equipped with a high vane wheel atomizer, the input temperature was 400°C and the output temperature was 150°C. The feed material was a solution of ferrous sulfate having a density of 38° to 42° Bé. The feed material was added to the drier at such a rate as to produce 44 pounds of dried product per hour. With the atomizer adjusted to a speed setting of 3.0 to 3.5, the particle size distribution of the product was as follows.

U.S. standard sieve screen	Percent by weight	Fischer sub-sieve sizer, microns
Retained on:		
100 mesh	2.8	(1)
140 mesh	9.6	>11
200 mesh	30.9	10.2
230 mesh	19.6	7.5
270 mesh	7.4	8.0
325 mesh	8.5	8.25
Passing 325 mesh	21.5	6.7
Total	100.3	

[1] Not measured.

The product contained 31.96% Fe^{++} and 11.86% water, both by weight. The bulk density was approximately 4.3 pounds per gallon loose and 5.2 pounds per gallon tapped.

Example 2: Procedure for Testing the Stability of Iron Enriched Flour – The following procedure was used to test the stability of flour enriched with various iron-fortifying agents. For these tests fresh flour was obtained from a mill immediately after milling and was stored at 0°C in double thickness polyethylene bags. An amount of the iron-fortifying agent to be tested equivalent to 11 milligrams Fe was added to 100 grams of flour and the sample was blended for 1 hour on a roller-mill. Each sample was then divided into two equal portions and stored in screw cap amber bottles, one at 0°C and the other at 50°C. The samples were examined for color and aroma after 1, 3, 5, 8, 11, 15, 19, 24 and 29 days of storage. For control purposes a sample of unenriched flour was included in both series of tests. Each test sample stored at 50°C was compared with a control sample of the same flour which had been stored at 0°C as well as with a control sample of unenriched flour which had similarly been stored at 50°C. A test sample was considered to have failed as soon as any off-color or off-aroma was detected by a trained observer.

Example 3: Results of Stability Test – The stability of flour fortified with the spray dried ferrous sulfate was compared with flour fortified with ordinary commercial dried ferrous sulfate, USP, which is in the form of a crystalline powder, and with reduced iron powder. The spray dried ferrous sulfate was tested both as it comes from the chamber of the spray drier and after being sieved to remove particles smaller than 230 mesh. Likewise, the commercial ferrous sulfate was tested as is and after it had been sieved to remove particles larger than 100 mesh and smaller than 200 mesh. The reduced iron powder was of a USP grade that is used for enriching flour. The results obtained were as follows.

Fortifying agent	Odor	Color
	Failure of flour sample at—	
Ferrous sulfate dried USP (crystalline powder "as is").	4 days	11 days.
Ferrous sulfate dried USP sieved (−100+200 mesh).	do	Do.
Ferrous sulfate spray-dried microspheres (from chamber as is).	25 days	Approx. 30 days.
Ferrous sulfate spray-dried microspheres sieved (+230 mesh).	Approx. 30 days.	Do.
Reduced iron	do	Do.
None	do	Do.

In other tests the bioavailability of iron from spray dried ferrous sulfate was found to be as great or greater than from ordinary commercial grades of ferrous sulfate crystalline powder.

RICE AND POTATO PRODUCTS

RICE ENRICHMENT

Nutrient Absorption Process

A special process is used by *T. Satake; U.S. Patent 3,674,514; July 4, 1972* to prepare parboiled rice with improved nutritional characteristics. Nutritious substances are allowed to penetrate inwardly into the endosperm in the process which requires cracking the brown rice in advance of the moistening step. The brown rice treated in this manner is moistened in a shorter time.

More specifically, a system is used where the cracks of brown rice enwrapped in chaffs are produced by subjecting the rice to a blast heated to a high temperature or chilled and by inducing the sudden change of temperature in the rice.

Figure 3.1a is a view schematically showing the devices employed in a system, in which cracks are produced in rough rice while it vertically flows, and Figure 3.1b is a view similar in which cracks are produced in the rough rice while it is horizontally agitated.

In Figure 3.1a, an elevator **1** for loading rough rice is provided with an outlet opening **2** which connects to an upper inlet opening of a device **3** for providing rough rice with cracks, in which the rice flows vertically. Inside the device **3**, is the blast supplying outlet of a blower **4** which supplies into the device through the outlet a blast which may either be heated to a high temperature or chilled. At the bottom of the device **3**, there is provided an outlet opening **5**, from which the rough rice subjected to the blast comes out. When a heated blast is employed, the temperature thereof may preferably be around 150°C. Due to the heated or chilled blast applied onto the rough rice while it is descending through the device **3**, the temperature of the rice suddenly changes and the brown rice enwrapped in chaffs cracks.

The rough rice thus cracked of its brown rice and brought out from the device through its outlet opening **5** is received by a belt conveyer **6** which is provided below the opening. The rice carried by the conveyer is released therefrom at one of its horizontal ends and is supplied to the inlet opening **8** of a loading elevator **7**, the upper outlet opening **9** of which opens to the upper inlet openings of a plurality of tanks **10**, in which the rice is moistened.

These tanks may be kept under pressure, if required. Outlet openings of a water supply

FIGURE 3.1: SYSTEM FOR PREPARING PARBOILED ROUGH RICE

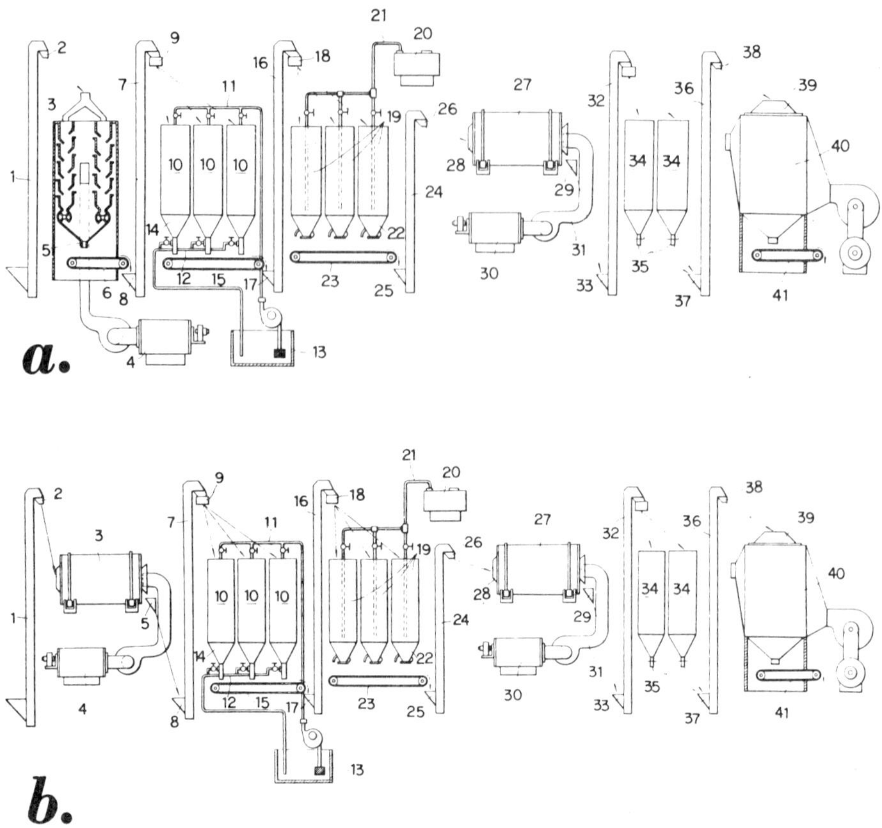

Source: T. Satake; U.S. Patent 3,674,514; July 4, 1972

tube **11** are provided at the upper inlet openings of the tanks **10**, and an exhaust tube **12** is connected at one of its ends to the bottom of the tanks and at the other end to a water supply tank **13**. Water circulates by means of a pump through the water supply tube **11**, tanks **10**, exhaust tube **12** and water supply tank **13**. At the bottom of the tanks **10**, there are provided outlet openings **14** which are normally closed and which serve to release the moistened rough rice when they are opened.

In the above constructions, the brown rice cracked and enwrapped in chaffs is stored in the tanks **10**, under water and becomes properly moistened within about two to three hours. When the tanks are kept pressurized, the time required for moistening the rice can be shortened to about 20 to 30 minutes.

A belt conveyer **15** located below the outlet opening **14** of the tanks **10**, and operatively connected to a loading opening **17** of another elevator **16** provided with an outlet opening **18** carries the moistened rough rice to steaming tanks **19**. Heated steam is introduced into the steaming tanks **19** by pipes **21** having one end connected to a boiler **20** and the other end connected to the tanks **19**, whereby the moistened brown rice enwrapped in chaffs is steamed and becomes parboiled rice. It should be noted that during steaming, the endosperm of the rice becomes melted within the chaffs and the cracks thereof are thereby eliminated. The water content of the rice at this stage is about 35 to 45%. The rice thus processed is further transferred to a drier **27** by means of a conveyer **23** located below the exhaust openings **22**, of tanks **19** and operatively connected with an elevator **24** having a loading opening **25** and outlet opening **26**.

The drier **27** is of a horizontal agitating type which can dry articles without their lumping together, and hence is most suited for drying moistened articles such as moistened and steamed rough rice, though the time for which it can hold the articles to be dried is comparatively short compared to a drier of other types.

In order to shorten the time necessary for drying the moistened and steamed rice, and subsequently to make it possible to utilize pipe **31** for a heated blast producing device **27** is connected to the drier **27**, so that a blast heated to 100° to 400°C may be introduced into the drier. The drier **27** is provided with an inlet opening **28** which connects to the outlet opening **26** of the elevator **24**, and an outlet opening **29** from which the rice dried in the drier is brought out. The rice is further transferred, by means of another elevator **32** having a loading opening **33** which is connected to the outlet opening **29** of drier **27** and outlet openings, into tempering tanks **34, 34** in which the rice released from these outlet openings into the tanks is stored for a predetermined time so that the water content of rice can be properly adjusted.

Numeral **40** indicates a secondary drier, in which the rice tempered in the tanks **34, 34** and transferred thereinto by means of an elevator **36** is further dried. The elevator **36** is provided with a loading opening **37** which is connected with exhaust openings **35** of tanks **34, 34** and outlet opening **38** which is connected to upper inlet opening **39** of the secondary drier **40**. Numeral **41** indicates a belt conveyer which is provided to the secondary drier for bringing out the dried rice therefrom.

The system illustrated in Figure 3.1b is substantially similar, except that the device indicated by the numeral **3** for providing rough rice with cracks is of a horizontal rotating type, in which the rice flows horizontally. The portions of Figure 3.1b which are identical or correspond to those of Figure 3.1a are denoted by the corresponding numerals of Figure 3.1a.

Thiamine Fortification

It has been known that the vitamin B_1 content of the polished rice is fortified by soaking the rice in an aqueous solution containing a water-soluble salt of thiamine. However, the enriched rice has the disadvantage that a considerable amount of thiamine is lost when the rice is washed with water, because the thiamine is impregnated into the rice tissue in the form of its soluble salts. For the purpose of avoiding such loss of thiamine, it has been attempted to glutinize the surface layer of the rice by steaming the soaked rice, however, this produces an unsatisfactory result. It also gives the rice a tendency to crack into small pieces.

An alkaline treatment is used by *I. Chibata and H. Ito; U.S. Patent 3,573,926; April 6, 1971; assigned to Tanabe Seiyaku Co., Ltd., Japan* after soaking rice in the thiamine solution to minimize losses of the thiamine salt.

The alkaline treatment is carried out by contacting the rice with an aqueous alkaline solution having a pH of higher than about 8.5 throughout the process. For this purpose, aqueous solutions of an alkaline compound such as ammonia, sodium hydroxide, sodium carbonate, etc., are used. The suitable pH range of the alkaline solution is about from 8.5 to 9.2. Although there is a tendency for the more concentrated alkaline solution to result in less washing loss of thiamine, the use of the highly concentrated alkaline solution caused either the deterioration of thiamine or coloring the rice to pale yellow.

An adequate volume of the alkaline solution is 40 to 50 milliliters per kilogram of rice. Under the abovementioned conditions, the rice is stirred for a few minutes in an alkaline solution and the solution drained off from the rice. Then, the moistened rice obtained by the foregoing process is dried in the usual manner.

In an alternative process, the alkaline treatment can be carried out by exposing the rice to ammonia gas. The enriched rice thus prepared shows much less washing loss of thiamine. Moreover, the rice has the same appearance as when untreated, and shows no difference in taste, smell, etc. In order to prevent the washing loss of thiamine, more effective results can be obtained by allowing the rice treated by this method to be subjected to further steaming treatment until the surface layer of the rice is glutinized.

Polished rice is enriched by the following four methods and the comparison of the washing loss to the respective samples is conducted by the testing method stated below.

Sample A is prepared by soaking the rice in an equal amount of 0.3% aqueous solution of thiamine (nitrate) at room temperature for two hours and drying in air at 70°C. Rice prepared in the same thiamine solution as Sample A is exposed to steam for 3 minutes and dried in air at 70°C to form Sample B. Rice prepared in the same thiamine solution as Sample A is moistened with a small amount of 0.8% aqueous solution of sodium hydroxide for 3 minutes and dried in air at 70°C forming Sample C. Sample D is obtained when rice in the same thiamine solution as Sample A is moistened with a small amount of 0.8% aqueous solution of sodium hydroxide for 3 minutes, exposed to steam for 3 minutes and dried in air at 70°C.

Five grams of each sample is moderately stirred with 50 ml of distilled water in an Erlenmeyer flask of 100 ml volume. After allowing to stand for 10 minutes at room temperature, the rice is filtered on a filter paper. The rice and the filter paper are washed with distilled water until the total volume of the filtrate and the washings become 100 ml. The washing loss of thiamine is calculated from the contents of thiamine in the solution obtained by the foregoing procedures. The results are shown in the following table.

Enriched Rice	Washing Loss of Thiamine, percent
Sample A	30
Sample B	16
Sample C	10
Sample D	7

Example 1: 3.0 g of thiamine nitrate is dissolved in one liter of water and one kg of the rice is soaked in the solution for 2 hours at room temperature with moderate stirring. After draining off the solution, the rice is stirred with 40 ml of 0.8% aqueous solution of sodium hydroxide for 3 minutes. Then, the rice is steamed for 3 minutes and dried in air at 70°C. Thus, enriched rice containing 1.5 mg/g of thiamine calculated as its hydrochloride is obtained. The washing loss assayed by the method is 6.8%.

Example 2: 3.6 g of thiamine hydrochloride is dissolved in one liter of water and one kg of the rice is soaked in the solution for 2 hours at room temperature with moderate stirring. After draining off the solution, the rice is stirred with 50 ml of 3% aqueous ammonia for 3 minutes. Then, the rice is steamed for 3 minutes and dried in air at 70°C. Thus, enriched rice containing 1.5 mg/g of thiamine calculated as its hydrochloride is obtained. The washing loss assayed by the method is 7.0%.

Example 3: 6.0 g of thiamine nitrate is dissolved in two liters of water and two kg of the rice is soaked in the solution for 2 hours at room temperature with moderate stirring. After draining off the solution, the rice is stirred with 90 ml of 1.0% aqueous solution of sodium carbonate for 3 minutes. Then, the rice is steamed for 3 minutes and dried in air at 70°C. Thus, enriched rice containing 1.5 mg/g of thiamine calculated as its hydrochloride is obtained. The washing loss assayed by the method is 7.3%.

Example 4: 3.0 g of thiamine nitrate is dissolved in one liter of water and one kg of the rice is soaked in the solution for two hours at room temperature with moderate stirring. After draining off the solution, the rice is exposed to ammonia gas for 5 minutes until the rice shows yellowish color. Then, the rice is steamed for 3 minutes and dried in air at 70°C. Thus, enriched rice containing 1.5 mg/g of thiamine calculated as its hydrochloride is obtained. The washing loss assayed by the method is 8.1%.

Spray-Coating Method

Cereal grains such as rice and corn grits are frequently fortified with vitamins and minerals to restore the nutrients lost in the commercial milling process. The familiar polished white rice of commerce is generally enriched after milling with vitamin B_1, niacin, iron and phosphorus. The vitamins and minerals may be added to each grain of rice or a highly fortified premix may be diluted with the conventional polished white rice to produce a mixture which, in cooking, provides an enriched rice preparation.

Vitamin C does not occur naturally in a cereal grain such as rice but it is frequently desirable to incorporate this vitamin to overcome nutritional deficiencies particularly in those people for whom rice forms a substantial part of their diet. Vitamin C is only sparingly soluble and is partially destroyed in the acid medium by means of which the thiamine and niacin are conventionally introduced into the rice. Therefore, it is not feasible to add vitamin C to rice by this method.

The well established process of *M. Mateles and P.F. Widmer; U.S. Patent 2,755,521; Dec. 25, 1956; assigned to Hoffman-La Roche Inc.* includes vitamin C as part of the vitamin fortification. This is accomplished by adding the vitamin C to the grain in the form of a dry powder. This may conveniently be effected by dusting the cereal grain, previously enriched with thiamine and niacin and covered with a protective coating, with a solid powder containing vitamin C and talc.

The vitamin C dusted grain is then covered again with an additional layer of protective coating. Since the vitamin C content of enriched rice or corn grits is subject to washing losses, i.e., certain amounts of the vitamin are lost by solution in the water when the rice or corn grits are water rinsed, it was found that washing losses of vitamin C are curtailed by incorporating an edible higher fatty acid such as stearic acid, palmitic acid or their salts in the dusting powder containing the vitamin C.

The fatty acid is also used in the solid form. The proportion of fatty acid used in the dusting powder may vary within broad limits, but an amount within the range of about 0.5 to 2 parts by weight of higher fatty acid per part of vitamin C is preferred.

The milled cereal grain is first impregnated with a solution containing thiamine and niacin, and, after drying, the grain is coated with a layer of protective material. While the protective coating is still somewhat tacky, the enriched foodstuff is dusted with a powder containing a mixture of vitamin C, a higher fatty acid and talc, covered with an additional layer of protective coating and dusted again with talc. The minerals may be introduced with the vitamin C if desired. The protective coating in each case contains a corn protein or prolamine such as zein or another edible, water-repellent film.

An adhesive agent such as abietic acid and a plasticizer such as stearic acid may also be incorporated in the protective coating. The adhesive agent and the plasticizer added to the protective coating are independent of and perform a different function from the higher fatty acid which is included with the vitamin C in order to curtail washing losses. In this manner, each grain of milled cereal grain may be vitaminized or a highly fortified premix may be formed by enriching only a portion of the milled product with concentrated amounts of the vitamins and minerals.

Example: A vitamin solution containing 190 g thiamine, and 2,400 g niacin in a solution of 1,400 g of sulfuric acid and 2,700 cc of water was sprayed onto 430 pounds of rice in a rotating tumbler. The vitamin enriched rice was then dried for one hour. One half of a coating solution comprising 2,000 g zein, 1,625 g Neo-Fat No. 1-65 (a commercially available fatty acid mixture containing 90% stearic acid, 6% palmitic acid and 4% oleic acid) and 13.75 g abietic acid in a solution of 500 cc of water and 13,750 cc of isopropanol was sprayed onto the dried vitaminized rice. A vitamin C powder containing 5,450 g vitamin C fine powder, 4,000 g stearic acid and 5,450 g talc was screened through a 20-mesh screen and then dusted onto the rice from movable hoppers in the rotating tumbler while the protective coating was still tacky. The rice was dried for approximately one hour. The remaining half of the coating solution was added and then the rice was dusted with 7,500 g talc. The rice was finally dried for one hour.

The product assayed 9,800 mg vitamin C, 340 mg thiamine, and 4,800 mg niacin per pound of rice. When tested for loss of vitamin C by water washing, the washing loss was 2.4%.

B_{12}-Rich Fractions from Rice

The by-products from rice and bran production are utilized in the process of *S. Tatara; U.S. Patent 3,751,261; August 7, 1973* to produce edible solids rich in vitamin B_{12}.

The germ-containing endodermis of vegetable seeds such as rice, wheat, soybean and the like, after removal of the exodermis on thrashing and polishing the wastes of soybean processing, rice-bran, wheat-bran, bean-curds husk and core medulla crashes of corn after thrashing, draff (bagasse) of sugar cane and the like are ground and mixed. To the mixture is added green leaf powders, and an aqueous solution containing ammonium complex salts of organic oxygenated acids such as tartaric acid, malic acid, citric acid, etc., of essential minerals such as iron, calcium, magnesium and trace amounts of cobalt and a small amount of inorganic salts, and an aqueous solution containing hot milk added with a small amount of potassium triiodide, i.e., iodine containing milk.

Unsaturated fats and proteins are added and combined with iodine, blended, steamed and sterilized by heating. It is mixed again after cooling with seed microorganisms belonging to one or more species of Aspergillus and Saccharomyces, made into pellets and incubated for a specified time interval at an appropriate temperature to effect partial digestion.

The mixture is dehydrated immediately by using a hot air or infrared ray-heated vacuum dryer to give edible powders which can be utilized alone or processed on addition of sugar into granules, chocolate forms or honey-candy drops as highly nutrient and healthy foods containing various essential enzymes and vitamins as well as protein. The process is explained by the following formulation:

(1)(a)	Fresh rice-bran containing germs	27 kg
(b)	Fresh wheat-bran containing germs	25 kg
(c)	Crushed powders of corn cores	20 kg
(d)	Dried powders of corn leaves	3 g
(2)(a)	Ferric ammonium citrate	250 g
(b)	Calcium lactate	150 g
(c)	Magnesium malate	50 g
(d)	Cobalt ammonium tartrate	5 g
(e)	Ferric nitrate or ferric sulfate	60 g
(f)	Fresh water	25 l
(3)(a)	Fresh milk	1 l
(b)	Potassium triiodide	1.5 g
(c)	Ammonium malate	100 g
(d)	Fresh water	50 l
(4)(a)	*Aspergillus oryzae*	100 g
(b)	*Aspergillus flavus*	50 g
(c)	*Zygosaccharomyces soya*	50 g

According to the above formulation, the raw materials cited in (1) are freshly air-dried and mixed thoroughly. The minerals in (2) are dissolved in about 5 liters, of fresh water (f) successively in the order of (a), (b), (c), (d), and (e), made alkaline by adding a small amount of ammonia, finally made up to 50 liters.

The starting materials described in (3) are prepared as follows: To 1 liter of fresh milk is added 0.5 liter of warm water, heated to maintain at about 60°C, and then a solution of about 1.5 g of potassium triiodide dissolved is about 100 cc of water is added gradually with stirring which results in a milky white solution after iodination, after cooling on standing, it is mixed with a solution of 100 g of ammonium malate dissolved in 1 liter of water separately, made up to 25 liters with fresh water, and finally added with 50 liters of the solution prepared in (2), with stirring to give 75 liters of the processing solution.

The solution is adjusted at pH 7 by adding either 5% citric acid solution or 10% aqueous ammonia, blended and kneaded on adding gradually into the raw materials of (1), made into lumps after thorough impregnation, steam-sterilized in an autoclave, and allowed to cool down to below 35°C with continuous blending, and then mixed well with the microbes given in (4).

After incubating at about 37°C for approximately one day and night in a suitable incubator to afford a milky white Koji-like mass with well grown mycelia, the mixture is dried and dehydrated as rapidly as possible under 50°C by using a hot air dryer, or an infrared ray heated vacuum dryer. The dried matter is finely divided by a ball mill or an impulse grinder to give edible powders having the nutriments shown in the following table.

Moisture	6.46 %
Crude protein	17.74 %
Crude fats and oils	21.52 %
Sugars	34.33 %
Coarse fibers	9.04 %
Crude ashes	10.91 %
Calorific value per 100 g	401
Vitamin E	24.2 mg%
Vitamin B_6	1.51 μg%
Vitamin B_{12}	0.25 mg%
Calcium	149.59 mg%
Sodium	1.18 mg%
Phosphorus	2.34 mg%
Iron	113.84 mg%
Vitamin B_1	4.32 mg%
Vitamin B_2	0.64 mg%
Nicotinic acid	25.3 μg%
Folic acid	33.9 mg%
Pantothenic acid	5.06 μg%
Sucrose	2.23 %
Glucose	3.77 %
Fructose	0.40 %

The obtained product can be used directly or after processing with sugar syrup, honey or malt dextrose into granules, chocolate-like matter, or candy style healthy foods as it is compatible with those sugars.

Ornithine-Aspartate Fortification

The unique advantages of using ornithine in combination with aspartate is given by *K. Hoshino and A. Kanemitsu; U.S. Patent 3,669,689; June 13, 1972; assigned to Kyowa Hakko Kogyo Co., Ltd., Japan.*

Many alimentary products are generally deficient in constituent amino acids, such as ornithine and aspartic acid. Ornithine, which is to be found in animal livers, performs, among other things, the important function of producing the indispensable amino acid, arginine. However, it cannot be found in the proteins of living bodies, except in livers. Aspartic acid not only takes part in ammonia neutralization together with ornithine, but is also considered to produce malic acid and to participate in the Krebs cycle.

The content of ornithine and aspartic acid in natural foods is extremely low, except in the case of animal blood and in liver. It is therefore desirable to enrich food with ornithine and/or aspartic acid, but the individual incorporation of these added amino acids prejudices the taste of the food in which they are incorporated. This defect is obviated by the incorporation of these desirable amino acids into alimentary products in combined form as an ornithine-aspartic acid salt or as a peptide of ornithine-aspartic acid. Not only does the taste of the food remain unimpaired but, on the contrary, the flavor is enhanced.

Ornithine has a strong alkalinity and when exposed to air in the uncombined form it absorbs carbon dioxide and is turned into an inactivated state. Hence, it has been used in the form of the hydrochloric acid salt when used as an additive in alcoholic beverages. However, when the hydrochloric acid salt is added to an alcoholic beverage such as sake, the constituent hydrochloric acid is liberated and the pH of the alcoholic beverage is lowered.

The liberated hydrochloric acid, with sodium ion previously existing in the sake or the like alcoholic beverage increases the salty taste excessively, Accordingly, when ornithine is added in the form of a salt or peptide of ornithine-aspartic acid, there is no deterioration of the taste, for example, the liberation of hydrochloric acid.

There has been a general recognition of the use of salts or peptides of amino acids. Although these nitrogen-containing compounds are known to furnish a special flavor to sake, the nature of the constituent amino acid is a subject requiring special consideration. In other words, some of the additive amino acids have a harmful effect on flavor even though they have the characteristic taste.

Rice is a food which is lacking in nitrogen constituents. Continued eating of rice sometimes results in dietary unbalance. The addition of the additive agent raises the nutritive value of rice. Simple mixing with polished white rice is not suitable because the additive may be washed out during the washing of the rice. It is therefore preferred to convert the additive agent into the form of a powder or tablet or granules, if desired in mixture with other substances, and to add the thus-converted material to the rice in the course of the cooking.

The other substances which may be mixed with the new additive can vary widely. A preferred composition for this purpose is a mixture of table salt, sodium glutamate and a salt or peptide of ornithine-aspartic acid, together with a binder such as starch, milk, sugar or crystalline cellulose. This mixture may be shaped, for example, into tablet form in order to produce a suitable commercial product. As for the amounts of these ingredients to be used as an additive for fortifying rice, about 0.1 to 1.5% by weight of salt, based on the weight of the rice, can be utilized. From about 10 to 30% by weight of the salt or peptide of ornithine-aspartic acid relative to the amount of table salt is preferred. The amounts of sodium glutamate, starch, milk sugar and/or crystalline cellulose are minor and optional.

The additives in the preparation of fortified rice, such as sodium glutamate, table salt, starch, etc., are used to improve the taste of cooked rice, to promote appetite and to aid in improving the absorption of the salt or peptide of ornithine-aspartic acid. The amount of the shaped product such as tablets or granules with respect to the cooked rice varies according to the content of the salt or peptide or ornithine-aspartic acid in the shaped product, but from about 0.01 to 0.5% by weight of the salt or peptide of ornithine-aspartic acid based on the amount of rice is suitable.

Reduced Amino Acid Losses

Various procedures have been proposed for increasing the nutritional value of grains. Nutrients used in these enrichment procedures include those of the vitamin B group, amino acids and inorganic substances. However, enrichment with vitamin B has been the only procedure used in practice, and enrichment with amino acids has not as yet been perfected. This is because amino acids were extremely expensive and, furthermore, the required intake amount cannot be obtained without also enriching the amount of vitamins by 100 fold or more. Moreover, even if the required amount of amino acids is supplied to the grain, such as in polished rice, it has been impossible to prevent the loss of amino acids by washing, a problem which does not occur with vitamins.

Resistance to losses of amino acids during washing is obtained in the process of *H. Mitsuda; U.S. Patent 3,623,886; November 30, 1971; assigned to Kyowa Hakko Kogyo Co., Ltd., Japan*.

Grains are soaked in an aqueous solution of amino acids or an acidic solution of the amino acids to permeate into the whole grains. Then, the water is removed. Next, the grains are soaked in or sprayed with a solution of dibenzoylthiamine, cetyl-B_1 or thiamine naphthalene-2,6-disulfonic acid, to make the thiamine substances permeate into the outer layer of the albumins of the grains. The grains are steamed to gelatinize the surface layer of the grains and then dried. Consequently, a gelatinized layer is formed on the surface layer of the grains. As a result of this treatment, the amino acid-enriched grains are resistant to the loss of amino acids, even by washing.

The enrichment of grains with amino acids such as lysine and threonine is nutritionally very important. For instance, it is well known that whole egg protein contains most of the desirable amounts of the essential amino acids as far as human nutrition is concerned. Polished rice protein is significantly deficient in three amino acids, i.e., lysine, threonine and isoleucine, as compared with whole egg protein. However, other amino acids are contained in polished rice in the preferred amounts with respect to human nutrition.

Furthermore, when a mixture of lysine and threonine is ingested by a human, a growth in height, an increase in weight and a reinforcement of physical strength have been noted by many researchers. Accordingly, especially in the case of polished rice, it is necessary to supply amino acids, particularly lysine and threonine. In the following examples, polished rice is used as the grain and lysine is used as the enriching amino acid.

(1) Permeation with Amino Acids: Polished rice is steeped in an aqueous solution of lysine or an acid solution and kept for a definite time (8 to 40 hours) and at a definite temperature (23° to 50°C). The lysine permeates into the whole polished rice. The water is then removed. The preferred amino acids for the enrichment procedure are lysine and threonine. The preferred concentration of amino acid is 10 to 40% by weight. The amount of amino acid used is in the ratio of 0.01 to 0.2% by weight of amino acid to 1 part by weight of polished rice.

(2) Treatment with Thiamine Compounds: (a) Benzoylthiamine – The polished rice obtained from step (1) is soaked in a solution of dibenzoylthiamine for a definite time (3 to 5 min). Water, acids such as acetic acid, hydrochloric acid, lactic acid and gluconic acid, or an organic solvent such as alcohol or acetone can be used to make up the solution. The concentration of this solution is 1 to 5% by weight. An amount of 0.5 to 2.0% by weight of dibenzoylthiamine with respect to the amount of polished rice is preferred.

Instead of the above described soaking step, organic solvent solutions of dibenzoylthiamine may be sprayed onto the rice. Spraying is carried out for approximately 5 to 10 minutes. With this procedure, the dibenzoylthiamine permeates the outer layer part of the albumins of the polished rice.

(b) Cetyl-B_1 – Instead of using dibenzoylthiamine, cetyl-B_1 can be used. Cetyl-B_1 is

thiamine dicetyl sulfate. Molecular formula: $C_{44}H_{86}O_{10}N_4S_3$. Molecular weight: 927.35. MP: 89° to 91°C. Solubility: Soluble in alcohol, hot water and hot acetone, insoluble in water.

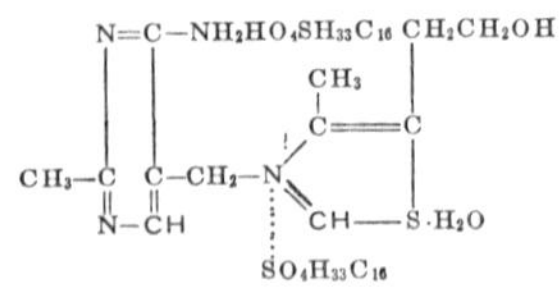

A lukewarm aqueous solution, an acid solution or an organic solvent solution of cetyl-B_1 can be used. Soaking is conducted for 5 to 10 minutes. The temperature of the lukewarm water, if used, is 60° to 70°C. Acids such as acetic acid, hydrochloric acid, lactic acid or gluconic acid and organic solvents such as acetone or alcohol are used in the soaking step. The concentration of cetyl-B_1 is 1 to 5% by weight. An amount of 0.5 to 2% by weight with respect to the amount of polished rice is used.

(c) Thiamine Compounds – Instead of dibenzoylthiamine or cetyl-B_1, the polished rice may be soaked in an organic solvent solution of thiamine naphthalene-2,6-disulfonic acid for a definite time (3 to 5 minutes). Instead of soaking, an organic solvent solution of thiamine naphthalene-2,6-disulfonic acid can be sprayed onto the polished rice and left as is for a definite time (5 to 10 minutes) so as to make the thiamine naphthalene-2,6-disulfonic acid permeate into the surface layer of the albumin. Acetone or an alcohol such as methanol or ethanol can be used as the organic solvent.

The concentration of thiamine compound is about 1 to 5% by weight. It is used in the ratio of 0.5 to 2.0% by weight with respect to the amount of polished rice. After any of the above described treating procedures (a) through (c) have been carried out, the water is drained from the rice.

(3) Steaming-Gelatinizing Procedures – Polished rice is steamed for a definite time (3 to 5 minutes), thereby gelatinizing the surface layer of the polished rice. Then, it is dried at a definite temperature (below 80°C). If an organic solvent or acidic solution has been used in step (2), it is removed by evaporation. As a result, a gelatinized layer is formed on the surface layer of the polished rice.

By means of these procedures, desired amino acids can be permeated into the grains. The permeated amino acid is substantially retained even when washing the rice in boiling water because of the two layers, a hard layer which is soluble in water and a gelatinized surface layer created by the permeation of the dibenzoylthiamine, cetyl-B_1 or thiamine compound into the outer layer of the grain albumins.

In order to observe this result, three samples, i.e., a nontreated polished rice enriched with lysine, a lysine-enriched polished rice which is simply gelatinized in the surface layer (modification to α-starch) and polished treated rice are steeped in water for 10 minutes. The amounts of lysine lost are 42%, 28%, 14% (dibenzoylthiamine and thiamine compound treatments) and 13% (cetyl-B_1), respectively, the last two figures reflecting the results obtained after treatment.

Moreover, grains enriched with an amino acid by this method are delicious, have no toxicity and present an appearance which is almost similar to the usual polished rice. Furthermore, the enriching amino acid is easily eluted by boiling the rice. When a mixture of amino acid-enriched grains and untreated grains are boiled in the usual way, the amino acid is uniformly distributed throughout the whole of the grains boiled. Therefore, the procedure of the process is quite useful for nutritional applications. Referring to the attached drawings, Figure 3.2a shows the concentration of lysine solution and the amounts of lysine

FIGURE 3.2: AMINO ACID ENRICHED RICE GRAINS

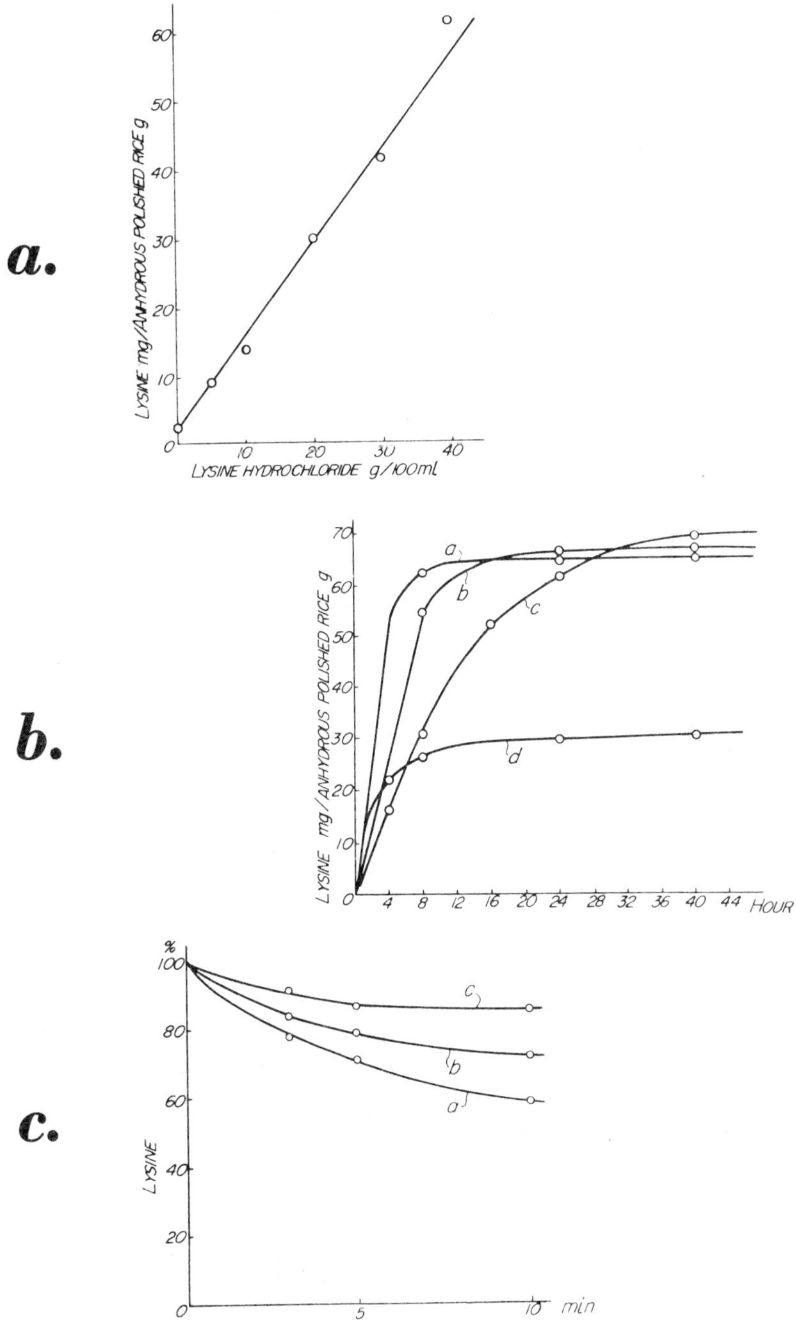

Source: H. Mitsuda; U.S. Patent 3,623,886; November 30, 1971

permeated into polished rice. Figure 3.2b shows the relationship between the amount of lysine permeated in polished rice and the time of soaking. In Figure 3.2b, curve **a** represents a 40% lysine hydrochloride solution at 50°C, curve **b** a 40% lysine hydrochloride solution at 37°C, curve **c** a 40% lysine hydrochloride solution at 23°C, and curve **d** a 20% lysine hydrochloride solution at 23°C.

Figure 3.2c shows that the loss of lysine by washing a rice enriched with lysine produced is small and that the new method is superior to methods known heretofore. In Figure 3.2c, curve **a** represents the loss of lysine in a nontreated product, curve **b** shows the loss after a modification of the product to α-starch and curve **c** shows the loss of lysine of a rice which has been treated by this process.

Example 1: (I) One kg of polished rice is soaked in 2 liters of an aqueous solution containing 40% lysine hydrochloride and is maintained for 40 hours at 23°C, the lysine thus permeating therein, and then the water is drained.

(II)(a) After the above described treatment (I) the polished rice is soaked in an ethanol solution containing 2% benzoylthiamine hydrochloride at 23°C for 5 minutes and is then removed therefrom. Dibenzoylthiamine ethanol solution is found to be permeated into an outer layer part of the albumins of the polished rice, and dibenzoylthiamine itself is present in the outer layer part of the albumins. Thereafter, the water is drained.

(b) Instead of the abovementioned treatment (a), polished rice is soaked in an aqueous solution containing 2% dibenzoylthiamine hydrochloride for 5 minutes at 23°C so as to make the dibenzoylthiamine permeate into an outer layer part of the albumins of the polished rice, and then, the water is drained.

(c) Instead of the above described treatments (a) and (b), polished rice is soaked in a lukewarm aqueous solution containing 2% cetyl-B_1 for 5 minutes at 60° to 70°C. Cetyl-B_1 is permeated into an outer part of the albumins of the rice and then, the water is drained.

(d) Instead of (a) through (c), polished rice is soaked in an ethanol solution containing 2% thiamine naphthalene-2,6-disulfonic acid for 5 minutes and then removed. Thiamine naphthalene-2,6-disulfonic acid-ethanol solution is found to be permeated into an outer layer part of the albumins of the polished rice, while thiamine naphthalene-2,6-disulfonic acid itself permeates into an outer layer part of the albumins.

(III) Polished rice, treated as described above, is steamed for 4 minutes so as to gelatinize a surface layer thereof and then dried with hot air at 80°C or less to remove the water and/or ethanol. A gelatinized layer is formed on a surface layer of the polished rice.

The lysine-enriched rice contains 12 to 13% moisture and 65 mg of lysine per gram of polished rice (the usual polished rice contains 2.2 mg of lysine per gram), and the appearance thereof is excellent. 1.4 kg of a mixture of the lysine-enriched rice and the usual polished rice (in the ratio of 1:66) is washed for 5 minutes and then boiled. The amounts of lysine lost by washing are 7% (benzoylthiamine organic solvent solution), 8% (dibenzoylthiamine aqueous solution), 6% (cetyl-B_1) and 7% (thiamine naphthalene-2,6-disulfonic acid).

Example 2: (I) One kg of polished rice is soaked in 2 liters of a 1% acetic acid solution containing 30% of lysine hydrochloride and 6% of threonine for 40 hours at 23°C, the lysine and threonine permeating into the whole polished rice. Then, the water is drained.

(II)(a) The polished rice from step (I), obtained after draining the water, is sprayed with 500 cc of an acetone solution containing 2% dibenzoylthiamine at 23°C and is left as is for 10 minutes so as to make the dibenzoylthiamine-acetone solution permeate into an outer layer part of the albumins of the polished rice. Dibenzoylthiamine is also found to be permeated into an outer layer part of the albumins.

(b) Instead of the abovementioned step (a), polished rice is soaked in a 1% acetic acid solution containing 2% of dibenzoylthiamine for 5 minutes at 23°C so as to make the dibenzoylthiamine permeate into an outer layer part of the albumin of the polished rice and then, the water is drained.

(c) Instead of using steps (a) and (b), polished rice is soaked in a 1% acetic acid solution containing 2% cetyl-B_1 for 5 minutes while maintaining the temperature at 23°C so as to make the cetyl-B_1 permeate into an outer layer part of the albumins of the polished rice. Then, the water is drained.

(d) Instead of (a) through (c), polished rice is sprayed with 500 cc of an acetone solution containing 2% thiamine naphthalene-2,6-disulfonic acid at 23°C and left as is for 10 minutes so as to make the thiamine naphthalene-2,6-disulfonic acid-acetone solution permeate into an outer layer part of the albumins of the polished rice. Thiamine naphthalene-2,6-disulfonic acid is also found to be permeated into an outer layer part of the albumins.

Polished rice treated by step (II) is steamed for 4 minutes and the surface layer of the polished rice is gelatinized. Thereafter, the polished rice is dried with hot air at 80°C or less to remove water, acetic acid and/or acetone. A gelatinized layer is formed on a surface layer of the polished rice.

The thus-obtained rice, enriched with lysine and threonine, contains 12 to 13% moisture, 32 mg of lysine and 18 mg of threonine per gram of polished rice (the usual polished rice contains 2.3 mg of threonine per gram). 1.4 kg of a mixture of the rice enriched with lysine and threonine and the usual polished rice (in the ratio of 1:37) is washed for 3 minutes and boiled. The total amount of lysine and threonine lost is the same as discussed in Example 1.

Example 3: (I) One kg of polished rice is soaked in 2 liters of an aqueous solution containing 40% lysine hydrochloride for 40 hours at 23°C so as to make the lysine permeate into the whole of the polished rice and then, the water is drained.

(II) The polished rice is soaked in ethanol containing 2% cetyl-B_1 at 23°C for 5 minutes and is then removed therefrom. Cetyl-B_1-ethanol solution is found to be permeated into an outer layer part of the albumin of the rice. Cetyl-B_1 is also permeated into an outer layer part of the albumins.

(III) The polished rice is steamed for 4 minutes and then, the surface layer of the polished rice is gelatinized. Thereafter, the polished rice is dried with hot air at 80°C or less to remove the ethanol. A gelatinized layer is formed on the surface layer of the polished rice.

The thus-obtained lysine-enriched rice contains 12 to 13% moisture and 65 mg of lysine per gram of polished rice. The appearance is excellent. 1.4 kg of a mixture of the thus-obtained rice enriched with lysine and the usual polished rice (in the ratio of 1:66) is washed for 3 minutes and then boiled. The amount of lysine lost by washing is 6%.

RICE-SOY MIXTURES

It was found that rice and soy could be combined to provide a balanced amino acid pattern, such that exists, for example, in cow's milk. *A.C. Benjamin; U.S. Patent 2,991,181; July 4, 1961; assigned to Salada-Shirriff-Horsey, Inc.* found that a suitable amino acid pattern can best be obtained by keeping within certain critical limits. For example, in the rice-soy cereal product the percentage of rice to soy should be between 86% rice to 14% soy and 72% rice to 28% soy with an optimum ratio of 79% rice to 21% soy. While a variation above or below the ratios stated would not be harmful to the product or to the child eating the cereal, it definitely would tend to throw the amino acid pattern out of balance and to prevent the obtaining of the results secured with the preferred pattern.

As an explanation, it might be noted that rice flour is low in certain of the amino acids while high in others of the eight amino acids generally considered essential to human nutrition. Soy flour is high in amino acids which are low in rice flour and is low in those amino acids which are abundant in rice flour. Thus a proper combination of the two flours will furnish a suitable balance of the amino acids when used within the ratios stated.

Since rice flour has a protein content of from 6 to 9% while soy flour has a protein content of from 50 to 52%, the higher the proportion of rice flour, the lower will be the protein level of the mixture, and the greater the proportion of soy flour, the higher the protein level. The following table shows the amino acid content of rice and soy flours expressed in grams of the eight amino acids per kilogram of protein:

TABLE 1: AMINO ACID CONTENT

	Grams per kilogram of protein	
	Rice	Soy
Iso-Leucine	52	53
Leucine	86	77
Lysine	38	63
Methionine	23	14
Phenylalanine	49	49
Threonine	39	40
Tryptophan	10	14
Valine	66	52

The moisture content of the cereal product is also fairly critical as it should not be greater than 11% nor less than 5%, with an optimum of 7%. If the moisture in the cereal product is below 5%, the product becomes stale and rancid on long storage and this tendency to become stale increases rapidly below the critical point of 5%. If the moisture content is above 11%, the growth of mold or bacteria is promoted, but near the 7% optimum, the cereal product may be stored for appreciable periods of time without becoming stale or showing mold growth.

Briefly, the process of making the cereal product and the cereal is as follows: Chosen flours as stated in the examples are mixed with water in a mixing tank, preferably but not necessarily with heating to a temperature of about 150°F. An amylase enzyme may then be slurried in water and added to the flours. The amylase, while optional, makes the process easier to control and does make the cereal somewhat smoother while being eaten. The advantage of heating the flour and water mixture before adding the enzyme is that this preheating makes it easier to handle the mix on the dryer.

This slurry is then heated to a temperature between 130° and 200°F to cause gelatinization of a part of the starch, which occurs between 130° and 175°F. The enzyme, preferably diastase although rhozyme or pangestin may be used, acts upon this gelatinized starch to change it into sugar, dextrins and polysaccharides. The exact temperature, which for example is 155°F with a rice-soy content of 18%, depends not only on the type of cereal used but on the quantity of solids in the slurry, the higher the percentage of solids the lower the temperature required.

The degree of breakdown of the starch is controlled by the quantity of amylase and the duration and temperature of the heating. The usual time required is from 10 to 20 minutes. The amylase is destroyed by heat at 190°F, thus stopping the reaction. Various mineral salts and vitamins may be added, either before or after the amylase enzyme is added. As stated, the cereal product can also be manufactured without using the amylase enzyme, as will be shown in one of the detailed examples, the presence or absence of the enzyme having no effect on the custard-like texture of the cereal.

The heated slurry is then pumped onto a drum dryer which may be either a single or double drum and which may be either atmospheric or vacuum jacketed. The product is here further cooked and dried. This further heating causes a gelatinization of the remaining starches of the flours so that they will more readily absorb water. With the 18% solid content just mentioned, the drums would be heated to about 290°F and be revolved at 5 or 6 rpm, the cooking requiring about 15 minutes, during which the temperature of the mixture never gets above 200°F, its boiling point. Subsequent drying in a thin layer needs only about 10 seconds after which the product is passed through a flaker and is reduced to discrete particles.

A special rennet mixture is prepared to be mixed with the cereal portion of the product, by first making an emulsion of sugar, gum arabic, and water and adding to this additional sugars together with rennet and a calcium salt such as the acetate. The sugars are all sucrose, never dextrose. It is preferred that the sugars shall be of at least three different grinds in order to obtain the proper particle size to insure that the sugars and the cereal portion shall remain homogeneous after being packaged. No. 2 industrial sucrose can be used but it is appreciably better to mix the sugars in proper ratio as for example: fine granulated, 41 parts; bakers' special, 44 parts; and pulverized (6x), 15 parts.

One part of the special rennet mixture may be mixed with four parts of the cereal portion of the product. The quantity of rennet for a given amount of the cereal product, however, is the same regardless of the ratio of the cereal product to the special rennet mixture used. After the cereal portion is mixed with the special rennet portion the product is in condition for packaging, usually in small boxes.

In using the cereal product, four level tablespoons of the product thus made is mixed with half a cup of milk which is at a temperature of about 110°F. The milk should be stirred for one minute after the cereal product is added and then stands for another minute or so to give the rennet plenty of time to set the milk. The cereal prepared in this manner exhibits none of the expected gumminess but surprisingly cuts sharp and clean, much as does a custard. The cereal possesses a fine, highly acceptable texture and an excellent body which does not dry out or become pasty on standing. The cereal remains homogeneous without the continued stirring required with earlier cereals.

Example 1: In this experiment, the rennet extract was mixed with a small quantity of water and applied directly to the cereal portion which consisted of:

	Parts
Rice flour	160
Soy flour	31
Dicalcium phosphate	7
Sodium chloride (common salt)	1.75
Iron pyrophosphate	0.25
Water	800
	1000.00

The ingredients were slurried with the water, heated to 150°F, pumped onto a double drum atmospheric dryer where they were cooked and dried. The calcium and iron salts were added to maintain a proper ratio of calcium to phosphorus, i.e., about 6:5. The sodium chloride was purely for flavor. After flaking the cereal portion was mixed with a special rennet mixture which was made as follows: 500 cc of rennet extract, 500 cc of water and 275 g of calcium acetate mixed together until the calcium acetate was dissolved. The resulting solution was then applied to 100 pounds of the cereal portion after which the excess moisture was removed by drying. The product was then in condition for packaging for the final purchaser or user.

The product thus made was tested by mixing with milk as stated in the general outline and the cereal was found to possess a light, creamy appearance, acceptable texture but a somewhat weak body. The protein level was a little low. The cereal exhibited none of the expected gumminess but surprisingly cut sharp and clean.

Example 2: In this experiment one part by weight of special rennet mixture was added to four parts of rice-soy mixture, the grain portion comprising:

	Parts
Rice flour	151
Soy flour (low fat)	40
Dicalcium phosphate	7
Sodium chloride (common salt)	1.75
Iron pyrophosphate	0.25
Water	800
	1000.00

The ingredients were slurried with the water, heated to 150°F, pumped into the "pinch" of a double drum atmospheric dryer where they were cooked and dried. After flaking the cereal portion was mixed with the special rennet mixture which was made as follows: 1596.3 g of finely granulated sugar and 798.1 g of gum arabic were mixed with 1025 g of water and blended for about one hour until smooth. To this emulsion was then added 92.399 pounds of sugar, to which was added 117.0 g of rennet powder and 936.2 g of calcium hypophosphite. All of these were mixed well until nicely blended, which required only about 5 minutes. The mixture was then dried, comminuted, and mixed with the cereal powder as stated. The product was then in condition for packaging for the final purchaser or user.

The product was tested by mixing with milk as stated in the general outline and the cereal was found to possess a light, creamy appearance, excellent texture and body which did not dry out or become pasty on standing. As in Example 1, the cereal exhibited none of the expected gumminess but surprisingly cut sharp and clean.

Example 3: In this experiment one part by weight of special rennet mixture was added to four parts of rice-soy mixture. The grain portion was similar to that of Example 2 but diastase was used.

The rice and soya flours were first slurried with the water and diastase was added in the ratio of 0.33 g per pound of the flours. The other ingredients were added, heated to 155°F, cooked and dried. The rennet mixture was then added as in Example 2.

The product thus made was tested by mixing with milk as stated in the general outline and the cereal was found to be quite similar to the cereal of Example 2 except that it was somewhat smoother.

Example 4: In this experiment one part by weight of special rennet mixture was added to twenty parts of rice-soy mixture. The grain portion was manufactured exactly the same as in Example 3. After flaking, the cereal protein was mixed with a special rennet mixture which was made as follows: 65.76 pounds of finely granulated sugar and 177.76 g of gum arabic, 666.7 g of rennet powder were mixed with 32.39 pounds (3.89 gal) of water and blended for about 30 minutes until smooth. This special rennet mixture was then mixed with the cereal powder as stated. After removal of excess moisture, the product was then in condition for packaging for the final purchaser or user.

The product thus made was tested by mixing with milk as stated in the general outline and the cereal was found to be the same as that in Example 3.

ENRICHED ARTIFICIAL RICE

Carboxymethylcellulose Binder

Low-priced and highly enriched artificial rice can only contain about 3 to 10% of enriching materials. Its taste is bad and the quality of cooked rice containing the artificial rice often becomes poor.

The method of *T. Yoshida, T. Sagara, T. Ojima, R. Takahashi and M. Takahashi; U.S. Patent 3,620,762; November 16, 1971; assigned to Ajinomoto Co., Inc., Japan* involves mixing rice powder with a suitable amount of enriching materials and, if desired a binder such as carboxymethylcellulose, kneading the mixture with enough water to make the water content 20 to 50%, heating the kneaded mixture with steam until the starch granules in the rice powder are partially gelatinized, forming the viscous intermediate product into grains on a suitable granulator, and drying the grains with warm air at a temperature below 80°C to reduce their water content to 5%.

Suitable raw materials for the production of artificial rice include natural rice of the type used for food but also broken rice or powdered rice that are not used for food. These materials are crushed, if desired, soaked in a 10% solution or a saturated solution of enriching materials such as amino acids and shaped into dry grains by a suitable method.

During soaking, the enriching materials pass through the cell membrane of the rice and, as the water content of the rice is reduced suddenly by drying the rice cells, become brittle and break on account of a kind of plasmolysis. Rice powder containing 10 to 15% of the enriching materials is obtained, through the content of enriching materials depends on the concentration of the solution in which the rice is soaked. Crushing is not necessary for rice powder; however, the process is very effective when natural rice grains are used.

Suitable enriching minerals include amino acids such as L-lysine, L-threonine, L- or DL-methionine and L- or DL-tryptophan, vitamins such as thiamine, niacin and pyridoxin, and minerals such as calcium phosphate, calcium carbonate, potassium iodide and ferrous sulfate. These enriching materials especially amino acids, may be added in an amount to constitute up to 50% of solid matter in the artificial rice. Such a high concentration has not been achieved previously.

Heating with steam is necessary to semigelatinize the starch granules in the rice powder and a temperature of 90° to 120°C is usually applied. If the heat treatment is more severe, the starch granules are expanded too much and break. However, the conditions for semigelatinizing the rice starch are observed as strictly as is necessary with other types of starch.

The starch product obtained by steam heating is an intermediate of high viscosity which coats the enriching materials. The process makes grain-forming easy and, prevents elution of enriching materials during washing with water.

The grains are formed on a granulator by known methods, for example the roller method (a press method) or the macaroni method. The resulting grains are dried in warm air up to 80°C for a suitable time until the water content is reduced to 5 to 15%.

The enriched artificial rice is white and has a lustrous surface, in spite of the high enrichment with amino acids and other nutrients, and the form of its grains is so similar to that of natural rice, that there is no significant difference from natural rice in appearance. When natural rice powder is used as main raw material, forming is very easily accomplished without loss and is economically advantageous as compared to the use of starch or gluten. The artificial rice may be coated if it is desired to be mixed with natural rice and cooked after soaking in water for a long time.

Example 1: In a solution containing 30% of L-lysine hydrochloride natural rice was soaked for 10 hours and dried, whereby the rice grains become very brittle. The L-lysine content of the rice after being crushed mechanically was 10.7%. To 700 g of the resulting rice powder containing L-lysine, 140 g of L-lysine hydrochloride, 100 g of L-threonine, 20 g of calcium carbonate and 150 ml of water were added and the mixture was kneaded for 15 minutes. The starch in the rice powder was semigelatinized by being steamed for 10 to 20 minutes, the steam material was formed into sheets (water content 27%) and formed into grains in a double roller granulator. This forming was done efficiently since the grains produced did not stick to the roller and did not break.

The grains were screened and dried to a water content of 5%. The obtained artificial rice could hardly be distinguished from natural rice in its appearance and color, it had a hardness of 15 g, and its taste was good enough even when cooked with natural rice. An alcohol solution containing 34% of shellac was sprinkled (30 ml of the alcohol solution and 5 to 15 g of talc were used at one time) over the grains in a pancoating machine, under warm air of 40° to 60°C until the quantity of shellac added amounted to 80 g.

The coated artificial rice had a whiteness of 75% and a hardness of 15 g. It contained 21% of L-lysine hydrochloride, 10% of L-threonine and 5% of water. The grains were rice shaped, their surface was smooth and had a desirable luster, and this rice could hardly be distinguished from natural rice in its appearance and taste even when cooked after being mixed with natural rice in an amount of about 1%.

Example 2: 750 g of powder of nonglutinous rice which had passed though a 100-mesh sieve and 210 g of L-lysine hydrochloride were put into a kneader and mixed well with stirring. After the mixture was kneaded with 250 ml of water for 15 minutes, its starch content was semigelatinized by heating at 100°C for 30 minutes, and it was mixed well with 80 g of a solution containing 10% of carboxymethylcellulose. The resulting mixture was put into a macaroni-form granulator and formed into rice-shaped grains which were dried to a water content below 5%.

The obtained rice grains could hardly be distinguished from natural rice in their appearance and had a hardness of 8 g. This artificial rice was coated by being sprayed with an alcohol solution containing 10% of shellac in a pancoating machine for 4 hours. When the coated artificial rice was washed with water and cooked after being mixed with natural rice in an amount of 1%, it lost very little of its amino acids upon washing, and the appearance and taste of the cooked rice could hardly be distinguished from natural rice.

Amino Acid with Starch Binder

In another process by *N. Katsuya, T. Sagara, R. Takahashi, T. Yoshida and T. Ojima; U.S. Patent 3,628,966; December 21, 1971; assigned to Ajinomoto Co., Inc., Japan* added starch is used as a binder for enriched artificial rice.

The first step is to mix amino acids with starch, knead the mixture with water and heat with steam; and the second step is to mix suitable binder such as gluten and starch (with vitamins and minerals, if necessary) with the product obtained in the first step and knead it with water. The product of the second step is shaped in a granulator.

In the first step of mixing amino acids with starch, kneading the mixture with water and heating it with steam, the crystals of the amino acids are covered and solidified by starch, which makes the mixing in the second step satisfactory and brings about good hardness and elasticity and prevents breaking of the artificial rice grains.

The starch used in the method, should be waterproof and have a low expansion rate such as is found in cross-linked starch (for example, distarch phosphate, and starch cross-linked with epichlorohydrin and amylose, for example, corn amylose). The ratio of amino acids and starch may be varied according to the desired enrichment. The amount of water used for kneading the amino acids and starch sould be 20 to 50% based on the weight of starch, which is enough to gelatinize the starch partly during heating with steam.

The heat treatment is performed for partly gelatinizing the starch at 90° to 120°C for 45 minutes to 1 minute. If the heat treatment is more severe, the starch granules are expanded and broken. The viscosity and elasticity of the starch and its ability to form grains are reduced as its tackiness increases, so that uniform grains of artificial rice cannot be obtained. If the heat treatment with steam is performed at lower temperatures for a shorter time, the starch granules do not gelatinize and do not adhere to nutrients. The term "partly gelatinized starch" or "semigelatinized starch" refers to an intermediate phase between the above two states, in which starch keeps its granule form and is strongly tacky at the same time.

In the second step, gluten and/or starch, and optional vitamins and minerals are added. Glutens are especially effective for maintaining the mechanical strength of the grains during boiling. Water is added as needed and the mixture is kneaded. The water content of the mixture at the time of forming grains is preferably 28 to 40%. Wheat gluten, corn gluten, and purified soybean protein are most suitable. After mixing in the second step, the mixture is formed into grains and dried.

The grains formed are dried in warm air of less than 80°C for a suitable time, until the water content is reduced to about 5%. It is necessary to coat the surface of the grains in order to avoid leaching of the nutrients during prolonged washing and soaking.

When cross-linked starch or amylose is used in the first step, dissolution of amino acids during washing and soaking can be avoided adequately even if a pinhole happens to be in the coating.

The artificial rice produced has adequate hardness and elasticity and a high content of nutrients. It resists washing and soaking for a long time. This artificial rice can be mixed with natural rice prior to cooking as usual. When cooking is finished, its grains are not broken and there is no difference between the artificial rice and the natural one in viscosity, elasticity and taste.

Example 1: To a mixture of 400 g of distarch phosphate, 140 g of L-lysine hydrochloride and 70 g of L-threonine, 300 ml of water was added and the mixture was heated at 100°C for 35 minutes with steam, whereby starch granules were semigelatinized. Then 90 g of cornstarch, 200 g of wheat gluten, 20 g of calcium carbonate and 150 ml of water were added, and the mixtures so obtained were rolled into sheets (water content 33%) which were converted to grains in a double roller granulator.

The grains were screened and only rice-shaped grains were put into a dryer to reduce their water content to 13%. An alcohol solution containing 34% of shellac was sprinkled (30 ml at one time) on the grains in a pancoating machine under warm air of 40° to 60°C until the added amount of shellac reached 80 g. The artificial rice thus produced was white, contained L-lysine hydrochloride 11%, L-threonine 5.5% and water 4%, had a hardness of 10 kg by hardness meter, and could hardly be distinguished from natural rice in appearance and taste.

For comparison purposes, 490 g of cornstarch, 140 g of L-lysine hydrochloride, 70 g of L-threonine, 200 g of wheat gluten, 20 g of calcium chloride and 300 ml of water were mixed and put into a granulator. The grains obtained were heated with steam at 100°C for 3 minutes. The grains were not uniform and adhered to one another so that their surface became rough when they were separated. Although the grains were coated according to the method mentioned above, they broke and lost their shape during cooking after soaking in water.

Example 2: To 460 g of amylose of cornstarch, 140 g of L-lysine hydrochloride and 70 g of L-threonine, 150 ml of water was added and mixed well. The mixture was heated with steam at 120°C for 10 minutes, and 130 g of corn gluten, 100 g of cornstarch, 20 g of calcium carbonate, 0.6 g of vitamin A, 0.1 g of vitamin B_1 hydrochloride and 200 g of water were mixed in thereafter. The product 1,000 g was finished by the same methods as in Example 1.

Example 3: To the mixture of 227 g of epichlorohydrin cross-linked starch, 333 g of L-lysine hydrochloride, 83.5 g of L-threonine and 83.5 g of DL-methionine, 170 ml of water was added and mixed well. The starch was semigelatinized by heat treatment with steam at 100°C for 35 minutes, and then 54 g of wheat gluten, 0.4 g of vitamin B_1 naphthalenedisulfonate, 0.1 g of vitamin B_2, 20 g of calcium carbonate and 8 g of ferric phosphate. Four hydrates were added and mixed well. The obtained mixture was put into a macaroni form granulator and formed to rice-shaped grains whose water content was reduced to 15% by drying with warm air at 70°C. The dried grains were coated by spraying with an alcohol solution containing 10% of shellac and 1% of ethylcellulose on an automatic coating machine under warm aeration at 40°C.

POTATO ENRICHMENT

Vitamin C Stabilized with Sulfite

Potato granules, a form of dehydrated instant mashed potatoes have been fortified with vitamin C and other vitamins for Armed Forces, school lunch programs, and other commercial uses for many years. Such potato granule products are about -80 mesh, have a moisture content of about 7%, and are packed in inert atmosphere of less than 2% oxygen in hermetically sealed tins. The vitamin supplement is added in approximately the same particle size as the potato granules. Such a product is known to have good storage stability in that it is not normally subject to discoloration or loss of vitamin C activity.

The bulk of dehydrated instant mashed products sold in retail markets, however, are large particles such as potato flakes, agglomerates, and Potato Buds. These products are conventionally air packed in containers which are permeable to both oxygen and moisture vapor and are susceptible to moisture absorption during storage at high relative humidity, often reaching a moisture level of 9% or higher.

When vitamin C in granular form is added to such products, segregation occurs, resulting in unequal vitamin distribution. When vitamin C is finely powdered to a particle size of about -200 mesh or finer, so the surfaces can be uniformly dusted, the dusted product turns pink and brown more rapidly upon storage. Adherence improves and discoloration increases as particle size decreases. The discoloration is more rapid as moisture content increases during storage. If the vitamin C is incorporated during processing, the dried product discoloration is accelerated in storage.

Currently there is emphasis on selling more nutritious dehydrated foods containing at least as much vitamin C as was present in the raw food prior to processing and in some cases with added vitamins to supply a significant part of the daily requirement.

A.O. Pulley; U.S. Patent 3,800,047; March 26, 1974; assigned to American Potato Company states that incorporation of a reducing agent, such as a sulfite salt in acceptable concentrations, along with the vitamin C has a slight beneficial effect but discoloration is not totally inhibited. The combination of a viscous but nonhydroscopic solution of a film-forming oxygen barrier, such as maltodextrin with vitamin C, is not effective in preventing discoloration.

However, it has been found that when a high solids mixture of vitamin C with sulfite salts (or other sources of active sulfur dioxide) and a film forming oxygen barrier such as light yellow corn dextrin is sprayed on the surface, neither pink nor brown discoloration occurs, even after long storage. In addition, vitamin C activity is preserved.

Furthermore, any discoloration of vitamin C normally resulting from the presence of even trace amounts of iron is obviated. High viscosity, but low solids mixtures of starches and gums, when incorporated with sulfites and vitamin C, are not effective. Other reducing agents, such as stannous chloride, likewise, are not effective. Complete protection against discoloration of vitamin C requires not only the combination of sulfite salts and viscous nontacky film-forming agents such as maltodextrin, but a concentration of maltodextrin in excess of about 10%. Although partial protection can be obtained at lower concentrations, undesirable yellow discoloration still occurs. When the concentration of maltodextrin (or its equivalent) is about 40%, discoloration is completely prevented. It appears that a solution of high viscosity and high solids is required for optimum protection of the vitamin C. It appears also from testing that useful components must not retrograde or form a gel and must not be sticky. When dextrins such as maltodextrin are applied as a water soluble coating in combination with vitamin C and sulfite salts, and dried as a film, it appears that oxygen is excluded, thereby protecting the vitamin C component against oxidation and discoloration.

Many starch-derived products appear to have these required characteristics. Thus, such products as British gum, 85-Fluidity thin boiling waxy maize starch and hypochlorite oxidized starch are useful. Other useful products include gelatin, gum arabic, gum tragacanth, carboxymethylcellulose and ethoxycellulose, each of which can be prepared in a concentrated solution which is moderately viscous.

Tests indicate that ascorbic acid (vitamin C), or soluble salts of ascorbic acid, such as sodium ascorbate, are equally protected by the process. In this application, the term vitamin C includes all such compounds with vitamin C activity.

Iron ions catalyze discoloration of vitamin C. Since small quantities of iron are present in potatoes and since iron is often added to supplement the nutritive value of foods, it is necessary to suppress this reaction. Tests have shown that the process was equally effective in stabilizing the color of vitamin C even in the presence of iron ions. It is well known that iron in the presence of vitamin C results in much greater discoloration than vitamin C alone.

By using a concentrated solution, the desired amount of vitamin C an be applied to products without incorporating sufficient moisture to require a subsequent drying step, although additional drying can be done if desired.

Example 1: Potato Buds, a dehydrated instant mashed potato product produced with a particle size of about -3+8 mesh was sprayed with various solutions to add vitamin C at a concentration of 80 to 160 mg/100 g Potato Buds. (A one-half cup serving of mashed potatoes made from the dehydrated product, fortified to contain 160 mg vitamin C/100 g, constitutes 100% of the minimum daily requirement of vitamin C for children 12 years of age and under, or ⅔ the minimum daily requirement for adults.)

Potato Buds, with solutions indicated in Table 1 below, were stored 4 months at 70°F at 70% relative humidity. Evaluations after 4 months storage indicated that a concentrated solution of maltodextrin was required in the presence of sulfite and vitamin C in order to get protection from discoloration. Sulfite alone with vitamin C decreased discoloration somewhat but was unacceptable. Sulfite with 10% maltodextrin resulted in slight yellowing which had marginal acceptability.

TABLE 1: COMPOSITION OF SPRAY SOLUTION

Test	Vitamin C (as sodium ascorbate)	Maltodextrin	Sodium Bisulfite	Appearance After Storage
1 (Control)	0	0	0	Light uniform color
2	8%	10%	5%	Yellow spots
3	16%	0	5%	Brown spots*
4	16%	0	0	Brown spots*
5	14%	40%	2.2%	Light uniform color

*Unacceptable

Example 2: A second series using the procedure of Example 1 was run varying the concentration of maltodextrin in the spray solution. Viscosities were run on spray solutions. Protection from discoloration was evaluated using an accelerated test in which moisture contents were adjusted in the range of 9 to 10% and the products were held in sealed bottles at 140°F for 40 hours.

TABLE 2

Test	Maltodextrin**	Vitamin C (as sodium ascorbate)	Sodium Bisulfite	Viscosity*	Protection
5	20%	20%	2.9	85.7	Partial
6	20%	14	2.2	84.6	Partial
7	30	20	2.9	204.3	Partial
8	30	14	2.2	145.5	Partial
9	40	20	2.9	1824.4	Complete
10	40	14	2.2	443.0	Complete

*Values are centipoises measured with Model LVT Brookfield Viscometer at 22°C, Spindle Nos. 1 and 2, 0.3 to 60 rpm.
**Dextrose Equivalent of 10.

This series shows that the 40% maltodextrin spray formulation was superior to 30 or 20% maltodextrin at all vitamin C and sulfite levels tested. Products sprayed with 20 to 30% maltodextrin formulations discolored somewhat and were less desirable in appearance at the end of the storage period. Viscosities are greatly affected by the vitamin C and sulfite in the solution. Viscosities for 20, 30, and 40% maltodextrin alone were 8.6, 14.6, and 121.4 cp respectively. The upper limit of viscosity is limited only by the ability to apply the solution as a spray.

Example 3: Tests were conducted in which commercial potato flakes, potato agglomerates, and dehydrated hash-brown strips were treated in the same manner as Test No. 10 of Example 2. The treated products were exposed to the accelerated test disclosed in Example 2. Protection from discoloration was complete in all cases.

The discoloration problem associated with vitamin C addition is not confined to potato products. This would indicate that the problem is a function of the vitamin C itself and not of the product tested. The synergistic mixture, when dry, apparently protects the vitamin C from oxidation or other changes which result in the common pink-brown color formation. This theory appears to be confirmed by tests in which pure sodium ascorbate was adjusted in moisture content to 9 to 10% and stored under the accelerated conditions outlined in Example 2. The product turned brown overnight when stored in a sealed bottle at 140°F. The synergistic mixture of Example 2, Test 10 when dried and then adjusted to 9 to 10% moisture, did not develop the brown color under the same storage conditions.

Vitamin C Stabilized with Sulfite-Antioxidant

In the preparation of dehydrated cooked mashed potatoes, it has been found desirable to add to the potato used to produce the end product an antioxidant to preserve the potato flavor, an emulsifier to improve the texture of the dehydrated product, and sulfites to further preserve flavor and to preserve color, and in particular to prevent browning of the end dehydrated product in storage.

However, attempts to simply incorporate additives, and in particular vitamin C, into the product with the antioxidant, emulsifier, and sulfites have resulted in the upsetting of the functioning, particularly of the emulsifier and antioxidants so that they are no longer capable of providing the high quality and consistent flavor and texture in the end product. It has not been practical to even restore to the dehydrated cooked mashed potato product the vitamin C lost in the dehydration process and for which potatoes may be relied upon.

Vitamin C is added to dehydrated cooked mashed potatoes by *T.F. Irmiter and G. Rubin; U.S. Patent 3,027,264; March 27, 1962; assigned to Salada-Shirriff-Horsey Ltd.*, to establish the loss of vitamin C in the production of dehydrated cooked mashed potatoes, 5 pounds of Russet Burbank potatoes were peeled and trimmed, then sliced into ½ inch slices and precooked in water for 20 minutes at 160°F. Following the precooking, the potatoes were then cooked in a pressure cooker or autoclave for 10 minutes at 15 psi and mashed.

Following cooking, the mashed potatoes were dried with a double drum drier with the drums operated at a steam pressure of 60 psi and a drum speed of 4 rpm. The resulting moisture content of the dried potato flakes was found to be 6.4%. The vitamin C content of the resulting dehydrated cooked mashed potatoes was found to be 12.3 mg/100 g dried flakes. Raw potatoes of the type used in the process were measured to determine their vitamin C level, and it was found that these potatoes contained 11.3 mg of vitamin C per 100 g potatoes.

If no losses of vitamin C occurred in the process to produce dehydrated cooked mashed potatoes, these potatoes in the dehydrated form would have a vitamin C level of 50.5 mg per 100 g of dehydrated potato instead of the actual vitamin level of 11.3 mg determined on measurement. It will therefore be seen that 78% of the natural vitamin C occurring in the raw potatoes was lost in processing to prepare the dehydrated cooked mashed potato product.

While on an experimental basis it was possible to restore the lost vitamin C by incorporating into the product a vitamin C additive. When it came to practical commercial manufacture where it is desirable to incorporate into the product an antioxidant, such as are commercially available and for example a mixture of butylated hydroxyanisole, butylated hydroxytoluene and corn oil to improve the flavor of the product, and an emulsifier such as glycerol monopalmitate to aid in the incorporation of the antioxidant, and to improve the product texture, it was found that the vitamin C could not be incorporated without impairing both the flavor and texture of the product.

The use of sulfites was also resorted to, to aid in the preservation of the flavor, but despite such additions, an acceptable product could not be obtained when the vitamin C was added into the product. It was found that the vitamin C appeared to have the effect of precipitating out the emulsifier with the result that the emulsifier could not be uniformly dispersed throughout the product during preparation.

Further, as a result of such random concentration of the emulsifier, it was found that the antioxidant also could not be uniformly dispersed throughout the product, with the result that here would occur random concentrations of the emulsifier and antioxidant which upset the flavor and left the product with a nonuniform texture. The addition of the sulfites, e.g., sodium sulfite and sodium bisulfite, failed to alleviate the problem.

Moreover it was found that although sufficient viatmin C was added to restore the natural vitamin C of the raw potatoes lost in the dehydrated product, the level of the vitamin C in the product decayed relatively rapidly in storage.

It was found that vitamin C could be added to dehydrated cooked mashed potatoes to restore the vitamin C level to the equal of that present in freshly dug potatoes without the problems of impairment of flavor and texture if the vitamin C was added to the potato mash prior to drying at a point separate from the emulsifiers and antioxidant, and more particularly it was found that if the vitamin C was mixed with the sulfites, not only could the vitamin C be isolated from effecting the apparent precipitation of the emulsifiers and antioxidant, but further the sulfites were found to retard the rate of vitamin C loss in storage so that the vitamin C level in the end product could be maintained throughout the maximum shelf life to which the product would be subjected.

In addition to the enhanced nutritional value occasioned by the incorporation of the vitamin C in the product, the product can be further enhanced if there is incorporated into the potato mash as an additive prior to drying a quantity of nonfat dry milk solids which result in an improvement in color in the product and which render the resulting potato flakes more opaque.

In addition to the ability to incorporate vitamin C into the product without adverse effects it was also found possible to incorporate other additives such as vitamin B_2 and niacin without adverse effect on the flavor, color and texture of the product providing the emulsifiers and antioxidant were delivered into the potato mash to a point separate from the sulfites, and the vitamin C when added was incorporated into the mash mixed with the sulfites.

It has also been found that fat soluble vitamins such as vitamin A, vitamin D, vitamin E and vitamin K can be successfuly incorporated providing they are mixed with the emulsifier prior to addition to the mash. A typical application of the process is illustrated by the following example.

Example: As before, five pounds of Russet Burbank potatoes were peeled and trimmed and then sliced into slices approximately one-half inch thick and blanched in water for 20 minutes at 160°F. Following precooking, they were cooked in the autoclave or pressure cooker for 6 to 10 minutes with the pressure cooker operated at a steam pressure of 15 psi. Following cooking, the potatoes were mashed, then there was added to the mash 17.5 cc of the following additives in the following proportions:

Emulsifier, glycerol monopalmitate	0.8488 g
Antioxidant	0.7377 g
Nonfat dry milk solids (instant milk powder)	1.6997 g
Sodium sulfite	0.5543 g
Sodium bisulfite	0.1794 g
Water	7.90 cc
Vitamin C	1.58 g

The emulsifier, antioxidant and nonfat dry milk solids were mixed into the mash separately from the sodium sulfite and sodium bisulfite into which was incorporated the vitamin C.

Following the mixing of the additives into the mash, the mash was again dried in a double drum atmospheric drier operating at a steam pressure of 60 psi and at a speed of 4 rpm. Following drying the moisture content of the resulting flakes was measured and was found to be 5.5%. The vitamin C content was also measured and was found to be 168.7 mg per 100 g of dried flakes.

To reconstitute the dried potato flakes into the table product, 86 g of flakes are added to one-half cup milk, one-half cup boiling water and one tablespoon butter, to produce four normal servings of potatoes. Therefore, translating the vitamin C content of the potato flakes into the required vitamin C level for one serving of fresh cooked mashed potatoes, the vitamin C level required in the flakes is 167 mg of 100 g of flakes.

In the example given above, the vitamin C content actually measured in the flakes was 168 mg per 100 g of dried flakes, so that it will be seen that the vitamin C level was fully restored.

Specific quantities set out in the above example are given for effecting 100% recovery of the natural occuring vitamin C normally lost in the processing of fresh potatoes to produce dehydrated cooked mashed potatoes. Should it be desired that additional vitamin C be incorporated or should something less than 100% recovery be acceptable, then the specific quantities and proportions of the additives may be varied accordingly.

In addition to the incorporation of the additives mentioned in the above example, it is also possible to introduce into the mash, other additives such as riboflavin (vitamin B_2), and niacin, and also as stated above one or more of the fat soluble vitamins such as vitamin A, vitamin D, vitamin E, and vitamin K, provided the latter are mixed with the emulsifier prior to its addition to the mash.

The sulfites (Na_2SO_3, $NaHSO_3$), whose function is to preserve flavor and color and to prevent browning on storage, especially at high temperature, appear to provide a protective mechanism enabling the requisite level of vitamin C to be maintained for the expected shelf life of the product.

Vitamin A Fortification

In the process of *F.H. Vahlsing, Jr.; U.S. Patent 3,423,213; January 21, 1969; assigned to Vahlsing, Inc.* beta-carotene or, provitamin A is added and stabilized on the surface of frozen potato products.

Beta-carotene is sometimes called provitamin A because beta-carotene can be converted in the digestive systems of animals into a substance which gives the activity of vitamin A. Thus, as a precursor of in vivo vitamin A, beta-carotene is often used as a substitute additive for vitamin A itself. It is also known, however, that the beta-carotene content of frozen foods, such as fruits and vegetables, in some unknown manner decreases markedly during storage during freezing temperatures. For example, the amount of added or naturally-contained beta-carotene in frozen foods may decrease by 25% or more after relatively short storage. For this reason it has been virtually impossible heretofore to achieve an effective long-term vitamin A fortification of frozen foods by addition of beta-carotene.

It was found that if beta-carotene is added to a frying oil for fried frozen potato products, then the frying of potatoes in such oil not only results in the beneficial pick-up of beta-carotene upon the surface of the potatoes, but this beta-carotene is also stabilized in some unknown manner so that it remains without diminution after the potato products have been stored in the frozen condition. No definite explanation or reasons have been found for this unique effect of a frying oil and a frying step in stabilizing beta-carotene against freezing temperature losses in potatoes.

The ordinary frozen french fried potato can be readily fortified with vitamin A by addition of beta-carotene to the frying oil and subsequent frying of the potatoes with such modified oil. Other products such as Potato Buds, which comprise potato shreds or particles reformed into larger compacted masses, can also be benefited by frying in the modified frying oil containing beta-carotene. In addition, potato chips and other fried potato products can be readily fortified with beta-carotene without the danger of unexplicable substantial losses of the additive during freeze storage.

In the process, cleaned white potatoes are first cut into pieces suitable for the particular product that is to be produced. For example, for frozen french fried potatoes, the potatoes may be cut into the so-called shoestring shape or the crinkle-cut shape and then treated in a frying oil containing beta-carotene. Potato Buds previously described above may also be reformed in known manner by compacting potato shreds or small particles in cylinders measuring approximtely ¾" x 1½" and these compacted masses may be fried in a modified frying oil.

The potato pieces can be fried in various hydrogenated vegetable oils such as hydrogenated cottonseed oil. The time of frying may vary from about 30 seconds to four minutes and the temperature of frying may vary from about 325° to about 385°F. Within these ranges the particular conditions that are employed will depend for the most part on the consumer for whom the product is intended. For example, where frozen french fried potatoes are to be prepared for restaurants or other commercial institutions, a short par-frying treatment will be given so that the commercial institution may complete frying according to its requirements. On the other hand, for the home consumer, frying in the manufacturing plant may be carried out to a much greater degree since the average consumer will usually use an oven or broiler to heat and brown the potatoes without additional frying.

This beta-carotene may be added as solids or, more conveniently, it may be added as a suspension of fine solids in a vegetable oil which is compatible with the oil that is to be used for frying. Another way of adding the beta-carotene is to use a partial solution and partial suspension of the beta-carotene solids in an edible oil medium, or in other cases a complete solution may be used since the beta-carotene is fat soluble.

As to the amount to be added, only a very small amount is necessary in order to fortify the potato product with a physiologically active and beneficial quantity of the beta-carotene which after ingestion is converted into vitamin A. In general, up to about 0.025% of beta-carotene can be added based on the weight of the frying oil that is to be used. Excellent results have been achieved with use of about 0.0010% to about 0.0020% of beta-carotene based on the weight of frying oil and such amounts are recommended. There is no actual lower limit of the amount of beta-carotene to be added since even the presence of small amounts will give some benefit.

In general, however, at the very least 0.0004% of beta-carotene should be added to the frying oil in order that a significant amount may be absorbed upon the surface of the potato solids which will give benefits in the form of vitamin A when the beta-carotene is converted in vivo.

If the frying oil is to be used over relatively long periods during large scale manufacturing operations, it is also recommended that antioxidants be included to stabilize the beta-carotene against high temperature degradation by oxidation. It is known that beta-carotene is susceptible to destruction by oxidation and this effect can be readily diminished or overcome by adding antioxidants to the frying oil. Excellent results have been achieved by adding small amounts of a mixture of butylated hydroxyanisole and propyl gallate.

However, any other antioxidants which can be tolerated in food applications such as butylated hydroxytoluene can be used. The amount of the antioxidant will be very small and in general will comprise up to 0.5% by weight based on the weight of the beta-carotene that is to be added. The following example illustrates the method.

Example: Strips of potatoes were cut from clean, white raw potatoes in elongated shapes which were suitable for manufacture of frozen french fried potatoes. The potato strips were submerged in a vat containing hot hydrogenated cottonseed oil having a temperature of approximately 385°F. To this oil had been previously added a suspension of beta-carotene dispersed in a peanut oil vehicle along with the antioxidants butylated hydroxyanisole and propyl gallate. The concentration of the beta-carotene in this suspension was 22% by weight and it was added to the cottonseed oil in the proportion of one pound per 15,000 pounds of the oil, thus giving a weight percent concentration of 0.0014% of beta-carotene based on the weight of the oil. The potato strips were fried in the oil for approximately 3 minutes and then removed from the vat. Excess oil was shaken off the potato strips and then the strips were immediately frozen at -20°F.

After storage for six months at 0°F, the original beta-carotene content of the potatoes remained substantially the same. Also a light yellow color in the originally frozen potatoes remained substantially the same over the same storage period.

Vitamin B_1 Fortification

A disadvantage in the manufacture of potato flakes is the reduction in the nutritional value of the fresh potatoes during peeling, cooking, and other operations up to and including dehydration. Attempts have been made to alleviate this problem by incorporating nutritional additives such as vitamins into the potatoes during processing. However, this presented still other problems since the flavor of the reconstituted potatoes was adversely affected by the addition of some vitamins. Moreover, it was discovered that the nutritional additives were unstable during processing and it was exceedingly difficult to obtain adequate levels of certain vitamins in the potato flakes without adding a relatively large excess of these vitamins.

Sulfites have a destructive effect on vitamin B_1 (thiamine) and the latter ingredient could not be incorporated into potato mash with conventional sulfite stabilizers without destruction of a very large proportion of the vitamin. Consequently, the preparation of potato flakes having a satisfactory flavor and texture upon reconstitution and retaining an acceptable level of vitamin B_1 as well as other nutritional values throughout processing would be a desirable advance.

A special process is used by *M. Pader and A.S. Hall; U.S. Patent 3,343,970; September 26, 1967; assigned to Lever Brothers Company* for preparing potato flakes in which vitamin B_1 and sulfites are incorporated.

Vitamins are added into the potato mash at a point during the processing of potatoes as far removed in point of time as possible from the addition of a solution containing sulfite stabilizers. By thoroughly blending the sulfites into the potato mash before adding the vitamins, the destructive effect of the sulfites upon vitamin B_1 is avoided in large measure, without adversely affecting the qualities of the potato flakes.

After peeling the raw potatoes such as by abrasion or lye treatment, they may be trimmed to remove blemishes and eyes and are then sliced into slabs of about ½ to ⅝ inch thickness. The slices are usually dipped in a preservative solution and may then be precooked by heating in water at a temperature of 160° to 165°F for about 20 minutes. If a precooking step is used, it is generally preferred to follow with cooling of the precooked slices to about room temperature or lower. This may be accomplished by using running water. Cooking is completed by the use of steam at atmospheric pressure for about 30 minutes. If desired, the cooked potatoes may be riced by forcing them through a perforated screen. The various additives are then blended into the potatoes. Since this blending step can also serve to mash the potatoes, the ricing step may be omitted. The mash is then dehydrated by application to a drum drier, and the product subsequently comminuted, screened and packaged.

In a continuous operation, an auger connecting the cooker and the drum drier may be used to transfer cooked potatoes from the former to the latter. The auger partially mashes or breaks up the cooked slices as they are conveyed from the cooker to the drier and additives are incorporated in the potatoes during this operation.

As stated above, it is absolutely necessary that the vitamins be added to the potatoes at a point as far as possible in time from the point of addition of the sulfite stabilizers. It is also highly desirable that emulsifiers, antioxidants and milk solids be incorporated separately from the point of addition of the vitamins.

In a batch operation, this result can be accomplished by adding the solution of sulfites to the potato mash in a mixer and thoroughly blending in these stabilizers. The emulsifier, antioxidant and milk solids can then be added separately or together in an aqueous slurry. After thoroughly mixing these ingredients in the mash, the vitamins are incorporated generally as an aqueous solution and intimately blended into the potato mash which is then dried.

In a continuous operation, the order of addition of the ingredients is unchanged and a maximum amount of time is allowed between the incorporation of sulfites and the addition of vitamins. In certain instances to be explained more fully in the working examples, it was necessary to incorporate the vitamins into the mash as the latter was applied to the drum drier.

The vitamins are added in the form of an aqueous solution. Individual vitamins can be incorporated if desired. However, it is preferred to add a mixture of water-soluble vitamins including B_1, B_2, C and niacin. The proportion of vitamins added will vary depending upon the ultimate levels desired in the final product.

The emulsifiers and antioxidants are those conventionally used in preparing dehydrated cooked mashed potatoes. Virtually any edible emulsifier can be added although monoglycerides are preferred. Examples include glyceryl monopalmitate and monostearate and monoglycerides of saturated monocarboxylic acids of about 12 to 20 carbon atoms.

Edible antioxidants conventionally used in preparing dehydrated foods include butylated hydroxyanisole, butylated hydroxytoluene, etc. Whole or skim milk solids may likewise be added. In a batch operation, the additives are added to the mash and thoroughly mixed prior to drum drying. In introducing the additives, the sulfite solution is first thoroughly mixed into the mash before adding the antioxidant, the milk powder and the edible emulsifier. The concentrated solution of vitamins is added last, after waiting the maximum practical time.

Whether a batch or continuous operation, the time delay between addition of sulfite(s) and of vitamin solution is dictated by practical considerations. In a batch operation, excessive holding of hot mashed potatoes prior to dehydration can result in deterioration of texture, flavor and color. The time lapse must not be so long as to cause such a defect. In a continuous operation, the same factors must be considered, in addition to rate of production, capacity of equipment, floor space available, etc. Thus, the shortest time delay is used that is compatible with obtaining the desired result. This time will vary with the characteristics of the starting potatoes, temperature and degree of mashing of the cooked potatoes when additives are introduced, and manner of introduction of the additives.

The drier in the continuous operation described in the following examples consisted of two single drums; operated independently. It was fed by a 25-foot auger connected to the discharge of the potato cooker. The auger formed a T in the drier housing dividing the mash to feed each drum. Since only one drum was used initially, only one side of the T was modified for the incorporation of additives. The sulfite solution was introduced at a point close to the cooker end of the auger allowing for thorough mixing before the introduction of the slurry and vitamin solution. Two inlet ports for the separate addition of slurry and vitamins were placed on the left lateral side of the T off the auger, at a point giving maximum separation between the point of vitamin addition and the sulfite introduction point.

The use of both drums of the drier necessitated relocation of the slurry inlet to a point 12 inches before the junction of the T. The vitamin solution addition point was moved to discharge directly above the mash exit feeding each drum. This change minimized fluctuations in the residual vitamin content of the finished product.

Example 1: In a pilot operation, Michigan Russet Rural potatoes were abrasion-peeled, cut into ½ inch slices and dipped in a 0.5% solution of citric acid and sodium bisulfite. The slices were precooked in water for 20 minutes at 160°F and cooled to 70°F for 20 minutes in flowing water. The slices were cooked in steam (atmospheric pressure) for 20 minutes and mashed in a Hobart mixer for one minute. A sulfite solution (30 ml) of 3 parts Na_2SO_3 and 1 part $NaHSO_3$ was added to 30 lb of mash and mixed in for 2 minutes. During the mixing 5.5 g of nonfat milk solids, 5.5 g of monoglyceride (Myverol 18:07) and 1.6 ml stabilizer (Tenox IV) were added.

Lastly, 25 ml of a solution of thiamine (1 mg/ml), riboflavin (0.4 mg/ml), niacin (12 mg/ml) and ascorbic acid (164 mg/ml) was added and mixed in the mash for 2 minutes. The mash was immediately applied to the drum and dried.

The drum drier used was a modified double-drum drier, each drum 18 inches long and 12 inches in diameter, heated by steam. Mash was applied to the nip of the two drums, which were set about ¼ inch apart. Only one drum was heated, the other serving as an applicator roll. Mash was rolled onto the heated drum by means of a hand-operated roller to ensure good dried potato sheet formation. The drums were rotated towards each other at about 1 rpm. The steam pressure was maintained at about 40 psig, and changed slightly as necessary to ensure the formation of a uniform sheet with not more than about 6 to 7% moisture. The sheet was removed from the drum by means of a doctor blade. The sheet was then comminuted coarsely. The comminuted dehydrated product (93 to 100 g) was reconstituted with hot water (1½ cups) and milk (½ cup) at about 160°F. The mashed potatoes had a desirable appearance and an excellent taste.

Example 2: In a one drum continuous operation, the sulfite solution addition point was in the cooked slice auger, close to the cooker and approximately 25 feet from the antioxidant-skim milk solids-monoglyceride slurry and vitamin addition points. The auger broke the soft slices and mixed the sulfite solution into the small pieces. It required about 30 to 60 seconds for the potatoes to reach the slurry and vitamin addition points. The distance between the addition points of sulfite and vitamins was the greatest allowable, so great that the mixing of the vitamins and slurry was accomplished mainly on the first and second applicator rolls of the drum drier.

The drum drier employed was 10 feet long and 3½ feet in diameter, and equipped with four rolls for application of mash. Potatoes and additives were fed at such a rate that flake (containing about 6% moisture) was obtained at a rate of approximately 300 lb/hr when the steam pressure was about 60 psig and drum speed 1.7 rpm. The sulfite solution consisted of 0.6 part Na_2SO_3 plus 0.2 part $NaHSO_3$ in 60 parts water. The additive solution (other than vitamins) contained 126 parts water, 2.8 parts distilled monoglyceride from hydrogenated cottonseed oil, 0.525 part antioxidant mixture (Tenox IV), and 2.8 parts of skim milk powder. The vitamin solution contained:

	Mg/Ml
Vitamin B_1	0.1
Vitamin B_2	0.04
Vitamin C	12
Niacin	1.2

The sulfite was added at a rate such as to give 200 to 400 ppm SO_2 in the finished flakes, and was varied according to analytical result as the run proceeded. The emulsifier-milk solids-antioxidant slurry was added at a rate such that the milk solids content of the flake was about 0.2%. The rate of feed of the vitamin solution was adjusted to provide approximately 1 mg of vitamin B_1/100 g of potato flakes.

The drum conditions were such as to optimally dry the mash without scorching the flake on one hand or incompletely dehydrating the mash on the other, and yet obtain a good production rate. Other conditions undoubtedly could be used within limits known to those skilled in the art.

The dehydrated potato sheet was comminuted to flakes by means of a mill with a ¼ inch mesh screen. The flakes when reconstituted as in Example 1 had excellent flavor, color, and texture (reconstitution ratio = 91 g flakes/2 cups liquid).

In this operation, the vitamin solution contained ascorbic acid and vitamins B_1, B_2, and C. The B_1 content was 0.1 mg/ml. The solution was metered in as indicated at such a rate as to provide a theoretical increase in B_1 content over the unfortified flake of approximately 1 mg/100 g potato flakes. In an identical run, vitamins were not added. Analysis of the flakes obtained in the two runs shows:

	mg/100 g Flakes
Vitamin B_1 added	1.19
Vitamin B_1 not added	0.27
Vitamin B_1 recovered	0.92

Thus, recovery of the added vitamin B_1 was about 92% when the vitamin was added as described.

Vitamin Flakes

Potato flakes, when compared with raw potatoes, are substantially reduced in nutritional value. Many attempts have been made to alleviate this problem by incorporating nutritional additives such as vitamins and minerals into the potatoes during processing. However, these methods of vitamin addition presented still other problems since many of the nutritional additives were unstable during the processing steps and some nutritional additives adversely affected the flavor and color of the reconstituted potatoes. Additionally, the finely ground minerals and vitamins readily segregated in the final product thereby causing great disparities of enrichment in portions removed from the package. Accordingly, large excesses of vitamins and minerals were required to obtain the original nutritional level of raw potatoes in all portions of the packaged product.

An improved process is used by *D.C. Pedersen and P.M. Sautier; U.S. Patent 3,833,739; September 3, 1974; assigned to The Pillsbury Company* to enrich dehydrated potato flakes with vitamins and minerals to any nutritional level.

Vitamin flakes are added to conventional potato flakes to obtain, preferably, the original nutritional level of raw potatoes. The vitamin flakes contain vitamins and minerals encapsulated in fat. The vitamin flakes are essentially nonsegregating in the potato flakes and substantially invisible to the eye.

More specifically, dehydrated flaked potatoes are produced by several commercial processes. In one such process, the cooked potatoes are reduced to a mash which is dried on a rotating metal drum dryer. The sheet of dried material is removed from the drum and broken into flakes. These flakes are then packed for shipment.

When potato flakes are manufactured, it has been found desirable to maintain the flake thickness in a monocellular layer having a thickness of from about 0.007 to 0.010 inch. Flakes having a thickness of less than 0.003 inch show excessive cell rupture and flakes measuring over the maximum dimension show increasing agglomeration of cells. The surface diameters of these flakes can and do vary widely but it has been found that a majority (over 60% and preferably over 80%) of the surface diameters should range from 1⁄16 to 1⁄4 inch for commercial purposes. Preferably, over 60% of the surface diameters should range from 1⁄8 to 1⁄4 inch. Because of the many and varied configurations of these flakes, surface diameter is used to indicate the longest surface dimension of the flake.

It has also been found desirable to reduce the final moisture content of the product to between 4 and 10%. Generally, the moisture is reduced by drying the potato mash on a heated drum dryer. If a moisture content of over 10% is present in the final product, storage stability of the potatoes may be adversely affected.

The potato flakes derived from any of the above processes have a packed density, i.e., bulk density after settling, ranging from 0.24 to 0.26 g/cc and, when white potatoes are used, the color is white to cream colored. There are generally no darkening effects due to the methods of processing.

Vitamin flakes are then added to the dehydrated potato flakes to, preferably, attain a nutritional value equal to freshly dug potatoes. The vitamin flakes have thicknesses ranging from about 0.007 to 0.020 inch. The surface diameters of the vitamin flakes, like those of the potato flakes, are of irregular configuration and the size distribution of these flakes conforms to the normal bell-shaped distribution curve. A majority (60%, and preferably 80%) of the vitamin flakes have surface diameters ranging from 1⁄16 to 1⁄4 inch. The density of these flakes ranges from about 0.45 g/cc to 0.55 g/cc. Vitamin flakes meeting the above size criteria are essentially nonsegregating in the dehydrated potato product. This nonsegregating feature is essential and critical to obtain proper distribution of vitamins and minerals in each portion of the reconstituted potato product.

In addition to the size criteria, the vitamin flakes must be free flowing at normal temperatures (below 95°F) to facilitate processing of the product. Additionally, this temperature stability insures that the flakes can be transported and stored without lumping. Another temperature related requirement for these vitamin flakes is that the vitamin flakes must not grease the potato flakes in the final product. The encapsulating agent, therefore, is a very important component of the vitamin flakes.

The encapsulating agents are water-insoluble substances, solid or semisolid at room temperature, consisting predominently of glyceryl esters of fatty acids. This definition includes monoglycerides, diglycerides and triglycerides which, for convenience, will be called fats. Fats used preferably are noncolored, that is, water white. This color is desirable because it is very similar to the color of the flaked potatoes and, therefore, the final vitamin flake is not readily discernible within the flaked potato.

The fat should also have a Wiley melting point within the range from 130° to 155°F. Wiley melting points are very difficult to obtain with monoglycerides because monoglycerides are soluble in alcohols. Therefore, the congeal point is used as the equivalent of the Wiley melting point. These melting characteristics of the fats are essential to insure that the vitamins and minerals are properly encapsulated while packaged and properly released during the final rehydration step. If the melting point of the fat is too low, the vitamins and minerals may be exposed in the package and degraded with concomitant color and flavor changes of the flaked potato product.

Additionally, the flakes may become greasy and adversely affect the rehydration characteristics of the potato flakes. If the melting point of the fat is too high, the vitamins and minerals may not be released from the vitamin flake during rehydration of the product and may contribute a waxy, grainy texture to the final product.

The following specific fats having melting points within the range of 110° to 165°F; capric monoglyceride, lauric monoglyceride, myristic monoglyceride, capric diglyceride, lauric diglyceride, myristic diglyceride, lauric triglyceride, myristic triglyceride, palmitic triglyceride and stearic triglyceride. Commercially available natural fats that meet the melting point criteria include soybean oil, cottonseed oil, safflower oil and tallow hydrogenated to a melting point of between 110° and 165°F.

One particularly preferred fat is a nondistilled, hard vegetable fat derived mixture of monoglycerides and diglycerides (DUR-EM 207). Its properties are set forth in Table 1.

TABLE 1

Form	Beaded solid.
Color	Cream to white.
Lovibond, color, max	5 Red.
Capillary melting point, °F	140 to 145.
Alpha-monoglyceride, min	52%.
Moisture, max	Trace.
Iodine value, max	2.
Free fatty acid, max	1.5%.
Free glycerine, max	3.0%.

Another preferred fat is a distilled monoglyceride (Myverol). Myverol is a hydrogenated vegetable oil having a congeal point of 158°F. Essentially all of the alkyl moieties are saturated. The alkyl chain length distribution is as follows:

Carbon Atoms	Percent
C_{14}	0.02
C_{16}	12.2
C_{18}	85.6
C_{20}	2.0

The vitamin flake contains from 60 to 70% by weight fat. This amount of fat is sufficient to properly encapsulate the vitamins and minerals without loading the potato product with excess fat.

The vitamins suitable for use are all nonreactive with the fat. Some oil-soluble vitamins can be used, such as vitamin A. However, in most applications it is preferred that only water-soluble vitamins be utilized in these vitamin flakes because of the tendency of the fat-soluble vitamins to remain at the surface of the vitamin chip where they may be degraded or where they may cause color or flavor degradation of the potato product.

The preferred water-soluble vitamins include vitamin B_1, (thiamine), vitamin B_2 (riboflavin), vitamin B_6 (pyridoxine hydrochloride, vitamin C (ascorbic acid), niacin, folacin, biotin and pantothenic acid. These vitamins can be used singly or in any combination within the vitamin flake.

The vitamins are utilized in granular form and generally have a particle size diameter of less than 0.018 inch. This size is required to obtain the proper thickness of vitamin flake without exposing the vitamins. Preferably, the vitamins are within a size range such that 70% of the particles are less than 0.003 inch. The following table further characterizes the preferred particle size of the vitamins.

TABLE 2: PREFERRED PARTICLE SIZE

Mesh Size (U.S. Bureau of Standards Sieve Size)	Min. Percent Through	Mesh Opening in inches
60	96	0.0098
200	70	0.0029
325	40	0.0017

From about 10 to about 50% of the vitamin flake is composed of the above vitamins. If substantially more vitamins are utilized in the flakes than specified, it is difficult to encapsulate the vitamins with the available amount of fat. If less vitamins are used in the flakes, the potato flakes carry an increased fat load and a substantial increase in calories.

The minerals utilized are also nonreactive with the fat and with the potato product and preferably are not oil soluble. Therefore, the minerals retain their integrity within the fat and are readily dispersed in the mashed potatoes during the preparatory step of rehydrating the potato flakes. The particle size of the minerals must also be less than 0.018 inch and preferably is such that 70% of the particles are less than 0.003 inch as described above.

Specific minerals which may be added to the vitamin flakes include calcium, iodine, phosphorus, iron and magnesium. These minerals are added as food grade salts such as dicalcium phosphate, tricalcium phosphate, potassium iodide and ferric orthophosphate.

The vitamin flakes can be manufactured by any suitable method. Generally, the fat is melted and the vitamins and minerals are added, with agitation. The molten mass is then flaked by any of the well-known methods. Preferably, the molten fat mixture is fed to the nip of a two-roll, chill roll. One roll is chilled and the other roll is heated. The fat is solidified on the chilled roll and scraped with a doctor blade. The continuous sheet obtained from the chill roll is rubbed through a mesh screen or cut to give pieces with a surface diameter of approximately ¼ inch.

The vitamin flakes are mixed with the potato flakes in amounts ranging up to 5% by weight of the potato flakes. To obtain proper distribution of the vitamin flakes in the potato flake product, the vitamin flakes are added by a proportional feeder into a stream of potato flakes in a continuous mixer. The final vitamin enriched potato product has the same appearance as ordinary dehydrated flaked potatoes. The vitamin flakes are substantially the same color and shape as the potato flakes, are practically invisible to the eye and do not segregate in the package. There are no color or odor negatives associated with the incorporation of these flakes with dehydrated potato flakes. It is, of course, possible to color the fat to emphasize the addition of vitamins and minerals to the final product.

Example 1: A distilled monoglyceride was added to a steam-jacketed kettle and heated to between 180° and 185°F until the monoglyceride was completely melted. The monoglyceride was a hydrogenated vegetable oil having a congeal point of 158°F. Essentially all of the alkyl moieties were saturated. The alkyl chain length distribution was 0.02%—C_{14}, 12.2%—C_{16}, 85.6%—C_{18}, 2.0%—C_{20}. A premix of vitamins and minerals, was prepared containing:

	Percent
Vitamin C (ascorbic acid)	86.6
Folacin	0.0482
Vitamin B_1 (thiamine)	0.3564
Niacin	2.9490
Vitamin B_6 (pyridoxine hydrochloride)	1.5380
Potassium iodide	0.0492
Vitamin B_2 (riboflavin)	0.0004
Dicalcium phosphate	8.4588
	100.0000

The mixture of vitamins and minerals had particle sizes within the following range: 96% through a 60-mesh screen, 70% through a 200-mesh screen, and 40% through a 325-screen. The premix was added with stirring to the molten monoglyceride and then pumped to a two-roll chill roll. One of the rolls was 8 inches in diameter and heated with 5 psig steam. The other roll was 4 feet in diameter and cooled with city water at 70°F. Both rolls were 10 feet long and were rotated at the same peripheral speed. The gap between the rolls was adjusted to 0.01 inch. A doctor blade was utilized to scrape the solidified film from the cooled roll.

The continuous film was passed through a breaker bar which divided the solidified film into pieces having a surface diameter of about ½ inch. These flakes were held at room temperature for 24 hours and then conveyed to a Urschel cutter. The pieces were cut into flakes; 60% of the flakes having a surface diameter between ⅛ inch and ¼ inch. The resultant vitamin flakes were fed continuously by a vibratory feeder into a conveyor line containing dehydrated potato flakes. The dehydrated potato flakes had thicknesses from about 0.007 to 0.010 inch and a moisture content ranging from 5.5 to 6 weight percent. Over 60% of the surface diameters were within the range of ⅛ to ¼ inch.

The potato flakes were conveyed at a rate of 99½ pounds per minute and the vitamin flakes were added to this conveyor and mixed with the potato flakes at the rate of ½ pound per minute. The bulk density of the resultant product ranged between 0.238 and 0.263 g/cc. The vitamin flakes were not visible to the untrained eye within the flaked potato product and the nutritional level of the resultant product was substantially equal to the nutritional level of freshly dug potatoes. A storage test was conducted and an analysis was made of vitamins. In Table 3 the initial vitamin content of the potato product is compared with potato flakes which were not fortified.

TABLE 3

Vitamin	Vitamin enriched potato product	Natural potato product
Vitamin B_2 (riboflavin)	0.139 mg./100 gms.	0.090 mg./100 gms.
Niacin	8.42 mg./100 gms. (Top A)	6.12 mg./100 gms.
	8.60 mg./100 gms. (Middle B)	
	7.56 mg./100 gms. (Bottom C)	
Vitamin B_1 (thiamine)	0.72 mg./100 gms.	0.10 mg./100 gms.
Vitamin C (ascorbic acid)	127 mg./100 gms.	14.7 mg./100 gms.

Storage tests were run at 6 weeks, 12 weeks and 30 weeks and in each case substantially all of the vitamins were available in nondegraded form.

TABLE 4*

Vitamin	Room temperature storage	100° F. storage
Vitamin C (ascorbic acid)	122 mg./100 gms.	117 mg./100 gms.
Vitamin B_1 (thiamine)	0.68 mg./100 gms.	0.62 mg./100 gms.
Vitamin B_2 (riboflavin)	0.127 mg./100 gms.	0.107 mg./100 gms.
Niacin	7.56 mg./100 gms.	7.36 mg./100 gms.

*6 week storage

TABLE 5*

Vitamin	Room temperature storage	100° F. storage
Vitamin C (ascorbic acid)	123 mg./100 gms.	113 mg./100 gms.
Vitamin B_1 (thiamine)	0.74 mg./100 gms.	0.61 mg./100 gms.
Niacin	7.52 mg./100 gms.	8.00 mg./100 gms.
Vitamin B_2 (riboflavin	0.124 mg./100 gms.	0.150 mg./100 gms.

*12 week storage

TABLE 6*

Vitamin	Room Temperature Storage
Vitamin C (ascorbic acid)	100 mg/100 g
Vitamin B_1 (thiamine)	0.73 mg/100 g
Vitamin B_2 (riboflavin)	0.157 mg/100 g
Niacin	7.10 mg/100 g

*30 week storage

As can be seen from the above tables, the encapsulated vitamins are not denatured over long periods of time even under rather severe storage conditions. In the product maintained at 100°F, the vitamin flakes did not grease out on the potato flakes and did not affect rehydration characteristics of the flakes.

Example 2: Dehydrated potato flakes having a thickness of between 0.007 and 0.010 inch and over 60% of the surface diameters in the range of 1/16 to 1/4 inch were made from the following components in percent by weight:

Potato flakes	98.4526
Dicalcium phosphate	0.6200
Mono- and diglycerides	0.5000
Citric acid	0.2000
Magnesium oxide	0.1400
Sodium bisulfite	0.0300
Calcium stearoyl-2-lactylate	0.0300
Sodium acid pyrophosphate	0.0200
Ferric orthophosphate	0.0055
BHA	0.0015
Vitamin B_2 (riboflavin)	0.0004
	100.0000

The final product contained about 5.5% by weight water (included in the potato flakes). The vitamin flakes were made by the process described in Example 1 and were of the same approximate size. The vitamin flakes contained the following components in percent by weight.

Monoglycerides (as described in Example 1)	61.00000
Ascorbic acid	33.77401
Folacin	0.01781
Niacinamide	1.15011
Thiamine mononitrate	0.13900
Pyridoxine HCl	0.59982
Dicalcium phosphate	3.31925
	100.00000

0.6 pound of vitamin flakes were added to 99.4 pounds of the potato flakes as described in Example 1.

HIGH PROTEIN POTATO SNACK

Snack-type food products have, in recent years, attained wide consumer acceptance, and continue to become more in demand at an ever increasing rate. These products are usually enjoyed as supplements to regular meals in the forms of chips, puffs, crackers, etc. The great majority of snack products feature taste appeal, and, for the most part, have high nutritional factors of carbohydrate and fat. In general, the conventional snack products enjoying wide distribution are notably lacking in protein content.

Potato chips and other French-fried potato snack products, such as potato sticks and potato strips have a special appeal and enjoy widespread consumer acceptance especially among teenagers. These products, especially potato chips, are usually sold as a salted deep-fat fried product.

Variants, such as synthetically colored or flavored potato chips are also available to a much limited extent. Although efforts have been made in the past to fortify potato snack products with proteins, within present knowledge, none of these products has enjoyed commercial success.

A unique method is used by *R.H. Waitman, M.H. Kelly and F. Hollis, Jr.; U.S. Patent 3,754,931; August 28, 1973; assigned to General Foods Corporation* for adherence of proteinaceous food particles to the surfaces of potato slices.

It has been found that adherence of the protein material serving as a coating can be successfully attained both before and during deep-fat frying by using a heat coagulable protein as the adhesive medium. A dilute aqueous solution, or sol, of heat coagulable protein has the synergetic function of providing a thin, tacky coating to the uncooked potato slice for adequately adhering the proteinaceous coating and, in addition, supplying a portion of the protein supplement to the potato structure.

When the uncooked potato slice is dipped, or otherwise coated, with the dilute aqueous solution of heat coagulable protein, a uniformly thin coating is effected which is adequate to adhere a uniformly thick layer of other dry proteinaceous material particles to the surface of the potato slice prior to the combination being deep-fat fried. Most importantly, during deep-fat frying, the thin coating of dilute aqueous solution of heat coagulable protein is capable of quickly coagulating under the temperature conditions of the deep-fat frying operation, and in so doing, further binds the protein material to the potato structure.

As distinct from any nonheat coagulable binding agent, the heat coagulable protein coagulates before the water in the potato has an opportunity to escape as steam during the deep-fat frying operation and force the proteinaceous coating from the surface of the potato structure. Since the heat coagulable protein binding medium is in the form of a uniformly very thin layer, the steam which does emanate from the potato chip during frying is capable of penetrating the thin layer and escaping through the interstices of the uniform but discontinuous coating of proteinaceous material. In effect, the heat coagulable protein binding medium is sufficiently tacky to adhere the proteinaceous material to the potato structure prior to deep-fat frying and, by coagulation, further binds the coating material during deep-fat frying, and still further, permits steam to escape from the potato without separating the coating from the potato structure. The net effect is a product with a crisp potato chip base or body upon which is coated a tightly adhering flavorable proteinaceous material, both potato base and coating material exhibiting a well-cooked deep-fat fried flavor.

Unlike batter coated potato chips wherein the thick impervious coating is separated from the chip body by the action of the escaping steam during deep-fat frying, the proteinaceous coating assures no separation and, in effect, produces a product wherein the coating may be considered integral with the potato base structure.

It has been found that potato slices having a thickness ranging from about 0.030 inch to about 0.060 inch are most suitable. As with conventional potato chip manufacture, the raw potato slices are subjected to a sulfite solution dip to inhibit surface darkening prior to deep-fat frying. This dip is especially effective if a delay is contemplated before the raw potato slices are to be deep-fat fried. The sulfited slices are then surface dried of excess water and immersed in an aqueous mixture of heat coagulable protein.

A dilute aqueous solution of egg white has been found to be the preferred heat coagulable protein material for applying to the potato slices to effect excellent adherence of the majority of coatings. Fresh egg white or reconstituted dry egg white to which is added to the mix about 1 to about 9 parts of water has been determined to produce the binding medium. This degree of dilution of the egg white assures a uniformly thin, tacky coating when applied to the potato slice; a coating which, as explained above, is distinguishable from a thick, impermeable batter coating in that, when coagulated by the heat of the deep-fat frying operation, the coating permits the steam to escape from the potato as it is being cooked and yet tightly binds the proteinaceous food supplement to the potato surface.

Also, the use of the above range of dilution of egg white limits the amount of egg white which is applied to the potato surface to the extent the egg white flavor is not a predominant factor in the overall taste of the finished product.

With the use of some proteinaceous coating materials, notably with fish, it has been discovered that the coating material inherently contains sufficient amounts of heat coagulable protein to satisfactorily adhere the supplemental protein coating to the potato slice without the use of egg white or any other binding material. It has been found that the heat coagulable protein, Myosin, occurs in fish in an amount adequate to satisfactorily adhere the fish protein supplement material to the potato slice.

The coatings of supplemental proteinaceous material which can be applied to the potato slice can be selected from many sources. To obtain the advantages of a significantly high protein content addition to the potato slice, it is, of course, necessary to add material having a high protein content. However, if the attribute of a significant amount of protein is not desired, nonproteinaceous material can, of course, be used. The critical factor, with all types of coating, is the capability of being adhered to the potato slice with a heat coagulable protein medium and that this adherence should be satisfactorily effective both before and after the heat coagulable protein is coagulated.

The most desirable appearing and most effective adherence of the coating occurs when the coating is in a shredded form. Although flakes and powdered forms of coatings can be successfully applied, the shredded form provides the optimum overall result. While the actual sieve size of the coating shredded material is not critical, it is essential for good adherence to the potato slice and for appearance purposes that a shred size ranging from about 8 mesh to about 20 mesh be used.

The coating is applied to the potato slice by any one of a number of conventional methods.

The coating may be cascaded over the potato slices or the slices dredged in a body of the coating material. The amount of protein added to the potato slice will, of course, depend upon the protein content of the coating material. In general, the protein added to the potato slice will range in amounts from about 10 to about 25%.

Example 1: Peeled raw potatoes were sliced to chips having a thickness ranging from about 0.035 inch to about 0.045 inch. The chips were surfaced washed in cold water, drained and then dredged in shredded, dry, salted codfish.

The coated chips were then deep-fat fried in a commercial frying oil at 350°F until substantially all of the water in the chips was removed, as evidenced by near cessation of bubbling in the fat. The cooked, fish-coated chips were removed from the fat, drained and cooled and then packaged. The finished chips had a tightly adhering coating of codfish shreds which provided a desirable delicate fish flavoring. Analysis showed a protein content of approximately 20% on a product weight basis.

Example 2: Raw potato slices prepared as in Example 1 were dipped in a 1:1, by weight, aqueous dilution of raw egg white and then allowed to drain. The egg white coated chips were then dredged in coarsely ground, well-fried bacon shreds.

The bacon coated chips were then deep-fat fried as in Example 1 resulting in chips having a unique, tightly adhering bacon shred coating which imparted a desirable bacon flavor to the product. Analysis showed the bacon coated chips to have a protein content of about 15%.

MILK AND DAIRY PRODUCTS

MULTIVITAMIN-ENRICHED MILK

In the enrichment of milk, most dairies introduce vitamins A and D into the product, but enrichment with riboflavin, niacin and thiamine hydrochloride has not received widespread acceptance heretofore because of the difficulty of incorporating these materials into the milk product without affecting the taste of the milk. This is due to the fact that no method had thus far been developed for solubilizing either riboflavin or niacin. Of these two vitamins, riboflavin is the most difficult to use for enrichment purposes because of its medicinal flavor. The undissolved vitamin particles tend to float rather than dissolve in the milk.

During drinking of the enriched milk, when the undissolved or imperfectly dispersed particles come in contact with the taste buds on the tongue, they impart a strong medicinal taste. Although niacin is generally classed as a water-soluble vitamin, it has been found to possess a pronounced electrostatic charge, which not only inhibits penetration of moisture into the vitamin particles but also causes the powder to agglomerate or roll up into tiny beadlets. These beadlets become suspended in the milk and impart a bitter and sour taste as they come in contact with the taste buds on the consumer's tongue.

Many milk processors have refused to provide full vitamin and mineral enrichment of their products because of the difficulty of dispersing the vitamins in the liquid and the unfavorable taste, especially after the milk has been permitted to stand for a few days before being consumed. Deterioration of the taste of the milk caused by the vitamin and enrichment materials has been found to progress very rapidly as the milk is permitted to stand and this effective shortening of the shelf life of the product has militated against widespread adoption of enrichment of milk and other foods with multivitamins.

Completely aside from the nutritional advantages of vitamin-fortified compositions, there is an economic advantage in the milk or food processor being able to introduce any desired mixture of vitamins into his formulations or milk because he can sell his product for an increased price greater than the actual cost of the multivitamins when he enriches the food with a full line of vitamins and minerals.

Most multivitamin enrichment compositions now used for fortifying milk and similar products involve a mixture of thiamine hydrochloride, riboflavin and niacin which are packaged as a dry mix and sold in predetermined weights for incorporation into relatively large amounts of the liquid material. In order to provide full enrichment of the milk, it is also conventional to include iron and iodine compounds with the vitamins. Usually the processor is instructed to introduce the package of dry vitamin and mineral fortification powder into several gallons of milk in a suitable container and then agitate vigorously for at least 10 minutes. The fortified milk was then poured into a large vat of several hundred gallons of milk and the milk further agitated to attempt to solubilize or completely disperse the vitamins and minerals in the liquid. Although the thiamine hydrochloride, which is relatively water-soluble, did disperse in the milk, the undissolved particles of riboflavin and niacin continued to float in the milk and therefore caused an off-flavor as they came in contact with the taste buds on the tongue. Also, a certain proportion of the particles of the water-insoluble vitamins tended to cling to the sides of the tank or to float on the top surface of the liquid because of the incompatibility of the vitamins with the milk.

In an effort to overcome the problem of nonuniform dispersion of the vitamins in the milk or similar products, the instructions have more recently been altered to state that the combination vitamin and mineral enrichment powder should be introduced into only a sufficient amount of milk, when the enrichment composition is added thereto, to produce a thin paste. The directions then call for addition of the paste to the vat of milk with continued agitation for an extended period. This procedure still did not overcome the problems noted.

A multivitamin enrichment composition which is readily soluble in milk is given by *H.J. Johnson; U.S. Patent 3,536,494; October 27, 1970.* Simple and rapid processing steps are given for the preparation of milk fortified with vitamins A, D, B_1, B_2, niacin. The enriched milk product is satisfactory in flavor and storage, and uniform in composition.

Briefly, the present method of processing normally water-insoluble vitamins to dissolve in milk or other water-containing products, involves adding one or more water-insoluble vitamins to a quantity of an organic, nontoxic, particulate carrier that is readily and highly soluble in the milk or food products, followed by addition of only a sufficient amount of water to the mixture to substantially saturate the composition. Next, the moist mass is subjected to a mixing operation to form a substantially homogenous product. The water is evaporated from the composition to produce a granular material and after proper screening, the enrichment composition may be packaged in suitable air and moisture-impervious packets for ultimate introduction into a predetermined volume of milk or other food product.

Processing of the riboflavin and niacin according to the operating conditions involves adding predetermined amounts of the vitamins to an organic, nontoxic, particulate, carbohydrate carrier such as lactose into a suitable receptacle. Although the thiamine hydrochloride is substantially water-soluble and presents no substantial problems with respect to dissolution in milk, it is convenient to incorporate this vitamin into the carbohydrate-containing carrier along with the normally water-insoluble riboflavin and niacin to simplify preparation of the final enrichment composition. The dry vitamins and the carbohydrate-containing carrier are thoroughly mixed and blended to obtain relatively uniform intermingling of the ingredients, whereupon, sufficient water is added to the mixture only to saturate the same.

The moist mass is then thoroughly mixed until an homogenous mixture is obtained with the vitamins being thoroughly penetrated by or embedded in the highly water-soluble carrier. The water is then removed from the homogenous mass by a conventional air-drying operation at a temperature not exceeding about 110°F to avoid altering the characteristics of the vitamins and producing a granular enrichment composition. Since only a very small amount of water is used in saturating the vitamins and the carrier with liquid, the granules resulting from drying of the mixture are of very soft texture requiring only screening of the material to produce a dry, free-flowing powder that may be introduced directly into a food product such as milk. Good results are obtained by passing the granules through a 60-mesh screen to assure relatively uniform particle size.

A dry vitamin A may also be added to the mixture described above. After mixing of the dry ingredients, the product is preweighed and individually packaged in laminated packets, as described. The amount of vitamin material in each packet is of a weight to fortify a prescribed volume of milk with vitamins of at least certain minimum concentrations. It has been found convenient to provide packages for fortifying either 100 or 500-gallon lots of milk. A specific example of a multivitamin food enrichment composition is as follows: add successively, 1.94 lb of thiamine hydrochloride, 3.78 lb of riboflavin, 18.92 lb of niacin, and 29.36 lb of lactose.

Mix the materials thoroughly in a dough mixer type of mixing apparatus. Five pounds of water is added to the dry ingredients to saturate the materials and the mixing is continued until homogenization of the material is obtained. The mixing step normally requires about 45 min. The product is removed from the mixing vessel and dried at an elevated temperature not exceeding about 110°F, whereupon the material is passed through a 60-mesh screen and preweighed and prepackaged as described. 2,000 packets may be weighed out of the formulation set forth above with each packet being designed to enrich 100 gallons of milk with at least 1 mg of thiamine, 2 mg riboflavin and 10 mg niacin per quart.

The quantities above provide, by weight, 3.6% of thiamine hydrocloride, 7.0% riboflavin, 35.0% niacin, and 54.4% lactose. The quantities of these vitamins may be varied in accordance with the vitamin formulation and fortification required or desired for the milk or other food product. In any event, however, the carrier should comprise at least about 40% by weight of the entire dry mixture to obtain proper dissolution of the enrichment composition in the milk. The multivitamin enrichment composition processed according to the procedure above, is fully soluble in milk without imparting any foreign flavor to the milk and without decreasing the shelf life. This is accomplished by securing complete solubility of all the vitamins and minerals.

Although the process described above relates particularly to simultaneous treatment of more than one water-insoluble vitamin material, it is to be recognized that the method is fully applicable to processing of only one vitamin and with the individual vitamin and carrier products being mixed and preweighed prior to packaging of the same in a single container. Also, the processing method described has application to the production of multivitamin or single-vitamin products which may be incorporated into capsules for administration to humans or animals as therapeutic agents.

The free-flowing characteristics of the multivitamin powders are especially useful in this respect during the step of filling the capsules with the powdered product. Since riboflavin, to a large extent, and niacin, to a somewhat lesser extent, are very difficult to introduce into small capsules or similar units because of the tacky and sluggish nature of the product, treatment with the new procedure results in a nonbridging material that will rapidly flow into the capsules prepared and filled in fully automated production equipment.

VITAMIN C-ENRICHED MILK

Milk usually contains bacteria which can be harmful to those consuming the milk or its products. For this reason, the milk is normally subjected to heat treatment for destroying the bacteria. This treatment can be carried out in the form of so-called pasteurization resulting only in the killing of certain types of bacteria, or in the form of a complete sterilization for killing all bacteria and microorganisms found in the milk.

As a result of heat treatment, the proteins contained in the milk usually are denaturated. The denaturation products consist, at least partially, of compounds containing so-called SH-groups, i.e., sulfurhydrogen compounds causing the milk to taste unpleasant and cooked. The amount of denaturation products depends, of course, on the extent of the heat treatment, for example on the temperature prevailing during the heat treatment and on the duration of the treatment. Milk products heated to a high temperature for a long time will undergo a more pronounced deterioration in taste than a product subjected to a less intensive heat treatment.

In certain cases the important ascorbic acid can be destroyed by the heat treatment of the milk, which in its turn results in a deficient amount of C vitamin in the final product.

When conversion of the ascorbic acid is caused by the heat treatment, or when the milk does not contain a sufficient amount of ascorbic acid, the milk can be subjected to vitaminization by adding to a suitable amount of new ascorbic acid.

H.E.W. Hansson; U.S. Patent 3,325,290; June 13, 1967; assigned to AB Tetra Pak, Sweden gives a method of vitaminizing milk with Vitamin C by addition of dehydroascorbic acid which reacts chemically with the denaturation products formed during the heat treatment, whereby ascorbic acid is reformed at the same time as the content of taste-deteriorating substances is reduced.

The milk to be treated is first subjected to the treatments which are usually applied in the processing of milk, i.e., homogenization, standardization etc. In order to obtain a sterile product, the milk must then be heated to a high temperature for such a long time that all bacteria and microorganisms found in the milk are made innocuous. By this treatment, it is true, the milk is made sterile but, as mentioned before, usually also part of the vitamins originally contained in the milk, such as for example ascorbic acid, is destroyed. Furthermore, proteins contained in the milk are denaturated which, for example, results in the formation of denaturation products containing so-called SH groups. These denaturation products which give rise to an unpleasent cooked taste, have a strong reducing effect, due to their content of SH groups, and therefore, readily react chemically with compounds rich in oxygen.

Dehydroascorbic acid which is an oxidized form of ascorbic acid, is reduced back to ascorbic acid when the compounds formed during the heat treatment and containing SH groups react chemically with the dehydroascorbic acid. Thus, by this treatment the two objects were achieved that the milk product was compensated for the C vitamin destruction caused by the heat treatment, and that at the same time the deterioration in taste resulting from the heat treatment was eliminated. The method may also be applied to albuminous food other than milk, for example to cheese, cream etc. which are subject to heat treatment with albumen denaturation and vitamin destruction.

ADDED WHEY FRACTIONS

Whey is an important by-product of the cheese manufacturing industry and although it contains valuable carbohydrate, protein and mineral constituents less than 40% of the annual production in the United States is processed into human and animal food. Consequently, of the more than twenty billion pounds of whey produced in the United States each year, an extremely large amount of it is discharged into streams or into sewage and waste disposal systems where, because of its high biological oxygen demand, it becomes a pollutant. In some areas where streams and waterways are overloaded with organic matter and sewage treatment units are inadequate or expensive to construct and maintain, the problem of whey disposal has become so critical that some cheese plants may be forced to discontinue operations.

Although many uses have been proposed, the utilization and disposal of whey is still a problem. By separating whey into fractions enriched in one or another of its components, better advantage can be taken of nutritional or functional properties of specific components such as protein. Thus, procedures such as electrodialysis, reverse osmosis, ultrafiltration, and selective crystallization of the lactose sugar component have been suggested and applied to the preparation of fractions having a higher percentage of a desired constituent than is found in the parent material.

The separation of solid dry powdery natural materials such as wheat flour into protein-rich and protein-poor fractions by fine-grinding followed by a classification step such as classification in a controlled stream of air, is well-known. The basis of such a separation is readily comprehended since the flour is derived from a heterogeneous substance, the wheat grain or berry, by dry milling steps that result in a mixture of particles with different character and composition.

One would not expect that similar types of separation could be effected on a solid derived from evaporation to dryness of solution containing a mixture of substances. It would be expected that the solid derived from such a solution would have the component substances too homogeneously dispersed in one another for differences between solid particles to be amenable to separation by air or other means of classification. This was found to be true when an attempt was made to separate dry skim milk by fine grinding and air classification. The dry skim milk could not be separated into protein-rich and protein-poor fractions.

However, when a sample of commercially dried whey was subjected to the same fine grinding and air classification process, it separated into a protein-enriched portion and a protein-deficient portion. Although one may speculate as to why the dried whey separated and the dried skim milk did not, the reason for the separation is not yet known. Dried whey is derived from the manufacture of cottage and other types of cheese. Whey from cottage cheese contains somewhat greater amounts of acid than that from other types of cheese.

Protein enrichment does not necessarily mean that the enriched fractions contain protein as the predominant component but only that they are significantly richer in protein, by several percentage points, than the starting material. It will be immediately apparent that the process can be used in conjunction with or in tandem with other separation processes, and that one may obtain a fraction containing a desired degree of protein enrichment by a stepwise procedure that is, the fine grinding of dried whey powder coupled with air classification of the ground powder.

A simple process is described by *I.A. Wolff and A.C. Stringfellow; U.S. Patent 3,784,114; January 8, 1974; assigned to the U.S. Secretary of Agriculture* to isolate and recover protein-enriched fractions from whey solids. Such fractions are useful to improve the nutritional quality of human foods.

Example 1: A commercial sweet whey product in the form of a dry powder was used. It contained 15.8% of crude protein as determined in a conventional way by the Kjeldahl procedure for percentage nitrogen and multiplying that nitrogen percentage by a factor of 6.38. The dried whey was ground by one pass through a pin mill operated at 9,000 rpm and then air classified into five, sized fractions with a standard air classifier. The protein-enriched fractions represented almost half the material and contained over twice as much protein as the remaining half, i.e., the coarse residue; 23.7 vs 10.4% (initially 15.8% protein, dry basis). The lactose showed a change of 21 percentage points between the extreme fractions with the coarse fraction (fraction 5, 56.8% yield) containing 76.6% lactose.

Example 2: Another sample of the commercial sweet whey product used in Example 1 was treated as in Example 1 except that it was ground by a single pass through a pin mill operated at 14,000 rpm. The results differed only slightly from those in Example 1.

Example 3: A commercial sample of nonfat dry milk was treated by the procedure used in Example 1. The original sample contained 38% crude protein on a dry basis. There was no change in protein content for the various fractions. The nonfat dry milk did not separate into protein-enriched and protein-deficient fractions.

Example 4: A sample of the commercially dried whey used in Example 1 was ground by two passes through the pin mill operated at a speed of 18,000 rpm and then the ground whey was air-classified in a standard air classifier. No separation of the ground whey into protein-enriched and lactose-enriched fractions was obtained. This example illustrates the limitations of the operating range of the process by showing that excessive grinding negates the process.

CASEIN AND WHEY PROTEINS

The preparation of a milk protein with a PER (Protein Efficiency Ratio) of 3.25 compared to casein with a PER of 2.50 is described by *M.E. Engel and A.D. Singleton; U.S. Patent*

3,361,567; January 2, 1968; assigned to National Dairy Products Corporation.

Casein, a protein, has been recovered from milk for many years and this has been accomplished by developing the acidity in milk to a point where the casein is coagulated and thereafter separated from whey. The acidification may be effected by the addition of mineral or organic acids and may also be effected by bacterial development. Unless the milk is specially treated, mere acidification of the milk will cause coagulation of casein, but will not cause precipitation of certain other proteins, referred to as whey proteins, from the milk. Accordingly, this whey protein substantially remains with the whey and will not be precipitated with the coagulated casein. Various attempts have been made to effect coprecipitation of casein and whey protein by treating the milk and by adding various materials to the milk.

Generally, these processes have not provided a coagulum or curd which can be readily handled in a continuous process and the disclosed processes require batch handling and processing of milk. The curd prepared from these processes is usually soft mush, or contains too many fines for satisfactory recovery except by uneconomical processes such as decantation or filtration. Furthermore these processes do not permit recovery of protein in high yield.

In an improved method to coprecipitate casein and whey protein the milk fat is first removed from milk to provide a skim milk, which is then subjected to a preliminary heat treatment to a temperature of 180°F for 10 minutes or more, for correspondingly diminished times at higher temperatures not to exceed 210°F. Overheating and/or high-temperature treatment of the skim milk in the preliminary heat treatment should be avoided.

The solids of the skim milk should then be increased to at least 10% solids and preferably to 30% solids. However, the solids should not exceed about 35% solids because such high solids result in difficulty in curd formation. The increased solids may be effected by adding nonfat milk solids to the skim milk or by concentrating the skim milk in an evaporator. The high-solids skim milk should then be subjected to an ultra high temperature-short time treating condition (UHTST) and, in this connection, the high-solids skim milk should be heated to at least 280°F for 15 seconds and preferably, to 285° to 295°F for 20 seconds.

The heat-treated skim milk should then be reduced in temperature to a temperature of 210° to 215°F and acidified by the addition of an acid under turbulent conditions so as to assure uniform distribution of the acid in the skim milk. The skim milk, after acid addition, forms into curd and whey with the whey protein coprecipitating with the casein to provide a handleable curd. Some compacting of the curd is desirable before further treatment. The curd and whey are then separated. The curd is washed to free most of the lactose from the curd which, if present at substantial levels, results in browning of the product. The curd is in such condition that it can be continuously roll-pressed on a conveyor and simultaneously and subsequently washed without substantial loss of solids and with high recovery of milk protein.

The curd is then ground in the presence of added water to a very small particle size and the resultant dispersion of milk protein particles is centrifuged from the aqueous phase. If desired, further addition of water, grinding and centrifugation may be used to further purify the milk protein. The milk protein is then ready for drying or further processing. In this connection, the protein may be dried in a tray dryer or it may be formed into a slurry and spray dried. Alternatively, a soluble milk protein may be prepared by dissolving the milk protein in sodium hydroxide in accord with known procedures. The solubilized milk protein can then be spray-dried.

It has been found that the preliminary heat treatment, prior to the ultra high temperature-short time treatment, is necessary for the achievement of a satisfactory curd for continuous processing. It was also found to be important for obtaining such curd that acidification be effected while the high-solids skim milk is at a high temperature. These conditions, in conjunction with the ultra high temperature-short time treatment of the milk and concentration of the milk provide milk protein in high yield with the casein and whey protein being coprecipitated.

As an example of the process and with reference to the flow sheet (Figure 4.1), skim milk was subjected to a preliminary heat treatment by heating it to 185°F for 20 minutes. This preconditions the skim milk for subsequent heat treatment. The milk was then fed to vacuum pans and concentrated to a solids content of 30%. In the flow sheet, the milk is shown to go next to cold storage, but this step is not necessary to the process.

FIGURE 4.1: PROCESS FOR THE MANUFACTURE OF CASEIN

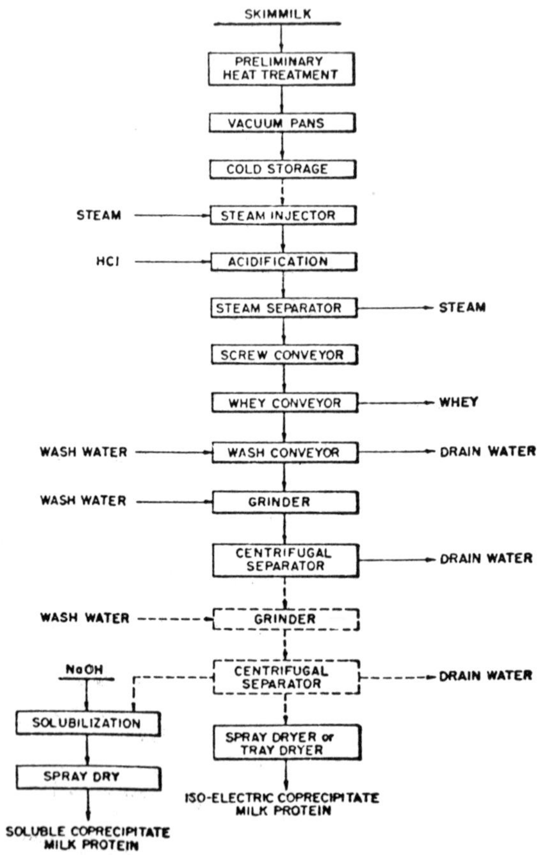

Source: M.E. Engel and A.D. Singleton; U.S. Patent 3,361,567; January 2, 1968

The concentrated skim milk was passed into a steam injector for UHTST treatment. The milk was raised in temperature to 290°F, held at such temperature for 20 seconds, and discharged to about atmospheric pressure conditions from the steam injector so that the milk was promptly cooled to about 212°F by flashing.

Flashing of the heat-treated milk was effected in a 1" sanitary pipe which included a U-shaped portion thereby effecting a high degree of turbulence. While in this turbulent condition and while at the elevated temperature, a preselected quantity of 11% hydrochloric acid was added to the milk. The high degree of turbulence effects uniform distribution of the acid throughout the heat-treated skim milk without localized high acid conditions.

Upon mixing of the acid and the heat-treated skim milk, casein and whey protein were coprecipitated to provide a mixture of coprecipitate curd, steam, and whey. Before attempting to separate the curd from the remaining constituents, the entire mixture is conducted into a separator, wherein liquid and vapor constituents are segregated. In the separator, steam was separated from the curd and whey with the whey passing out the bottom of the separator and steam being recovered from the top of the separator.

The steam separator consists of a vertical, open-ended 3" pipe, about 1 foot in length, and the mixture of curd, whey and flashing steam entered mid-point of the pipe. The mixture was directed tangentially into the pipe causing the curd and whey to spiral downwardly and the steam to escape from the top of the pipe. The curd and whey were passed into a hopper which fed a screw conveyor and which held the curd for a short period of time. During this hold period of about 10 seconds final precipitation of the milk protein occurred. The screw conveyor operated to physically agglomerate and compact the curd, which came out of the screw conveyor in the form of a rice-like granule.

The curd and whey were discharged from the screw conveyor onto a whey separator which comprised a riffle bottom trough, the trough being provided with a 40-mesh screen on its bottom. The removed whey had a pH of between about 4.5 and about 4.8 which is equivalent to a whey acidity of 0.6 to 0.8% measured in equivalent lactic acid level. The amount and concentration of acid added to the hot skim milk was selected to provide this pH level in the whey.

The curd discharged from the bottom trough had a solids content of about 25% and was fed onto an upwardly moving wash conveyor. The wash conveyor comprised a continuous belt over which was located a plurality of pressure rolls. A first roll was adjusted so as not to provide excessive pressure and to avoid compacting the curd into an impenetrable cake. The curd on the belt was washed with water so as to promptly cool the curd to about 90°F. Such prompt cooling is necessary to avoid undue browning of the milk protein. The curd was then washed again with water and pressed by a second roll on the continuous conveyor so as to substantially reduce the lactose level in the curd.

As illustrated in the drawing, the curd from the wash conveyor was conducted to a grinder wherein water was also added, and the curd was ground in a Morehouse mill to provide a fine dispersion of milk protein in water. The dispersion was then passed to a centrifugal separator where the milk protein was separated from the water. At this point, the protein comprised less than about 1% lactose. Optionally, as shown in dotted outline in the drawing, the milk protein may again be ground with added water to further reduce its lactose content, and centrifugally separated from the water.

As shown in the flow sheet, the resultant milk protein can be placed on a tray and dried. Alternately, it can be slurried in water and fed to a spray dryer. As an additional alternative, and as shown in the flow sheet, the milk protein slurry can be solubilized by the addition of sufficient sodium hydroxide to provide a pH of about 7.0, and then the resulting solution spray-dried to provide a soluble milk protein.

It has been found that a preliminary heat treatment at about 165°F for about 5 minutes, followed by concentration and heat treatment at 285°F for about 20 seconds does not give adequate harvestability to the milk protein. It also has been found that a preliminary heat treatment at about 185°F for about 20 minutes, followed by concentration and heat treatment at temperatures less than about 280°F does not give adequate harvestability. It has been found further that if the acid is not added at the high temperature, a satisfactory curd does not result.

The product has improved nutritional characteristics in comparison with a protein containing casein alone. This was demonstrated in a standard feeding test by feeding the product to rats and comparing their weight gain with the weight gain of rats which were fed a control diet comprising a commercially available casein. The weight gains per gram of protein consumed were tabulated as protein efficiency ratio, with a higher ratio indicating a higher nutritional value. The control casein resulted in a protein efficiency ratio of 2.50, whereas the new product resulted in a protein efficiency ratio of 3.25.

CALCIUM-FORTIFIED MILK PROTEIN

Protein concentrates and especially those prepared from milk have become important as additives to fortify the protein content of many foods. They have also found use in the pharmaceutical field. One form of concentrate is generally prepared by the simultaneous precipitation of the protein content from skim milk under acid conditions. The curd thus formed contains 25 to 30% total solids, of which about 85% is protein in the form of a mixture of casein and lactalbumin, together with a small amount of beta-lactoglobulin. This precipitate is sometimes referred to as a casein-lactalbumin fusion or coprecipitate. The major proportion of lactose and minerals originally present in the milk are carried off in the whey.

In the processes for coprecipitation of milk proteins, a proportion of the calcium originally present in the milk is lost. Additionally, the processes result in a decrease in density of the milk protein powder. This is often undesirable in that it results in increased transportation and handling difficulties, as well as unsuitability for certain uses.

A dried milk protein concentrate having increased density and calcium content is prepared in a process described by *M. Loewenstein; U.S. Patent 3,218,173; November 16, 1965; assigned to Crest Foods Co., Inc.*

It has been found that a milk protein product may be prepared having the above-stated desired properties by first removing the protein content from milk by precipitation. The precipitated protein curd is washed with water, separated from the whey, and broken up into very fine particles by passing through a high-speed mill such as a Fitzmill. Lime (CaO) is then added to the curd in slurry form and thoroughly mixed. The slurry is then mixed with calcium salt such as calcium carbonate ($CaCO_3$) and moderately heated. After the reaction is complete, the slurry is once again passed through a high-speed Fitzmill and subsequently spray-dried.

The best products are obtained by utilizing a protein precipitate in curd form obtained from milk, and especially skimmed milk. The precipitate products from skim milk contain a mixture of casein, lactalbumin, and a small amount of beta-lactoglobulin. Individual protein materials such as casein or lactalbumin may also be used.

Example: Coprecipitation — Into a suitable receptacle such as a conventional cheese vat are placed approximately 5,000 lb of freshly separated skimmed milk. The titratable acidity, as measured in terms of lactic acid, is adjusted downwardly to a point where the titratable acidity ranges between 0.04 and 0.11% by the addition of suitable alkalizing material such as sodium bicarbonate or soda ash, or a combination of the two materials. Other neutralizing agents which may be used either singly or in combination are NaOH, KOH, CaO, MgO, or other acid-neutralizing chemicals. Temperatures in the range of from 32° to about 145°F may be used, but neutralization should be carried out before the heat initiates precipitation of some of the whey proteins.

The neutralized milk is then heated to a temperature in the range of from about 190° to 194°F. Heating is discontinued and the milk brought to a quiescent state. A dilute acid is then added to accomplish the precipitation of the casein, lactalbumin, and beta-lactoglobulin, in the form of a homogeneous coprecipitate. The acid used for precipitation may be an acid such as sulfuric, hydrochloric, lactic, acetic, hydroxy-acetic, phosphoric, citric or tartaric. Hydrochloric acid is the preferred acid. Sufficient acid should be added to effect complete precipitation, with a resultant clear whey. The pH should be from about 4.3 to about 4.9. The titratable acidity of the clear whey should be from about 0.20 to about 0.35%.

The whey resulting from the process is withdrawn from the vat. This process may be accomplished by first allowing sufficient time for the whey to separate by gravity, and subsequently carefully removing the whey. The whey may also be removed by mechanical means such as strainers, filters, or centrifugal devices. The precipitated curd is washed to remove excess acid, whey, soluble salt, and other nonproteinaceous materials. Either hot or cold wash water may be used. Various additives for aiding the washing process may be used. The number of washings and the volume of wash water required are dependent upon the degree of purity desired in the finished protein. The protein may be broken into small particles during the washing operation to facilitate the complete removal of the nonproteinaceous material.

Example 2: Calcium Reaction with Wet Curd – The washed, purified, and precipitated protein curd as produced above in Example 1 may be further comminuted by a grinder or hammer mill, with the addition of cold water, to reduce the particle size sufficiently to form a slurry of heavy cream-like consistency. The curd particles should be 0.03" or less in diameter, and may contain a high proportion of moisture. The slurry thus formed is not a stable dispersion, but, upon standing, the solid phase may separate from the liquid phase.

While the slurry thus prepared is vigorously agitated, lime (CaO) NF powder is added in an amount of about 1% of the total solids present. The lime may alternatively first be dispersed in water and the dispersion added to the protein slurry. A proportion of 1 lb of lime to 100 lb of slurry solids is approximately equal to 150 grams of lime added to the slurry for each 1,000 lb of skim milk originally used, or 1 lb of lime to the slurry for 3,000 lb of milk. The addition of lime should be made while the slurry is at a temperature no greater than 120°F. The pH of the slurry at this point should be from about 7.0 to 7.4. In order to complete the reaction between the lime and the milk protein, it should be continued further for from 5 to 10 minutes.

After the reaction has been completed, calcium carbonate (USP precipitated powder) is added in an amount approximately equal to the amount of lime previously added. The amount may be varied where it is desired to modify the properties of the finished product. The addition of the calcium carbonate to the protein slurry should be preferably made at a temperature no greater than 120°F.

As a variation of the above process, the lime-containing slurry may be pumped from the reaction tank into a second reaction tank, no heat being added during this period. While the slurry is flowing into the second reaction tank, the calcium carbonate is added, either as a dry powder or as a liquid. In this case also, if the liquid form is used, agitation must be provided for in the supply tank to prevent the settling out of the insoluble calcium carbonate. Additionally, the slurry must be agitated vigorously as the calcium carbonate is added. The pH of the slurry, after the addition of both the lime and the calcium carbonate, should be in the range of from about 7.1 to about 7.5.

A sufficient reaction period should be provided after the addition of the calcium carbonate to permit the reaction with the protein to reach completion. The addition of both the calcium oxide and the calcium carbonate should be carried out at temperatures no greater than 120°F to prevent the formation of tough curds which would be difficult to dry as by spray drying.

When the reaction is complete, the protein slurry is heated to prepare it for drying. Heating may be accomplished either by the injection of live steam into the slurry, or by forcing the slurry into or through any of the conventional heat exchange devices such as tubular or plate heaters, jacketed vats, or kettles. Heating to a temperature prior to drying in the range of from 170° to 180°F gives excellent results, although either higher or lower temperature may be used dependent upon the bacteriological and physical properties required in the finished product. The slurry should be as concentrated as possible at the time of heating in order to facilitate handling through the heating and drying devices. After the slurry has been heated, but before drying, it may be treated in a high-speed comminuting machine such as a hammer mill in order to reduce the particles to a size suitable for drying, and to render the particle size of the dried product more uniform.

Drying may be accomplished either by spray drying, drum drying, or foam mat drying. Although a final moisture content of from about 3 to about 6% is generally desired, it may be varied from this range in either direction where such properties are desired in the finished product. In place of calcium carbonate, other edible calcium salts may be utilized for reaction with the protein in the second step of the present process. Among these are any of the calcium phosphates such as tricalcium phosphate, calcium gluconate, calcium tartrate, calcium citrate, calcium acetate, etc.

The final products have many advantages. Because they have low water-binding capacity, they are extremely useful for adding to baked goods to increase the protein content. Protein concentrates available on the market have water-binding capacities which are both high and variable. Consequently, it is often difficult to determine how much water need be added to the baking dough in order to obtain the proper consistency. Moreover, because of their high water-binding capacity, a large amount of water must be added to the dough in order to obtain a particular consistency. The presence of the excess water makes drying and baking considerably more difficult.

Because of the low water-binding capacity of the present protein precipitates, only a minimum amount of water is required to obtain the desired dough consistency. Moreover, the water-binding capacity from one batch to another is so uniform, that the problems of varying consistency are eliminated. The protein concentrates also have a higher calcium content than similar protein concentrates, since the calcium content is normally lost during processing. They are therefore extremely useful for the production of high-calcium-containing foods.

HYPOALLERGENIC MILK

Complex with Casein

The fact that the milk proteins beta-lactoglobulin, alpha-lactalbumin, and serum albumin are allergens for many persons is well-known. It has been proposed that this property be reduced by heating milk for a few minutes, but such processing is ineffective or undependable in correcting the milk for many sensitive people.

S. Saperstein; U.S. Patent 3,174,866; March 23, 1965; assigned to The Borden Company proposes that these allergenic proteins be heated in contact with casein until the beta-lactoglobulin becomes so complexed, i.e., associated by chemical combination or otherwise, with the casein that 90 to 99% or more of the beta-lactoglobulin will precipitate along with casein in the acidity of human gastric juice, as at a pH of about 4.5. When the beta-lactoglobulin has been so complexed, or modified, the other protein allergens, which are less difficult to correct, will also have been satisfactorily reduced in allergenic properties. The process includes maintaining the casein and beta-lactoglobulin at temperatures below that temperature which, if used for a long time, would cause release of the beta-lactoglobulin and its reactivation as an allergen.

As to materials, the beta-lactoglobulin, other proteins to be modified and the casein are suitably those of milk. In any case they are substantially fat-free and in aqueous dispersions as heated. Examples are skim milk and a reconstituted milk resulting from dispersion of skim milk solids in powder form in water to a solids concentration about that of fresh skim milk or higher up to about 15%. If it is attempted to use homogenized whole milk, the fat causes the effect to be unsatisfactory; the cause is considered to be the interference of the fat with the proper and necessary contact or action of the allergenic proteins with the casein during the heating step.

The proportions of casein to beta-lactoglobulin that occur in cows' milk is suitable for the process. The proportion may be varied, however, as for instance from about 25 up to 150 parts of casein for 1 part of the beta-lactoglobulin. With low amounts of casein, the proportion of the beta-lactoglobulin that precipitates at the isoelectric point for casein is reduced. Casein in amount above the maximum stated is unnecessary and introduces a taste variation that is objectionable for some food uses.

The pH during the heating, to correct the allergenic properties, is above that at which even slight curdling of the milk occurs and below that which if used would cause marked browning of the milk during heating at the temperature and for the long time used. The pH of the milk to be heated is within the range about 6.2 to 6.8 and usually 6.4 to 6.7. The pH drops slightly during the long extended heating as, for instance, by about 0.2 to 0.3 on the pH scale.

The time during which and the temperature at which the milk is subjected to heating to decrease the allergenic properties is critical as to the effect. The heating is continued until at least about 90% of the beta-lactoglobulin becomes precipitable with the casein at a pH of 4.5. This requires at least approximatley 2.5 hours and ordinarily not more than 5. Shorter periods of heating are not effective in making the beta-lactoglobulin precipitable with the casein at pH 4.5. Numerous tests have shown also that such precipitability is required for the desired inactivation step, that is, reduction of allergenic effect of the product when consumed orally. The temperature is within the range about 190° to 212°F and for best results 202° to 208°F.

Restriction of contact of the milk with oxygen during the inactivation heating improves the result. Introduction of air or other form of elemental oxygen is avoided as by not introducing oxygen purposely and not whipping in air by overly vigorous stirring.

Example 1: Pasteurized skim milk was heated in glass on a hot plate with constant but slow agitation at a temperature maintained within the range 202° to 208°F for a total of 3 hours. Samples were taken at intervals. Each sample was divided into two parts. One part of each was autoclaved for 15 minutes at 247°F in a glass container set in a closed retort, this heating simulating terminal sterilization that might be given to the product before actual use in the hospital or the home or during or after canning for marketing.

For test purposes, all samples were diluted with distilled water to a protein content of 1.5% (N x 6.25). The samples were then acidified to a pH of 4.4 to 4.6 with hydrochloric acid. The casein and other materials thus precipitated were removed by centrifuging and the clear centrifugate was mixed with dilute sodium hydroxide solution in amount to raise the pH to within the range 6.7 to 7.1. The additional insolubles so precipitated were again removed by centrifuging.

The separated whey, when tested by the addition of trichloroacetic acid, gave a positive test for protein. Other portions of the whey gave significant results in the precipitin test. In this test, two antisera specific for beta-lactoglobulin and alpha-lactalbumin, respectively, are introduced into the whey in test tubes and with a minimum of mixing. The formation or lack of formation of bands of precipitation at the zones of contact of the whey and the antisera in the test tubes is noted.

The wheys from all samples which had been heated at 202° to 208°F for 2.5 hours or longer but not autoclaved showed negative for the two proteins in this test. Samples heated less than 2 hours, even without autoclaving, showed positive. Wheys from all the samples which had been both heated at 202° to 208°F and then autoclaved for 15 minutes were positive for beta-lactoglobulin but negative for alpha-lactalbumin. This shows that the autoclaving at 247°F had released the beta-lactoglobulin.

Example 2: An infant milk formula using milk previously preheated to reduce activity of whey protein allergens was prepared as follows. Fresh raw fluid skimmed milk was heated in conventional steam-jacketed stainless steel kettles at 205° to 210°F for 165 minutes with constant agitation. The milk was then condensed by evaporation under vacuum.

The product in amount to provide skim solids, 444 lb, was pumped by a positive pump into a jacketed mixing kettle. Sufficient water was added to make the solids concentration of the milk 15 to 16%. The milk was then heated to 120°F and 498 lb of lactose was added. Then there were introduced in order and in aqueous solution potassium citrate, 17.37 lb; sodium citrate, 4 lb; citric acid, 2.5 lb. There were then added, in turn, and stirred in an oil mixture that had been warmed to 140°F and that consisted of 179.66 lb of palm oil, 89.83 lb of peanut oil, 89.83 lb of coconut oil and 5.28 lb of lecithin; 14.2 million (international) units of vitamin A, 2.3 million (international) units of vitamin D and 4.896 grams of vitamin E (alpha-tocopherol).

The resulting mixture was warmed to 160°F and hydrated ferrous sulfate, 194.9 grams, previously dissolved in water, then introduced. Additional nutirents were then added in amounts to make the mixture contain a total of 119 grams methionine, 1.96 grams thiamine hydrochloride, 2.94 grams riboflavin, 21.15 grams niacin, 1.96 grams pyridoxine hydrochloride, 11.75 grams calcium pantothenate, 4.9 mg vitamin B_{12} and 0.64 gram potassium iodide.

The mixture was then homogenized hot at a temperature of 165°F and at a total pressure of 2,500 psi with 1,000 psi on the second stage. The mixture was then cooled to 60°F and passed through a clarifier and a suspension containing 4 lb of calcium carbonate added to the clarified product. A solution containing 549 grams of ascorbic acid was then added to the mixture. The mixture was adjusted by adding water to a final solids content of 36 to 38%. The solution was passed through a tubular heater or other suitable equipment and heated to 200° to 205°F for 2 minutes. From the heater, the mixture was pumped through a nozzle into a spray box whereby drying was accomplished by conventional spray-drying techniques.

Example 3: Fresh raw fluid skimmed milk was heated in a conventional steam-jacketed stainless steel kettle at 205° to 210°F for 180 minutes with constant agitation. The milk was then condensed by evaporation under vacuum. The condensed product of total solids concentration approximately 17% was used in amount to provide skim solids, 8.45 lb. The milk was then heated to 130°F and 7.68 lb of lactose added. Then there was introduced a mixture of 3.4 lb palm oil, 1.7 lb peanut oil, and 1.7 lb coconut oil previously mixed with 226,000 IU of vitamin A palmitate, 37,000 IU of D-3 and 1.35 grams D-alpha-tocopherol acetate. To the mixture was then added 2 lb of lactose previously admixed with 4.53 grams of Sea Kem No. 2, i.e., carrageenin. The temperature was raised to 170°F and 0.2 lb of glyceryl monostearate (Myverol 18-06) was added.

To this mixture was then added 34.08 grams of calcium carbonate and 33.8 grams of potassium bicarbonate. There were then introduced, in aqueous solution, thiamine hydrochloride 54 mg, riboflavin 90 mg, niacin 400 mg, pyridoxine hydrochloride 44 mg, D-calcium pantothenate 217 mg, vitamin B_{12} 30 mg, ferrous sulfate hydrated 3.76 grams and potassium iodide 12 mg. The mixture was then homogenized at a temperature of 165°F and at a total pressure of 2,500 psi with 500 psi on the second stage. The mixture was then cooled to 60°F and 9.1 grams of ascorbic acid, previously dissolved in water, was added. Additional water was added to adjust the solids content to 25%.

The product was then pumped through a heated tube of 5/16" O.D. and 80 feet long at 300°F, 3½ seconds being required for the product to pass through the tube and about 2 of the 3½ seconds representing holding time at 300°F to give a sterilization equivalent of F_0 = 3 to 15 and suitably 8. The product temperature was then lowered to 165°F in a continuing tube and then passed through a homogenizing block in series with the outcoming line. The sterile product was thus homogenized at a pressure of 2,500 psi. The product was then cooled to 95°F by passage through a cooling coil and then aseptically canned in a Martin aseptic canner.

The method as illustrated by the examples reduces antigenicity of bovine serum albumin of milk and alpha-lactalbumin, two of the whey proteins which are potential and known allergens, to a level such that they cannot be detected by gross anaphylactic shock in guinea pigs or by the tube precipitin test described.

For the precipitin test here used, the milk is acidified to pH 4.5 with dilute hydrochloric acid or acetic acid and the precipitate that results is removed. The resulting whey is then neutralized by the addition of sodium hydroxide solution and all insoluble material that appears is removed, as by centrifugation. The resulting clear whey is placed in dilution tubes and diluted with 0.87% sodium chloride solution.

The highest concentration tested represents a dilution of the original milk protein (N x 6.25) of approximately 1:66. When tested by the conventional tube precipitin methods, using a high titer specific antiserum, no visible precipitin bands are obtained at the highest concentrations of the proteins, for either bovine serum albumin of milk or for alpha-lactalbumin. The antisera used are capable of detecting the specific proteins within 10 minutes when these proteins are present at a concentration of from 1:10 up to 1:200,000.

When an antiserum of equal potency but of kind specific for beta-lactoglobulin is used in the test, the test becomes positive only after not less than 90 minutes in the 1:66 dilution and is usually negative at all higher dilutions. Gross anaphylactic shock tests on guinea pigs presensitized to the specific proteins, when such animals are challenged with the neutralized whey made as described above are negative, even with doses as high as 1 ml administered intravenously whereas control animals give positive responses in such a test to as little as 1 microgram of the specific protein. Previously available milk products, for which low allergenicity is claimed, give positive precipitin tests and positive anaphylactic shock tests for both alpha-lactalbumin and beta-lactoglobulin.

Example 4: The procedure and materials of Example 1 are used except that the components mixed and processed were of kind and in amount to make the final dried powder of the following composition.

Components	Weight for Approximately 100 lb of Powder
Skim milk solids	33.3 lb
Lactose	37.3 lb
Palm oil	13.5 lb
Peanut oil	6.75 lb
Coconut oil	6.75 lb
Lecithin	0.4 lb
Sodium citrate	0.3 lb
Potassium citrate	1.3 lb
Citric acid	0.19 lb
Methionine	8.9 grams
Vitamin A palmitate, 1 million units/g	1.1 grams
Vitamin D_2, 1 million units/g	0.17 grams
D-alpha-tocopherol acetate, 1,360 units/g	1.35 grams
Ascorbic acid	36 grams
Thiamin hydrochloride	146 mg
Riboflavin	158 mg
Niacin	1,464 mg
Pyridoxine hydrochloride	146 mg
D-calcium panthothenate	880 mg
Vitamin B_{12}, 3,000 μg/g	120 mg
Ferrous sulfate·$7H_2O$	14.6 grams
Potassium iodide	48 mg

Lactalbumin- and Lactoglobulin-Free Drink

A simulated milk which does not contain lactalbumin and/or lactoglobulin is described by

P.P. Noznick and C.W. Tatter; U.S. Patent 3,397,992; August 20, 1968; assigned to Beatrice Foods Co. The drink is prepared from a polyglycerol ester of a higher fatty acid, a vegetable fat, colloidal cellulose and water. Preferably dextrins (carbohydrates) are present and it is also preferable, except in the case of the preparation of a nonprotein dietetic drink to include a protein other than lactalbumin or lactoglobulin. Normally carboxymethylcellulose (usually as the sodium salt) is also included. The colloidal cellulose is Avicel RC which is composed of 92% microcrystalline cellulose and 8% sodium carboxymethylcellulose. Because of the fine particle size of the Avicel RC there is no problem in drinking the imitation milk. The preferred emulsifying agent is decaglycerol decastearate but there can also be used other diglycerol to triconto glycerol esters of fatty acids.

The term fat includes both liquid and solid fats. Among the suitable edible fats which can be used are cottonseed oil, coconut oil, safflower oil, corn oil, soybean oil, peanut oil, hydrogenated cottonseed oil, hydrogenated soybean oil, hydrogenated peanut oil, olive oil, hydrogenated olive oil, hydrogenated coconut oil, e.g., hydrogenated to a melting point of 92°F. There can also be used animal fats such as butterfat and lard.

As the carbohydrate, there can be used corn syrups. The DE equivalent has a profound effect on the emulsion stability in that the greater the amount of monosaccharide present the harder it is to properly emulsify the composition. Hence the use of corn syrups with very low monosaccharide content is preferred. Illustrative of preferred corn syrups are Frodex 15 (corn syrup with a DE of 15) and Mor-rex 1913 and 1918 both of which have a low monosaccharide content. Additional flavoring and color agents can be added to simulate fresh milk but they are not essential and do not contribute to stability of the emulsion.

It has also been found that emulsion stability can be improved by adding a citrate-phosphate buffer. Thus there can be used citric acid-tetrasodium pyrophosphate or citric acid-disodium phosphate. Preferably sodium chloride is included with these buffers. The buffers (and sodium chloride) also improve the overall flavor of the product and make it less bland or dry to the taste. The citric acid can be used in an amount of 0.01 to 0.2% of the mixture, the sodium phosphate component can be used in an amount of 0.05 to 0.3% of the mixture and the salt can be used in an amount of 0.01 to 0.2% of the mixture.

When a protein is used it is preferably sodium caseinate although there can also be used soybean protein, calcium ammonium and potassium caseinates, low-viscosity gelatin, fish protein, wheat gluten or corn gluten. Next to sodium caseinate the preferred protein is soybean protein. The most convenient method for preparing the product is to dissolve the emulsifier, e.g., decaglycerol decastearate, in the melted fat. The remaining ingredients are dispersed in water and heated, e.g., to 150° to 160°F. Then the fat and aqueous materials are combined and pasteurized, e.g., at 155° to 160°F for 30 minutes. The product is then homogenized, e.g., at 2,000 to 3,000 psi, and chilled over a surface cooler, i.e., chilled immediately to about 40°F and shipped as such.

The proportions of materials are as follows: fat 1 to 5%, preferably 3 to 3.5%; polyglycerol esters 0.02 to 1.0%, preferably 0.05 to 0.3%; colloidal cellulose 0.2 to 2.0%, preferably 0.3 to 1.0%; protein 1 to 4% (can be omitted); corn syrup solids 1 to 10% (can be omitted); carboxymethylcellulose 0.02 to 0.15% (can be omitted); the balance is water. The water is usually 82.5 to 94.5% if protein is present and 85 to 95% if the protein is omitted. Unless otherwise indicated all parts and percentages are by weight.

Example 1:

	Percent
92° coconut oil	3.4
Decaglycerol tristearate	0.1
Frodex 24 (corn dextrin solids DE 24)	5.0
Avicel RC	0.4
Sodium caseinate	3.5
Water	87.6

The decaglycerol tristearate was dissolved in the melted coconut oil. The Frodex 24, Avicel RC and sodium caseinate were dispersed in water and heated to 155°F. The decaglycerol tristearate in the coconut oil was then added and the mixture pasteurized at 155° to 160°F for 30 minutes, homogenized at 2,500 psi and chilled over a surface cooler to 40°F. The product was similar to milk in appearance and was a satisfactory milk substitute for people who cannot eat lactalbumin and lactoglobulin.

Example 2:

	Percent
92° coconut oil	3.15
Decaglycerol decastearate	0.20
Corn products 1918 corn syrup solids	5.5
Sodium caseinate	2.8
Avicel RC	0.5
Water	87.85

The emulsion was prepared in the same manner as in Example 1.

Example 3:

	Percent
92° coconut oil	3.10
10-19-S (decaglycerol decastearate)	0.2
Sodium caseinate	2.8
Frodex 24	5.0
Avicel RC	0.5
Citric acid	0.03
Tetrasodium pyrophosphate	0.07
Salt	0.025
Water	88.275

The emulsion was prepared in the same manner as in Example 1. The citric acid, tetrasodium pyrophosphate and salt were added to the water dispersion prior to adding the coconut oil.

AMINO ACID FORTIFIED MILK

Cow's milk is known to be unsuited as the sole or principal source of human nourishment whereas human milk can supply all nutrients required by humans, particularly infants. One of the most important differences between cow's milk and human milk resides in the distribution of constituent amino acids in bovine and human milk proteins. Cow's milk protein is deficient in methionine and tryptophan as compared to human milk protein, and the nutritional value of cow's milk can be substantially enhanced by adding free methionine and free tryptophan. These amino acids are very well tolerated in the amounts required, and are readily added to fluid milk, milk concentrates, such as evaporated milk or condensed milk, and to dry nilk prepared from whole milk or from skim milk.

T. Hino, K. Hayashi and N. Iizuka; U.S. Patent 3,415,655; December 10, 1968; assigned to Ajinomoto Co., Inc., Japan further discuss the problem and give a process to eliminate the undesirable effects of addition of methionine and tryptophan to cow's milk.

Methionine and tryptophan, however, impair the taste and odor of milk products, and fresh or reconstituted cow's milk containing an amount of tryptophan and/or methionine substantially corresponding to the deficiency of the milk protein in these amino acids as compared to human milk is unfit for general human consumption for this reason. Amounts of methionine and tryptophan which do not deleteriously affect the taste and aroma of cow's milk or cow's milk products are too small to have significant nutritional value, while amounts of these amino acids which are truly beneficial not only impart their own characteristic taste and odor to the product but they also suppress the characteristic taste of milk which is found even in evaporated milk or in milk reconstituted from dry skim milk, and makes the liquid product readily distinguishable from cow's milk to which tryptophan and methionine are not added.

The amount of methionine necessary for making up the deficiency of cow's milk as compared to human milk is approximately 50 to 100 mg of methionine per gram of protein nitrogen in the cow's milk, or per 6.39 grams of cow's milk protein. The corresponding amount of tryptophan is between 10 and 20 mg. While these values are characteristic of normal cow's milk and normal human milk, differences somewhat outside these limits may exist, and may require corresponding additions to cow's milk for compensation.

It was found that the undesirable effects of methionine and/or tryptophan on the taste and odor of cow's milk and milk products, can be avoided by mixing to such milk products monosodium glutamate and at least one amino acid of the group consisting of glycine, alanine, serine and threonine. The improvement thus mainly resides in an enriched milk product which contains the protein of natural cow's milk, at least one of the amino acids methionine and tryptophan in an amount substantially corresponding to the deficiency of the cow's milk protein in these amino acids, monosodium glutamate, and at least one of the four amino acids of the group of glycerine, alanine, serine and threonine.

A milk product indistinguishable from a control sample not containing any additives is obtained when the glutamate and one or more members of the group of the four mentioned amino acids are present in about equal amounts of 10 to 20% of the amount of the indispensable amino acid or acids. The effects of various amino acids and amino acid salts on the taste of fresh cow's milk reinforced with 0.05% methionine and 0.01% tryptophan were as follows. The reinforced milk had the strong odor and bad aftertaste characteristic of methionine. It was somewhat bitter, as is characteristic of tryptophan, and had little milk taste and aroma. The milk taste was more noticeable with the addition of 0.006% alanine, glycine, serine or threonine, but the methionine odor was still too strong to make the milk palatable.

Proline produced only a slight improvement as to milk taste and methionine odor, and imparted a sweetish taste of its own when added in an amount of 0.006%. Phenylalanine, leucine, isoleucine, and valine were also useless, and tended to reduce the milk taste in plain cow's milk. The hydrochlorides of arginine and histidine did not produce any improvement in the taste of reinforced milk. Monosodium aspartate in an amount of 0.006% restored the milk taste to a small extent but could not restore the sweetness of ordinary cow's milk. 0.006% monosodium glutamate fully restored the milk taste and completely obliterated the methionine odor but the treated milk was somewhat lacking in sweetness. The flavor and aroma of the reinforced milk could not be distinguished from that of the original cow's milk after addition of 0.006% each of monosodium glutamate and glycine.

The other three amino acids of the group are similarly capable of restoring sweetness to milk which contains methionine, tryptophan, and enough monosodium glutamate to overcome the odor and other taste effects of the indispensible amino acids added. The sweetening effect of alanine is approximately equal to that of 1.2 weights glycine, and the glycine equivalents of serine and threonine are 0.8 and 0.44 weights, respectively.

0.002% of each of monosodium glutamate and glycine could not satisfactorily overcome the methionine odor nor restore the milk taste in the opinion of a panel of trained tasters. They considered 0.005% each of monosodium glutamate and glycine fully to restore the original's cow's milk aroma and taste. At higher concentrations of glutamate and glycine, more tasters objected to the increasing prevalence of the taste of the addition agents. 0.01% was still considered good by most tasters. The objectors almost prevailed at 0.02%, and a reinforced milk containing 0.05% each of monosodium glutamate and glycine was almost unanimously rejected.

The results were similar when the glycine was replaced partly or entirely by alanine, serine and/or threonine in amounts having comparable sweetening effects. Replacement of the sweet-tasting amino acids by more conventional sweetening agents such as glucose and sucrose in any proportion, or by artificial sweeteners, resulted in a product which was readily distinguished by a majority of the panel members from original cow's milk. The specific sweetening effect of the amino acids mentioned above is believed to be related to the fact that they buffer the liquid milk product to the slightly alkaline to neutral pH value of ordinary cow's milk in a manner not possible with carbohydrates or the conventional synthetic sweetening agents.

Example 1: Commercial dry milk containing 16% protein by weight was intimately mixed with 0.3% DL-methionine and 0.06% L-tryptophan. Samples of a 15% solution of the mixture in warm water were submitted to a 40-member panel for comparison with a control solution free from added methionine and tryptophan. Thirty panel members declared the two samples to be substantially different in taste and aroma. Of the 30 panel members asked for a preference between the two samples, 25 preferred the control sample.

The original dry mixture was further mixed with 0.036% monosodium glutamate, and the panel was again asked to determine the difference between a 15% solution in water and a control sample made up from the original dry milk. There was some preference for the control samples, and recognition of a difference by some panel members but the panel was too small to make the results statistically conclusive.

The addition of 0.018% glycine and 0.018% DL-alanine to the mixture of dry milk with methionine and tryptophan produced a liquid sample which was considered different from the control sample by a statistically significant number of panel members but the results of a preference test were not statistically conclusive. There were some complaints about a strange odor in the methionine-bearing mixture.

A sample of dry milk, methionine and tryptophan, when further mixed with 0.036% monosodium glutamate, 0.018% DL-alanine and 0.018% glycine could not be distinguished from a control sample by a 40-member panel, and there was no statement of significant preference.

Example 2: Commercial fluid cow's milk containing 2.9% crude protein was partly mixed with 0.05% DL-methionine and 0.01% tryptophan. A sample of this material was considered different from an untreated milk sample by 40 panel members at the 1% significance level. 27 members of a 30-member panel preferred the control to the treated sample (significant on the 0.1% level). Those panel members who preferred the control objected to the methionine odor and the lack of milk taste in the treated sample.

When the treated milk was additionally mixed with 0.005% monosodium glutamate, 0.004% glycine and 0.004% DL-alanine, only eleven members of a panel of forty members found a difference between treated sample and control, and the eleven were about evenly split as to their preference for the sample or the control.

In a third test, the glycine and alanine of the preceding test were replaced by 0.002% each of glycine, DL-alanine, L-serine, and L-threonine. The results were approximately the same. Several panel members indicated that the taste of the treated sample was somewhat stronger than that of the control.

Example 3: Evaporated milk containing 6.8% crude protein was mixed with 0.1% DL-methionine and 0.02% L-tryptophan, and two batches of this mixture and corresponding batches of untreated evaporated milk were diluted with about 2½ parts water. The difference between a sample of one treated batch and the control was recognized by 29 of the 40 panel members, and 25 of the 29 recognizing a difference preferred the control.

The other treated batch was mixed with 0.01% monosodium glutamate, 0.04% DL-serine and 0.04% DL-threonine, based on the weight of the original evaporated milk. Of the forty panel members questioned, fourteen claimed to notice a difference between the treated sample and the control, but the 14 noticing a difference were about equally divided as to their preference between the sample and the control.

SIMULATED MILK

Soy-Sesame-Coconut Formula

Because of allergies, many individuals cannot include cow's milk in their diets and frequently their physicians will direct them to use soybean milk which is presently the only acceptable substitute. Soybean milk is usually prepared from soybeans which have been finely ground to provide a powder which is dispersible in water. The substitute soybean milk conventionally marketed has a dark unattractive appearance and is vastly different in flavor from cow's milk and is often found unpalatable by those accustomed to cow's milk. Conventional soybean milk is nutritionally less acceptable than cow's milk in several ways. Soybean milk contains only approximately half of the essential amino acid, methionine, which is found in a like quantity of cow's milk.

Soybeans contain a toxic factor which is antithyroid in activity and an enzyme which interferes with protein digestion. This toxic factor and objectionable enzyme are conventionally destroyed by light roasting of the soybean but unfortunately the detoxification procedure also destroys beneficial enzymes of the soybean including the highly valuable phosphatase enzyme. Other nutritional materials are to some extent damaged and lessened in value by the toasting procedure. Soybean milk contains less than one-fourth of the calcium content of milk, which shortcoming is especially undesirable for a growing child. Soybean powder now employed for making of simulated milk is inherently inadequate in antioxidant properties and deteriorates in the presence of air.

A simulated cow's milk utilizing a special blend of vegetable proteins is described by *G.W. Johnson; U.S. Patent 3,386,833; June 4, 1968.* The composition contains three principal ingredients: soybean flour, sesame seed flour, and coconut meal. The three components are ground to extremely fine powder with the soybean being provided preferably in four parts by weight to three parts by weight of the sesame seed and two parts by weight of the coconut meal. Somewhat varying amounts of the three components may be used in the composition. The soy and sesame seed flour are best employed for a superior milk respectively in the range of 3 to 5 parts by weight per 1 to 2 parts by weight of the coconut meal. The coconut meal is prepared from dried coconut meat (copra).

A small amount of lecithin, usually in the range of 0.5 to 2% by weight of the three principal ingredients, is incorporated as a water-dispersing agent for the three ingredients. The powdered mixture may be placed in liquid suspension by stirring in water, preferably employing a Waring-type blender. The product will remain in suspension for a long period of time without the aid of an emulsifying or suspending agent. However, a superior product is had by incorporating a small stabilizing amount of an edible emulsifying agent such as extracts of Irish moss, (*Chondrus crispus*) or other edible alginate-base material. The composition may be enriched by incorporating vitamins or other desired ingredients.

In the preparation of the liquid simulated milk, 6 to 9 teaspoons (approximately 4.9 ml per teaspoon) of the powdered composition are added to one quart of water and thoroughly mixed in a blender. The precise amount of powdered composition used may be varied somewhat to suit the taste of the user. In the plant manufacture of the composition it is advantageous to first mix the finely ground powder composition with water to make the simulated cow's milk and then spray-dry to obtain a product which may be placed by the consumer more readily into a liquid form.

Cow's milk and conventional soybean milk have high protein contents of about 3.4% compared with human breast milk of about 1.4% protein. This higher protein content of cow's and soybean milks is frequently not acceptable to infants and it becomes necessary to dilute the milk where it is being substituted for human milk and there is thus a resultant lowering of other desirable nutritional values. The simulated milk contains a significantly less protein content of approximately 1.5% which more nearly approximates that of human milk than either cow's milk or soybean milk. The lower protein content is attributable to the significantly less protein found in sesame seed flour and coconut meal than in undiluted soybean flour.

A principal shortcoming of soybean milk apart from its relative unpalatability has been its relatively low methionine (an essential sulfur amino acid) content, being approximately 60% that of cow's milk. If it should be that the diet of an individual is deficient in betaine and choline, the principal sources of the methyl groups needed by normal metabolism of the body, methionine is called on to supply the needed methyl groups, thus leading to a further deficiency of this essential sulfur amino acid. This simulated milk composition has a high methionine content, considerably more than that of conventional soybean milk. Sesame seed flour contains over twice the amount of methionine found in soybean flour. Cow's milk is deficient in iron. The product has approximately six times as much iron as found in cow's milk.

Soybean milk heretofore used as a cow's milk substitute is produced from a raw material which is necessarily toasted to eliminate the toxic factor characteristic of soybeans. The simulated milk composition includes as principal ingredients the raw and untreated sesame seed flour and coconut meal which contain their full natural enzyme complement and other valuable nutritional elements undamaged.

A principal objection to the use of soybean milk as a substitute for cow's milk is its exceedingly low calcium content. Soybean milk characteristically contains approximately one-fifth of the calcium found in cow's milk. The simulated milk composition contains significantly more calcium than is found in conventional soybean milk. Sesame seeds contain approximately six times as much calcium as is characteristic of soybeans. Because of this the composition of the process contains a much higher calcium content than that of conventional soy milk.

The fat of cow's milk is predominantly saturated and this is objectionable for the consumer who is on a low-cholesterol diet. The simulated milk composition is made of vegetable materials having their fats principally in the unsaturated form. The simulated milk composition contains a relatively high proportion of the two important amino acids, namely, lycine and methionine, which have been demonstrated as being capable of initiating full growth in cases of severe malnutrition. Soybeans, sesame seed, and coconut all contain high amounts of both lycine and methionine and this makes the product of the process especially attractive for use in countries having low protein resources.

Soybean flour commonly used heretofore in the production of a simulated cow's milk has an objectionably short keeping quality, primarily because of the absence of a suitable antioxidant material to protect the unsaturated fats of the soy bean against oxidation. It has been found that sesame seed flour intimately mixed with the soybean flour and coconut meal protects the latter two materials effectively for long periods of time against oxidation. An unrefrigerated and unsealed powder composition at the end of eight months, when added to water and blended therewith, showed but slight change in flavor. A soybean composition containing no sesame seed under similar handling was rancid.

Whey and Peptone Proteins

A potentially cheap source of protein is tank water—an aqueous liquid from steam rendering of fat which includes a soluble protein material. This soluble protein is a heat hydrolysate of collagen coming from skin and fat cells. After lard is drawn off, and tank water is screened, the screened tank water is concentrated, for example, in a triple effect evaporator, to 60 to 70% solids. This concentrate is known as stick. Very few plants have wet rendering facilities so stick may be considered to be generally unavailable at the present time. However, it has been estimated that for the entire pork industry, there should be 500 million pounds of collagen available per year (dry solids basis). If economic considerations were attractive enough, it could be steam-processed for utilization.

Hydrolysate of collagen produced from steam rendering of fat is notably deficient in tryptophan, threonine, methionine plus cystine and isoleucine. On the other hand, whey protein supplies generous amounts of these amino acids. Equal amounts of whey protein and peptone protein would appear to give an amino profile similar to cow's milk.

The details of the process to produce an imitation milk utilizing these protein materials are given in the following by *F.C. Olson; U.S. Patent 3,778,514; December 11, 1973; assigned to Oscar Mayer & Co., Inc.*

Example 1:

Imitation Milk Product, Ingredients	Pounds
Lactose-reduced whey concentrate (40% solids)	90
Collagen hydrolysate (40% solids)	10
	100

Whey concentrate is prepared in a conventional method by low-temperature concentration of centrifuged cheddar cheese whey to 35 to 40% solids. It is cooled to 40°F and permitted to stand. Lactose crystallizes in the standing liquor and is separated by centrifuging. The effluent is adjusted to 40% solids to provide the lactose-reduced whey concentrate for use as an ingredient in this example.

The collagen hydrolysate ingredient is prepared in accordance with the following procedure: U.S.D.A. inspected and passed edible fatty connective tissue is used as a starting material. This material includes fatty trimmings with and without attached skin. The material is placed in a pressure vessel and heated by direct injection of steam at 265°F for 2.5 hours. The contents of the vessel are then permitted to settle for 1.5 hours. A rendered fat layer accumulates at the top and is drawn off separately. The lower aqueous phase is separated, filtered, and concentrated. The resulting material is approximately 49.2% protein (dry solids basis). The amino nitrogen in the concentrate is between 4 and 8% of the total nitrogen present.

Whey concentrate and the collagen hydrolysate are thoroughly mixed. The mixture is heated to 130°F, pasteurized by high-temperature, short-time treatment, and spray-dried. The product analysis (product basis) was as follows:

	Percent
Protein	33.6
Lactose	42.0
Ash	19.1
Moisture	4.2

The resulting nutritious protein product provides an amino acid profile between that of soybean meal and casein. The thiamine, riboflavin, niacin and pantothenic acid and vitamin B_{12} are believed to be double the level observed in milk.

Example 2:

Protein Additive, Ingredients	Pounds
Whey protein concentrate (40% solids, 18% protein)	70
Collagen hydrolysate (40% solids, 36% protein)	10
Soy flour (8% moisture, 47% protein)	20
Water	50

The above ingredients were thoroughly mixed and spray-dried. The yield was 52.5 lb. The resulting product was dispersible in water although not soluble. The protein content was 48.7% and was of good nutritive value. The resulting product was high in vitamin B complex. It can be used as a nutritive additive to wheat or corn flour to increase the protein value of native bread in foreign countries. It can also be used as an imitation milk.

As illustrated in the examples, the protein content of the whey-collagen hydrolysate mixture can be increased by addition of appropriate vegetable protein concentrates. For example, soy flour can be used. This combination's first deficiency would be in methionine plus cystine, but otherwise, the mixture would be equivalent to cow's milk. Also, wheat germ is considered to be a good vegetable protein, and, it can be combined with the whey-collagen hydrolysate mixture to provide a nutritional mixture whose first deficiency would probably be isoleucine.

Further, the whey-collagen hydrolysate mixture can be combined with sesame flour. Sesame protein is generally known to provide rather high methionine plus cystine but a low lysine content. The whey, collagen hydrolysate, sesame protein combination would provide a high nutritive value of protein. The protein content of some yeasts is about 45%, and this protein has good nutritive value. Although brewers' yeast is rather bitter, it can be debittered. Yeast grown from waste products of wood pulping, e.g., Torula yeast, also provides high-nutritive-value protein. Debittered brewers' yeast, or Torula yeast can be used in the whey-collagen-hydrolysate combination to give a protein amino acid profile which is believed to be equivalent to cow's milk.

NUTRITIOUS BEVERAGES

PROTEIN DISPERSION PROCESS

In a description of a process for a protein containing beverage *A.L. Gordon; U.S. Patent 3,773,961; November 20, 1973; assigned to Kraftco Corporation* gives some background on the problems of formulating and processing such a product.

From a nutritional standpoint, soft drink beverages contain carbohydrates, which supply energy, but contain no protein. It is possible to add protein to soft drink beverages which have a pH above the isoelectric pH of the protein used. Most proteins, such as casein and soy protein, are markedly insoluble at the isoelectric point of the protein. The isoelectric point for most proteins is in the range of about pH 4 to 5. It is well known to provide casein and other proteins in soft drink beverages when the soft drink beverage has a pH above about 5.

However, substantial problems are encountered in providing a protein source in soft drink beverages when the pH of the soft drink beverage is below the isoelectric point. It is difficult to reduce the pH of a soft drink beverage containing protein below the isoelectric point and avoid coagulation of the protein. It is, furthermore, difficult to provide a stabilized soft drink beverage containing protein when the pH of the beverage is maintained below the isoelectric point. Also, when a protein dispersion is maintained for a substantial period of time at a pH below the isoelectric point for the protein, there is a tendency for the protein to gel or thicken. A special process is used to prepare an acidic aqueous beverage containing protein.

A protein source is blended with water to provide a protein dispersion. The protein source is provided at a level sufficient to provide from 1 to 8% protein in the dispersion. Thereafter, the protein dispersion is acidified by the addition of an edible acid to the dispersion. The edible acid is rapidly added to the protein dispersion, while the protein dispersion is maintained under agitating conditions. The edible acid is added to a level and rate sufficient to reduce the pH of the protein dispersion to below the isoelectric point of the protein source as rapidly as possible. In general, the isoelectric point for the majority of protein sources is from pH 4 to 5.

It is desirable to reduce the pH of the protein dispersion to below about 3.5 and preferably to a pH within the range of from 2.0 to 3.3. Any edible acid may be used to acidify the protein dispersion. Suitable edible acids include phosphoric acid, hydrochloric acid, malic acid, citric acid, lactic acid, adipic acid, ascorbic acid, carbonic acid, acetic acid, tartaric acid, fumaric acid and mixtures. Preferred edible acids for reasons of economy and flavor are phosphoric acid, citric acid and malic acid. A system for reducing the pH of the protein dispersion is a mixture of phosphoric acid and malic acid. The mixture provides a particularly desirable flavor and mouthfeel in the finished beverage product. The mixture should have from 1 to 3 normal equivalents of malic acid per normal equivalent of phosphoric acid.

An aqueous, flavored base is then prepared. The aqueous flavored base generally includes water, flavoring material and may include a sweetening agent. Natural or artificial flavoring materials may be used to provide the aqueous base with any suitable flavor, such as lemon, lime, orange, cherry, strawberry, grape, cola and other fruits and vegetable flavors. The sweetening agent, when used, may be a natural or artificial sweetening material, such as sucrose, fructose, invert sugars, lactose, dextrose, corn syrup solids, saccharin, cyclamates, dipeptides and mixtures.

The acidity of the aqueous flavored base is maintained at a level such that when the base is mixed with the protein dispersion, the acidity of the mixture will be below the isoelectric point for the protein material used. Generally, the acidity of the aqueous flavored base will be in the range of from 2.5 to 4.0 pH, and the pH of the mixture will be in the range of from 2.0 to 3.3. The level of use of the sweetening agent and the flavoring materials are sufficient to provide the desired levels of flavor and sweetness in the final beverage product.

In general, the protein dispersion and the aqueous flavored base are combined at levels which are sufficient to provide from 2 to 6 grams of protein per 8 ounces of the beverage product. Higher levels of protein may result in protein gelation or protein precipitation. Lower levels of protein may be used, but a less significant nutritional benefit is derived when the protein level in the beverage is below about 2 grams of protein per 8 ounces of beverage product. To achieve the desired level of protein, from 150 to 300 grams of aqueous flavored base is combined with from 150 to 300 grams of protein dispersion.

It has been found that the protein dispersion and the aqueous flavored base should be combined in particular sequence and under particular conditions. In this connection, if the pH of the aqueous flavored base is higher than the pH of the protein dispersion the aqueous flavored base is added to the protein dispersion. If the protein dispersion is added to the aqueous flavored base, precipitation occurs. In general, the aqueous flavored base may be added as rapidly as possible to the protein dispersion, so long as agitating conditions are maintained. However, if the pH of the aqueous flavored base is equal to or lower than that of the protein dispersion the sequence and rate of addition is immaterial.

It has also been determined that the temperature of the mixture that results from blending the aqueous flavored base with the protein dispersion should be within a predetermined range. In general, the temperature of the mixture should be from 30° to 60°C. When the temperature of the mixture is below 30°C, precipitation of the protein may occur. When the temperature of the mixture is above 60°C, gelling of the protein may occur. It is preferred that the temperature of the mixture be from 35° to 50°C. The temperature of the protein dispersion should be above 40°C at the time of the blending to prevent any gelation of the protein prior to blending. In general, it is preferred that the temperature of the protein dispersion be within the range of from 40° to 80°C. The temperature of the aqueous flavored base may be at a level which results in a mixture temperature within the range of from 30° to 60°C. Preferably the temperature of the aqueous flavored base is from 10° to 50°C prior to blending.

The pH of the mixture that results by blending the aqueous flavored base and the protein dispersion must be below the isoelectric point for the protein used. For most proteins the pH of the mixture is preferably from 2.5 to 3.2. However, lower pH values can be used without disturbing the stability of the mixture but the acid flavor below about 2.5 pH is considered too pronounced. Higher pH values for the mixture up to about pH 4.0 may be used with certain protein materials, such as the serum protein of milk. However, at pH values above about 3.2 for most protein materials the mixture becomes destabilized and the protein precipitates. The protein beverage may be carbonated if desired. In this connection, a protein beverage may be prepared which is concentrated with respect to protein and flavor levels. Carbonated water is then added. In an alternative method for preparing a carbonated beverage, a flavored protein beverage is prepared having the desired level of protein and flavor. Thereafter, carbon dioxide gas is dispersed in the flavored protein beverage under suitable conditions of temperature and pressure.

Example 1: Casein is used as the proteinaceous material in a lemon-lime flavored protein beverage. For each six ounces of the lemon-lime flavored protein beverage to be prepared, 3 grams of casein are dispersed in 28.4 ml of water. The casein dispersion, while being continuously stirred, is heated to a temperature of 75°C. 2.5 ml of 6 N H_3PO_4 is added to the casein dispersion to provide a pH in the dispersion of 2.5. An aqueous, lemon-lime flavored base is prepared having the following composition per gallon of base.

Ingredients	Amount
Sucrose	6 lb
Anhydrous citric acid	10 g
Fluid lemon-lime flavor concentrate	2 oz
Sodium benzoate	7.75 g
Sodium citrate	5.0 g
Color	0.25 g

The pH of the flavored base is 4.2. The temperature of the flavored base is adjusted to 35°C and 1 gal of the flavored base is then added to 1 gal of the acidified protein dispersion to provide 2 gal of a lemon-lime flavored beverage. The temperature of the acidified casein dispersion is 58°C at the start of the addition of the lemon-lime flavored base to the casein dispersion. The temperature of the mixture is 43°C and the pH of the mixture is 2.9. After the addition of the flavored base to the casein dispersion, a stabilized lemon-lime flavored beverage is provided having 3 g of protein for each 6 oz of beverage.

Example 2: A lemon-lime flavored beverage product is prepared as in the procedure of Example 1, with the exception that milk serum protein is used as the proteinaceous material. The milk serum protein is obtained by treating whey from a cheesemaking operation by ultrafiltration so as to remove lactose and salts from the whey and provide a whey product with 35 to 80% milk serum protein on a dry solids basis. The beverage obtained is stable for extended periods of time at room temperature and is stable indefinitely under refrigeration conditions.

Example 3: A lemon-lime flavored beverage product is prepared as in the procedure of Example 1, with the exception that soy flour is used as the proteinaceous material. The soy flour used has a soy protein level of about 55% on a dry solids basis. Six grams of the soy flour are used for each 6 oz of beverage product so as to provide 3 grams of soy protein in the beverage product. The beverage product obtained has a desirable lemon-lime flavor and is stable for extended periods of time at room temperature.

PHYTASE TREATMENT

In the enzymatic reaction of *E.M. McCabe; U.S. Patent 3,733,207; May 15, 1973; assigned to Societe d'Assistance Technique pour Produits Nestle SA, Switzerland* a soluble protein fraction is prepared for use in beverages.

The soy protein starting material (isolate, soy flour or meal) may first be suspended in water or dilute alkali, for example at pH 7.0 to 8.5. The solids concentration should not exceed about 15% by weight as at higher levels the solutions become very viscous and hence less easy to handle. An acid is then added to adjust the pH value of the suspension to about 4.6 to 6.0. The enzyme is most active at a pH of 5.0 to 5.5 but it is preferred to work at pH values of 5.5 to 6.0 as in this range a greater proportion of the protein present is dissolved. Phytase is then added in an amount corresponding to 0.01% by weight of the protein present, and enzymic reaction is carried out at 50° to 55°C, this being the temperature of optimum activity of the enzyme. The reaction time should be sufficient to secure breakdown of substantially all the phosphate linkages present in the protein. Depending on the activity and purity of the enzyme, as well as the reaction conditions, the total time may be up to about 36 hours. Thereafter, the pH of the reaction mixture is adjusted to a value of about 4.6, and the resultant precipitate, containing unreacted material, is separated as by filtration or centrifugation. Hydrochloric acid is preferred for pH adjustment, but any other nontoxic acid may also be used.

The pH of the recovered clear solution is then adjusted to a value of 5.0 to 5.4, preferably about 5.2. A precipitate is again formed, and it may be recovered as previously. This precipitate consists of the protein fraction soluble at low pH values (below 4.6), and it may be formulated in various edible compositions. It is particularly suitable for incorporation in beverages such as soft drinks which may be carbonated, for example at a level of 2.0 to 2.5% by weight. The desired protein fraction may be further purified by washing, molecular sieving or gel filtration to remove traces of salts formed during the process and to eliminate off flavors. The product has a bland flavor and is soluble in water as acidic solutions at pH values below 4.6.

Example 1: Ten parts of soy protein isolate are dissolved in 100 parts of water to which sufficient sodium hydroxide has been added to raise the pH to about 8. Hydrochloric acid is then added to lower the pH to 6 and 0.1 part of a phytase-containing preparation (obtained from wheat meal by the method described by F.G. Peers, *Biochem. J.,* 53, 102 (1953) are added. The temperature is maintained at 50° to 55°C for 24 hours, which is generally sufficient for the phytase to break down the phosphate linkages in the protein.

Upon completion of the reaction the pH of the medium is adjusted to 4.6 by addition of hydrochloric acid. The resultant precipitate is removed by centrifugation and discarded, whereas sodium hydroxide is added to the supernatant to raise the pH to 5.2 where a precipitation occurs. The precipitate thus formed, consisting of the protein fraction soluble at pH values of 4.6 and below, is recovered. It may be further purified by washing, reprecipitation or gel filtration to remove traces of salts formed during neutralization. The protein fraction may also be dried, giving a white powder with a bland taste, which is suitable for incorporation in different foods and beverages.

Example 2: A lemon-lime flavored soft drink is prepared from the following ingredients:

	Parts
Sugar	10.0
Protein fraction of Example 1	2.0
Citric acid	0.1
Sodium benzoate	0.05
Lemon-lime flavor	0.05
Carbonated water to give 100 parts.	

The resulting beverage (pH 2.8) is clear and has an excellent mouthfeel, with no beany or paint soy flavor.

Example 3: A cola-flavored soft drink is prepared from:

	Parts
Sugar	9.0
Protein fraction of Example 1	2.5
Phosphoric acid	0.15
Sodium benzoate	0.05

(continued)

	Parts
Cola flavor	0.05
Coloring	Trace
Carbonated water to give 100 parts.	

The resulting beverage (pH 2.6) has excellent mouthfeel and no off-flavor associated with soybeans.

Example 4: A still beverage (pH 4.5) with added vitamins and minerals is prepared from the following ingredients:

	Parts
Sugar	10.0
Protein fraction of Example 1	2.0
Lemon-lime flavor	0.05
Sodium benzoate	0.05
Coloring	Trace
Water to give 100 parts.	

The following quantities of vitamins and minerals are then added to each 100 ml of the prepared beverage:

Vitamins:	
B_1	0.3 mg
B_2	0.4 mg
B_6	0.5 mg
B_{12}	0.8 mcg
Niacin	3.0 mg
Potassium iodide	1.0 mg
Calcium (as calcium phosphate)	15.0 mg
Phosphorus (as calcium phosphate)	15.0 mg
Iron (as ferric ammonium citrate)	1.0 mg

ACID ACTIVE ENZYME PROCESS

Simple proteins, e.g., globulins, are insoluble in aqueous solutions and even the relatively soluble secondary and primary digestion products of simple proteins, e.g., proteoses and peptones, do not dissolve enough to give clear solutions. Additionally, proteins vary in their solubility characteristics. Generally, proteins are least soluble at their isoelectric point which is the pH at which proteins are electrically neutral. As the pH is changed from the isoelectric point to more acid or more alkaline values, the protein becomes relatively more soluble. This is true of the generally insoluble plant globulins (simple proteins) and of the relatively soluble peptones and proteoses, etc. (primary and secondary split-products).

The increased solubility at acid or alkaline pH is generally thought to be caused by the formation of proteinates or protein-salts which are soluble salts of proteins formed by the reaction of the ionizable groups in the protein molecule with ions produced by the base or acid, respectively. However, it will be apparent that since the cationic content of the protein is generally low, it is not possible to obtain substantially complete solubility of the protein by adjusting the pH. It is further apparent that extremely acid or extremely basic solutions are not adapted for consumption. Thus, even though conversion of plant globulins to lower derivatives generally increases the protein solubility and even though decreasing pH further increases solubility, it has not been possible to produce sparkling clear solutions of proteins at pH in the range of 2 to 3 since the increased solubility is not adequate and since proteins are difficult to dissolve due to physical parameters.

In the process of *A. Pour-el and T.C. Swenson; U.S. Patents 3,713,843; January 30, 1973 and 3,741,771; June 26, 1973; both assigned to Archer Daniels Midland Co.* insoluble plant proteins are digested with acid active enzyme systems and freeze dried to form a soluble proteinaceous powder.

The process provides a method for treating the simple proteins (globulins) with enzymes to digest such simple proteins and form the primary and secondary split-product derivatives. These derivatives are proteins of a less complex nature and their digested products, which upon digestion by elementary enzymes, will yield amino acids adapted for assimilation in the body. The treatment facilitates solutions of proteins in beverages, particularly carbonated beverages. Since the insoluble proteins are split and since the pH is strongly acidic, the solubilizing tendencies of these treatments are combined with the increased solubility rendered by this process. By conducting the proteolysis at acid pH, it is possible to digest only the insoluble or dispersed protein, thus avoiding the necessity of treating the already soluble protein at the given pH.

Before specifically describing the process steps, the range of materials adapted to be treated and to be formed should be illustrated. The proteins which can be treated are all plant globulins, i.e., simple proteins which yield only alpha-amino acids or their derivatives upon hydrolysis. The simple proteins are more particularly those vegetable proteins, such as glycinin, which are derived from soybeans. However, the process is adapted to treat all globulins derived from plants. Plants, of course, include not only vegetables but cereals such as corn, wheat, rye, malt and fruit kernels, etc. The enzymes are proteinases, i.e., those which are active in breaking the peptide bond, particularly those which are active at acid pH of 2 to 5 such as fungal proteinases, pepsin, or any other acid active enzyme.

The digestion medium is generally an aqueous dispersion of protein containing 3 to 6% protein in colloidal dispersion. Any suitable acid may be added to the aqueous dispersion to provide the acidic pH required for digestion. An example is a combination of 6 N hydrochloric acid and concentrated phosphoric acid. The addition of the acid active enzyme is performed by stirring approximately 0.1 to 1% based on the protein or 0.003 to 0.03% based on the total slurry of the enzyme and allowing the reaction medium to settle.

During digestion, quiescent conditions are maintained. Since the pH gradually rises during digestion, it is sometimes necessary to add acid to maintain a low pH. This pH adjustment is performed by periodically adding concentrated phosphoric acid or other acids to adjust the pH; adjusting to pH 2.7 is generally adequate. After sufficient time has elapsed for the colloidal protein to be digested and for the larger solids to settle out of solution, the digested protein slurry is cooled and centrifuged and the solid residue is discarded. The soluble material is dried and gives a powdery product which is adapted for forming sparkling clear solutions at pH of approximately 2.7 to 4.75.

Example: Approximately 100 grams of Nutrisoy 7-B (a commercially available soy protein flour) is extracted by mixing with about 20 times its weight in water at 96° to 100°F for 1 hour with constant agitation. The resulting slurry is allowed to settle and is decanted. The solid residue is extracted by mixing with approximately 10 times its weight of water for 1 hour at 96° to 100°F with no agitation. The resulting slurry is centrifuged and filtered and the solid residue discarded. The supernatent liquid from the first extraction and the second extraction are combined and hydrochloric acid is added to bring the pH of the liquid to 4.75. This liquid is centrifuged (filtering may be substituted here) and the supernatant liquid is discarded.

The solid residue is washed twice with 10 times its weight in water and the washings are discarded. The washed solid residue is used to make a 3% solids aqueous slurry to which is added a mixture of 6 N hydrochloric acid and concentrated phosphoric acid to bring the pH to 3.5.

To the acid slurry is added 0.03% by weight based on the weight of the slurry of acid fungal proteinase with stirring. After settling the medium is left for a period of 2 days where the pH is periodically adjusted to about 2.7 to 4 by addition of concentrated phosphoric acid. The digestion medium is maintained in a generally quiescent state during the digestion period except during the acid additions and the originally present colloidal dispersion gradually is converted into a sparkling clear phase. After the 2 days, the digested protein slurry is cooled to about 15°C, centrifuged, and the solid residue is discarded while the supernatant liquid is converted by means of freeze drying into a powdered product. Upon adding the powdered product to water and adjusting the pH of from 2.7 to 4.75, the proteinaceous material is substantially completely dissolved and forms a sparkling clear liquid when quantities up to 20% by weight of the soluble proteinaceous material are used.

During the process, it is desirable to avoid allowing the pH to rise above a point at which solubility occurs, this generally corresponding to the maximum pH at which the digested protein may be used, i.e., 4.75. If the reaction is allowed to proceed to a point where the pH reaches 4.75, the resulting product has a bad taste and is incompletely soluble. One explanation is that the colloidal solubility of soy protein is lowest at a pH of approximately 4.75 and since it is desired to keep the colloidal sized protein dispersed for maximum digestion, it is desirable to avoid a pH of low solubility. It may be desirable to maintain the pH of the reaction medium at approximately the pH which will be present in the beverage. For example, if a beverage is intended to be produced having a pH of 2.7 it is sufficient to maintain the pH of the reaction or digestion medium at approximately 2.7.

Another aspect of the process relates to the quiescent condition of the digestion medium during the digestion period. Initially, agitation is provided to mix the enzyme thoroughly throughout the dispersion; however, the dispersion should be allowed to settle and should be maintained in a relatively quiescent state during the entire digestion period, except that some disruption occurs when acid is added. The material which is larger than colloidal size will thus be allowed to settle to the bottom and will be separated from the colloidal proteinaceous material which is intended to be treated by the process.

A suitable time for digestion is on the order of 1 to 2 days, for example, 15 to 20 hours. The time of digestion depends on the activity of the enzyme used and the temperature of digestion; generally long time periods over 4 days, for example, produce a bitter product with specific enzymes and are to be avoided, if possible. The maximum efficiency is attained when the digestion medium is maintained at approximately 40°C. However, other temperatures may be utilized with a somewhat reduced efficiency when the enzyme used dictates other temperatures. Cooling after digestion is performed so as to precipitate insoluble material and avoid such precipitation from beverages formed from the protein when the beverages are refrigerated by the consumer.

When a carbonated beverage is formed using the permanently soluble protein (PSP) a suitable formulation is prepared as follows. Flavored carbonated protein drinks were prepared using PSP produced by a variety of proteolytic enzymes. The following formulation was used: 1.4 g cool ade (any flavor), 62.5 g sugar, 500.0 g carbonated H_2O, and 15.0 g PSP.

The PSP can be added either before or after addition of the carbonated water. The mixture should be stirred gently, to prevent excessive foaming, until all the PSP is dissolved. The pH of the final solution will be between 2.5 to 3.0. A noncarbonated beverage is an orange drink prepared as follows: 100 g reconstituted orange juice and 3 g PSP. The ingredients are combined in either order. The mixture is stirred gently until all PSP is dissolved. The final pH is between 3.0 and 4.0.

HYDROLYSIS FOR SOLUBLE PROTEIN

Solvent extracted soybean meal is the starting material in the process of *P.J. Magnino, Jr. and C.W. Frederiksen; U.S. Patent 3,645,745; February 29, 1972; assigned to Ralston Purina Company* to produce a soluble protein product suitable for beverage use.

A slurry is formed with defatted soybean flakes and water. The slurry is agitated for a sufficient time to extract the soluble protein and carbohydrate materials contained within the physical, cellular structure of the original flakes. The slurry consists of the suspended solid, cellular seed material with a major portion of the protein and carbohydrate materials in solution. The slurry may be adjusted to an alkaline pH to increase the speed of removal of the soluble materials. The aqueous slurry may be formed at a pH of from 6.8 to 10.5.

As an alternative procedure, the protein and carbohydrate materials may be dissolved out of the flakes by placing the flakes into an acid solution having a pH of from 1.2 to 2.5. The liquid portion is separated from the solid cellular material as completely as is possible.

The liquid portion is adjusted to an alkaline pH range of 11.8 or to an acid pH in the range of from 1 to 2. The liquid is then heated to a temperature of from 140° to 180°F for a time period of from 20 to 120 minutes. The pH is adjusted to the above described alkaline range with suitable chemicals such as sodium hydroxide, calcium hydroxide, and ammonium hydroxide; however, sodium hydroxide is preferred. It has been found that the best results are obtained when the chemical is present in liquid portion in an amount from 6 to 11% by weight based on the estimated dry weight of the protein present in the liquid. The acid range may be obtained by the addition of suitable acids such as phosphoric, hydrochloric, sulfuric and sulfurous.

It is essential that the above ranges of time and temperature be observed since insufficient hydrolysis will not provide the necessary solubility properties of the process. On the other hand, if excessive hydrolysis takes place, the protein will be reduced to short chain peptides or its amino acids and some of the amino acids destroyed. Amino acid destruction would drastically reduce the nutritional value of the protein as well as being detrimental to the functional properties of the product. Where desired, the protein may be precipitated from the solution immediately after it is extracted and separated from the solid flake material and prior to the subsequent pH adjustment in the range as described above, i.e., from 1 to 2 or 10 to 12. The protein may then be dissolved in the pH range set forth above, reprecipitated and further processed.

The pH-adjusted, heated, liquid is subsequently adjusted with an appropriate edible acid to the isoelectric point of the protein at which point the protein is precipitated out of solution. Acids that may be used in such precipitation include suitable food acids such as phosphoric, hydrochloric, sulfuric and sulfurous. The precipitation is carried out at a pH of from 4 to 5 and preferably at about 4.5. The soluble extracted carbohydrates remain in solution.

The solid precipitated protein material is separated from the liquid as completely as practicable and is washed sufficiently to remove substantially all of the soluble carbohydrates that may remain in the protein material. The protein material is then suspended in water to obtain an appropriate solids content and the slurry is adjusted to a pH of about 6.8 to obtain a dispersible product. If an acid product is desired, the pH is adjusted to about 3. The slurry in either of the pH ranges may be subjected to pressure and dynamic instantaneous heating at an elevated temperature.

The most satisfactory way of achieving this is to pass the slurry through a device commonly known as a Jet Cooker. For example, it includes adjacent jet nozzle orifices, normally concentric, through which the slurry and the pressurized steam used as a heating agent are ejected at high velocities in intersecting flow patterns, so that each tiny bit of slurry is instantly and dynamically heated by the steam while simultaneously being subjected to severe physical forces at the nozzle, such physical working being largely of a shearing nature and to expose new surface areas of the protein to promote both the volatilization of the undesirable flavor components and also provide for the desirable physical alteration of the protein molecules.

The physical working of each tiny portion is believed to expose the undesirable flavor components to physical action, and this physical action of working together with the elevated temperature heat treatment is believed to weaken and/or loosen the tenacious attraction between the undesirable components and the complex protein molecules, to an extent where the undesirable flavor components can be removed with the flashed off vapors in a subsequent vacuum chamber treatment. It is preferred that the jet cooking be combined with a subsequent vacuum chamber treatment to obtain the desirable properties of the product. This physical working and heating results in substantially eliminating the characteristic soy flavor where soybeans are employed, but more importantly, assists in physically altering the protein molecule to obtain the desirable solubility and viscosity properties of the product. The temperature range to which the protein slurry is heated and jet cooked for the desired results is 220° to 400°F, although the temperature should not be in the lower region of this range unless the product is subsequently vented into a high vacuum chamber after being held under pressure in a special holding chamber.

Normally, the temperature should be 285° to 320°F, for best results. The slurry is held at such temperatures for a period of from 5 to 30 seconds. The heated slurry is then immediately released to atmospheric pressure or to a vacuum chamber to flash off the volatile, undesirable flavor components and subsequently the evaporatively cooled material is dried and reduced to the desired particle size. Drying may be accomplished by conventional means such as roller drying, vacuum drying, freeze drying and spray drying. It is important to note that the above described treatment produces a product which is nearly sterile, and therefore, highly desirable for use in producing food products. It is to be understood that, where desirable, the precipitated protein material may be subjected to more than one jet cooking treatment in order to improve the desirable properties of the product.

In an alternate method, a soluble protein is extracted from defatted oleaginous seed materials such as soybeans in the manner as previously described. The solids and liquid are separated and the liquid is adjusted to a pH of from 10 to 12. The liquid is directly passed to a jet cooker where it is instantaneously heated to elevated temperatures in the range of from 220° to 350°F, passed to a vacuum chamber and the pH adjusted to the isoelectric point of the protein to precipitate the protein. The precipitate and liquid are separated and the precipitate is adjusted to a neutral or acid pH. The precipitate is again subjected to jet cooking and the vacuum chamber and subsequently converted to a resulting dry, particulate protein product.

The extraction of the soluble proteins and carbohydrates from the flake is best accomplished in the slurry by having the water and flakes present in a ratio in the range from 8:1 to 20:1 (water:flakes). The preferred ranges of the ratio of water:flakes provides a highly efficient and economical means of extraction.

The products have been described as having either acid or neutral properties. The acid products are soluble in acid mediums such as in soft drinks, citric juices and other foods having an acid pH. The neutral products are soluble in neutral or slightly alkaline mediums such as in coffee whiteners, imitation milk, imitation ice cream, whipped toppings. Both the acid and neutral products provide the property of being soluble in a clear medium under suitable pH conditions and once dissolved, to retain the clear property of the mediums as opposed to becoming cloudy or murky. More importantly, the product provides a soluble source of protein for producing protein enriched food products.

Example 1: (1) Protein Cured – Defatted soybean flakes were slurried in water at a temperature of 85°F with a water to flake ratio of 9:1. To the slurry was added 1.6% of calcium hydroxide and 1.0% of sodium sulfite, the percentages being based on the total weight of the flakes. The slurry was agitated for a period of 30 minutes and subsequently the liquid was separated from the solid materials in a centrifuge. The solid materials were reslurried in water at a ratio of 6 parts of water to 1 part of flakes and this slurry was agitated for a period of 10 minutes, after which time the solid and liquid materials were separated, as before, in a centrifuge. The two liquid portions or extracts were combined.

To the combined liquid portions, which contain soluble proteins and carbohydrates, was added 7.5% sodium hydroxide, the percentage being based on the estimated dry weight of protein present. The liquid portion has a pH of about 11.5 after the addition of sodium hydroxide. The liquid was then heated to 160°F and that temperature maintained for a period of 40 minutes. After heating, phosphoric acid was added to the liquid to adjust the pH from 11.5 to the isoelectric point of the protein, i.e., pH of about 4.5. The dissolved protein was precipitated out of solution and the precipitated curd was concentrated and washed in a centrifuge to remove the carbohydrate materials as completely as possible. The washed protein curd was diluted in a ratio of 1:1 by weight with fresh water and concentrated to about 17% by weight solids.

(2) Neutral Soluble Protein Product – The concentrated curd of (1) above was diluted to about 15% by weight solids and the pH adjusted to about 6.8 with sodium hydroxide. The neutralized protein curd was then passed to a Jet Cooker where the temperature of the curd was raised instantaneously to a temperature of 310°F and held at that temperature for a period of about 5 seconds. The curd was then injected into a vacuum chamber under 27" of mercury vacuum. The heated curd was passed from the vacuum chamber to a spray dryer under a pressure of 4,000 psi and dried at an exhaust temperature of 180°F. The resulting product was a finely divided particulate neutral protein product. Several tests were conducted to compare the properties of the product with other similar products which have been produced by different methods. The products compared are listed as (A), (B), (C) and (D):

(A) isolated soy protein which has not been hydrolyzed but has been jet cooked and available on the market
(B) isolated soy protein which has been hydrolyzed but not jet cooked
(C) product of Example 1, isolated soy protein which has been hydrolyzed and jet cooked
(D) product of (C) which has been jet cooked a second time before spray drying.

Solubility Index – Disperse 4 grams of the protein product in 100 ml of water with agitation in a Waring Blender Model No. 1120, for about 90 seconds. Centrifuge 50 ml of the mixture for about 5 minutes at about 1,000 rpm. Remove all but about 5 ml of the supernatant liquid, add water to make volume up to 50 ml, shake gently and centrifuge again for 5 minutes at 1,000 rpm. The solubility index is expressed in terms of the amount of residue, expressed in milliliters, remaining undissolved. That is to say, the more the residue, the less soluble and less desirable the product.

Viscosity 1 – A 12% protein solution was prepared by mixing 30 grams of protein with 220 ml of water. The mixture was placed in an open beaker and heated in a hot water bath, with agitation, for about 30 minutes at a temperature of about 165°F. The mixture was cooled to 75° and 40°F and the viscosity measured with an LVT Brookfield viscometer using a No. 1 spindle at 60 rpm.

Viscosity 2 – A 12% protein solution was prepared by mixing 30 grams of protein with 220 ml of water. The mixture was sealed in a metal can and retorted without agitation at 250°F for 15 minutes. The product was cooled to about 75°F, the container opened and the mixture tested for viscosity with an LVT Brookfield viscometer using a No. 1 spindle at 60 rpm. The products of (C) and (D), consistently show better physical properties than similar products that are currently available.

Example 2: The procedure of Example 1 is substantially repeated. The concentrated curd was diluted to about 15% by weight solids and the pH adjusted to about 3.5 with an acid mixture comprising phosphoric, citric and malic acids. The acid curd was passed to a Jet Cooker where the temperature was raised instantaneously to about 310°F, held at that temperature for about 5 seconds and subsequently injected into a vacuum chamber under 27" of mercury vacuum. The curd was then passed from the vacuum chamber to a spray dryer under a pressure of 4,000 psi, the dryer having an exhaust temperature of about 180°F. The resulting product was a finely divided, particulate, acid product.

The properties of the acid product were tested according to the procedures set forth in Example 1. Product (E) represents the product made in accordance with the procedures of Example 2 and product (F) corresponds to product (A) except that the product has a pH of 3.5.

The concentrated curd of Example 1 was diluted to about 15% by weight solids and the pH adjusted to about 6.8 with sodium hydroxide. The neutralized protein curd was then passed to a Jet Cooker where the temperature of the curd was raised instantaneously to a temperature of 310°F and held at that temperature for a period of about 5 seconds. The curd was then injected into a vacuum chamber under 27" of mercury vacuum. The heated curd was passed from the vacuum chamber to a spray dryer under a pressure of 4,000 psi and dried at an exhaust temperature of 180°F. The resulting product was a finely divided, particulate, neutral, protein product having viscosity and solubility properties substantially the same as those products produced with the procedures of Example 1.

The concentrated curd produced in Example 1 is diluted to about 15% by weight solids and the pH adjusted to about 3.5 with an acid mixture containing phosphoric, citric and malic acids. The acid curd is passed to a Jet Cooker where the temperature is raised instantaneously to about 310°F, held at that temperature for about 5 seconds and subsequently injected into a vacuum chamber under 27" of mercury vacuum. The curd is then passed from the vacuum chamber to a spray dryer under a pressure of 4,000 psi, the dryer having an exhaust temperature of about 180°F. The resulting product is a finely divided, particulate, acid product.

CEREAL GRAIN ENZYME TREATMENT

Cereal grains are treated in an enzymatic process to produce carbohydrate, proteins and other nutrients in a more soluble form. The process is given by *W.F. Allen and J.J. Thompson; U.S. Patent 3,157,513; November 17, 1964; assigned to Kellogg Company.*

In carrying out the process, grain material in finely divided condition is digested, in the form of an aqueous slurry, first with a proteolytic enzyme so as to convert in excess of 50% of the protein content of the cereal grain treated to soluble and colloidal partially digested proteins. From this liquid phase, the bran and starch residue solids are readily separated, and the high protein content liquid recovered as a product of the process. The proteins in this digested liquid constitute from 70 up to 90% of the protein content of the raw grain or flour treated. The liquid is then heated to inactivate the enzyme content.

In the process, an upper sludge layer is separated from the recovered granular starch, which sludge layer contains some protein and starch material. This sludge layer can then be subjected to gelatinization by heat and then digested in the presence of amylolytic enzyme in aqueous solution to convert the starch content to sugars, following which the solution is heated to inactivate its enzyme content. If desired, this step of treating the sludge separated from the starch fraction can be simultaneously treated with added proteolytic enzyme so as to further solubilize and convert any protein content.

As another alternative, the sludge layer separated from the starch fraction can be mixed with the digested protein-containing liquid prior to heating and the mixture then heated to gelatinize starch content, following which it is then treated with amylolytic enzyme to convert the gelatinized starch content to sugars and then subjected to heat treatment to inactivate all enzyme content.

The recovered digested liquid containing the soluble and colloidal solids may be used directly in food products, or spray dried or vacuum dried before adding to other food ingredients without resulting in a sandy or coarse texture and without settling out and clogging the processing equipment. The recovered protein material is bland in taste as well as smooth in texture, and hence can be used in a wide variety of food and flavors and in the formulation of different food products where more easily digested protein as well as maltose and dextrose sugars enhance the food product. The recovered liquid material can be used in instant high protein food beverages, and aid in promoting smoothness and in giving a stable, creamy texture to the drink and to supply drink, with, for example, from 30 to 60% of its protein content.

Example 1: (1) Two liters of water heated to a temperature of 95°F and having a pH of 9.73 resulting from addition thereto of calcium hydroxide was put into a 1-gal size Waring Blender. (In lieu of the calcium hydroxide other substances such as sodium hydroxide, carbonate or bicarbonate, may be added as well as calcium hydroxide or oxide and the like basic substances to provide a pH during the process of from 6 to 10.) While running at a low speed, 1 kg of second clear grade hard wheat flour (containing 10.9% moisture, 15.3% protein) was slowly added. The resulting slurry was homogenized in the blending device at slow speed for about 2 minutes, at which time the pH was reduced to 6.18 and the temperature became elevated to 134°F.

(2) The slurry from the previous step was allowed to stand for about 18 minutes at a temperature of 130° to 134°F and the pH adjusted to 6.63 by adding 1.0 gram of dry calcium hydroxide to the slurry while slowly homogenizing.

(3) The slurry was cooled to 120°F at which time 0.50 gram (0.05%) of proteolytic enzyme (Takamine #HT proteolytic 200 *Bacillus subtilis* proteinase) was added after dissolving in 2 to 3 ml of water. Digestion was continued for 4 hours at 125° ± 1°F with constant stirring.

(4) The partially digested slurry was immediately centrifuged to separate and recover the starch and bran from the supernatant liquid containing the digested protein.

(5) The supernatant protein liquid separated in step (4) was centrifuged again. The sludge thrown down was put into a 1-liter stainless steel beaker and the separated digested protein liquid recovered.

(6) The digested liquid (designated as liquid A) was then heated slowly to the boiling point with constant stirring and continued for 5 minutes to inactivate the enzyme and destroy the raw wheat flavor and recovered as a product of the process.

(7) Bran was removed from the separated starch by suspending the mixture recovered in step (4) in 600 ml of cold tap water and then putting through a #230 USS screen on a vibrator. Bran remaining on the screen was washed with 6 small portions of tap water from a wash bottle.

(8) Starch slurry with bran wash water passing through screen was centrifuged.

(9) The top sludge layer on the granular starch fraction was carefully removed and added to that sludge in the beaker [step (5)]. About 300 ml of starch wash liquid was also put into this beaker.

(10) The washed and separated bran and starch recovered respectively from steps (7) and (8) were air dried on stainless steel trays, and then dried for about 2 hours at 200° to 212°F before being weighed and analyzed.

(11) Starch content in the sludge or mixture fraction [step (9)] was gelatinized by heating slowly to 159°F with constant stirring.

(12) Then 0.10 gram of amylolytic enzyme (Takamine #HT-440, *Bacillus subtilis* amylase) was added (in solution). The mixture was digested, stirring continuously, for 80 minutes at 158° to 160°F.

(13) It was then cooled to 135°F and 0.02 gram of Takamine HT proteolytic 200 enzyme was added (in solution). The digestion was continued for 3 hours at 125° to 130°F.

(14) Temperature was then increased to 170°F and held for 15 minutes. Heat was increased over a 30 minute period to 212°F to inactivate the enzymes and destroy the raw wheat flavor. This fraction was designated as digested liquid B.

Yields and Analyses

Constituent	Yield, g	Moisture, %	Protein (N x 5.7), Grams in Constituents	pH
Bran	33.0	3.06	1.80	-
Starch	291.5	6.20	6.21	-

(continued)

Constituent	Yield, g	Moisture, %	Protein (N x 5.7), Grams in Constituents	pH
Digested liquid A	2,134.2	80.81	111.83	6.53
Digested liquid B	564.0	72.63	28.76	6.39
Starch wash liquid (remaining)	410.0	-	3.03	6.97
Total			151.63	

Nitrogen recovered in digested liquids A and B = 91.89%.*
Nitrogen recovered in all constituents = 99.10%.*
[*Total protein (N x 5.7) in flour used was 153.0 grams.]

The following is an example of the preparation of a high protein enzyme digested liquid from whole wheat.

Example 2: (1) One kilogram of unground soft white winter wheat (moisture 11.48%, protein 9.47%) was put into a 2-liter beaker and washed with 500 ml of cold tap water. Water was drained off immediately and the wheat was covered and allowed to temper about 30 minutes.

(2) Two liters of calcium hydroxide treated hot water (pH 10.06, temperature 95°F) were put into the large (1 gallon) Waring Blender. The washed wheat was added slowly to the water in the blender while running at slow speed. After all the wheat had been added, the blender was run at high speed for 2 minutes to shear and finely subdivide the wheat and to form a slurry thereof with the water. The pH of slurry was 6.40 and the temperature was 106°F.

(3) The pH of the slurry was adjusted to 7.25 by slowly adding 1.024 grams of dry $Ca(OH)_2$ while running the blender at slow speed. The slurry was then homogenized at high speed for about 3 minutes until a temperature of 125°F was reached.

(4) Immediately 0.50 gram of proteolytic enzyme dissolved in water, as in Example 1, was added and the slurry (pH 6.75) was digested for 2 hours and 15 minutes at 126° to 131°F with constant stirring. More $Ca(OH)_2$ (0.334 gram) was added during the digestion to raise the pH to 6.81.

(5) Bran was removed from the digested slurry by putting the liquid through a #230 USS screen on a vibrator in 4 portions, and washing it thereon with warm (110° to 120°F) tap water.

(6) Granular starch was separated from the above filtrate slurry by centrifugation. The supernatant digested liquid, and a top sludge layer, further separated from the starch, were put into a 4-liter stainless steel beaker.

(7) Starch was washed once by suspending in about 1 liter of distilled water and then centrifuging again. The top sludge layer on the starch was again scraped off and added to the main portion in the 4-liter beaker (step 6).

(8) The 4-liter beaker and contents were heated in a salt water bath with constant stirring to 160°F to gelatinize the starch content. pH of mixture was 6.90.

(9) Liquid was cooled to 150°F and 0.30 gram of Takamine HT-440 amylolytic enzyme (*Bacillus subtilis* amylase) added (in solution). Digestion was continued for 2 hours at 146° to 154°F with constant stirring.

(10) The temperature was then gradually increased to 207°F over a 1 hour and 45 minute period to inactivate the enzyme and destroy the raw wheat taste.

(11) Digested liquid was cooled, weighed, and sample taken for analyses. The separated and washed bran and starch were air dried on stainless steel trays, then dried for about

2 hours at 200° to 212°F before being weighed and analyzed.

Yields and Analyses

Constituent	Yield, g	Moisture, %	Protein (N x 5.7), Grams in Constituents	pH
Bran	133.3	3.95	14.84	–
Starch	458.3	3.25	3.67	–
Digested liquid	1,993.3	92.18	69.37	6.66
Starch wash liquid	710.0	97.35	4.08	6.90
Total			91.96	

Nitrogen recovered in digested liquid = 73.25%.*
Nitrogen recovered in all constituents = 97.11%.*

[*Total protein (N x 5.7) in wheat used was 94.7 grams]

Several experiments were conducted with wheat flour using the same general procedure described in Example 2 [omitting step (1)]. The removal of bran on the #230 screen was considerably easier with whole wheat because of the larger size of the bran particles. The addition of a proteolytic enzyme for the second or starch sludge digestion period (as in Example 1) is recommended for obtaining higher digested protein in the liquid. The following is an example of a formulation for an instant high protein beverage containing enzyme digested liquid from wheat, the formula being plain and without flavor.

Example 3:

Ingredients	Weight Percent
Nonfat milk solids	40.00
Brewers' yeast	1.00
Digested liquid A (Example 1)	37.11*
Digested liquid B (Example 1)	14.39*
L-lysine-HCl	1.00
Corn oil	5.00
Wheat germ oil	1.00
Sodium alginate	0.40
Ascorbic acid	0.10
Total	100.00

*Moisture-free basis.

The ingredients are mixed and a drink prepared therefrom as follows. (1) The digested liquids A and B were weighed directly in the small Waring Blender bowl. Then all dry ingredients were added to the liquid and homogenized at high speed for about 30 seconds. The oils were then added while homogenizing at high speed. (2) An 8-ounce drink was prepared by weighing out 167 grams (56.7 grams or 2 ounces on a dry basis) of the emulsified mix and then adding 0.12 ml of vanilla extract. The mix was diluted with 69.6 grams of cold water to make 8 ounces and homogenized for about 30 seconds. A second drink was prepared by merely stirring rapidly with a spoon. This food beverage had a very smooth creamy consistency and color, resembling egg nog. It did not have any grainy or sandy feel in the mouth and the emulsion was stable for at least 30 minutes. The taste was bland, with no objectional aftertaste. Several flavors could be used with this formula including vanilla, butterscotch, maple nut, egg nog, etc. A chocolate drink would require a modified formula.

Analysis of beverage before dilution: total solids, 34.43%, protein (N x 6.25), 10.38% or 30.15% on moisture-free basis, and pH 6.29.

DEAMIDIZED GLUTEN ADDITIVE

Most beverages, such as sodas, noncarbonated fruit-flavored drinks, simulated fruit juices, preferred by young people, are nutritionally unbalanced. They are high in sugar but lacking in proteins. One way of aiding the battle against malnutrition would be to raise the protein content of these drinks. Gluten, a rich source of protein and readily obtainable from cereal grains, has been considered as an additive for this purpose. Unfortunately, gluten is not compatible with the acidic aqueous media that make up such beverages.

If unmodified (native) gluten is added to an acidic beverage, a cloudy suspension forms. Very little, if any, of the gluten is dissolved. As a result, the appearance of the beverage is impaired and when imbibed has an undesirable effect on the palate because of its gritty texture. Also, if the beverage is allowed to settle and only the clear liquid imbibed, the consumer will obtain none of the benefit of the added protein. Another item to be considered is that native gluten develops undesirable astringency when contacted with an acidic aqueous medium such as an acidic beverage. The taste sensation is developed far back in the mouth and is described by some as a throat-burning, puckering, or bitter sensation.

In the process of *J.W. Finley; U.S. Patent 3,770,452; November 6, 1973; assigned to the U.S. Secretary of Agriculture* deamidified gluten is used to fortify carbonated and noncarbonated beverages. The starting material is vital gluten which contains approximately 50% gliadin and 50% glutenin. It is prepared by making a dough of wheat flour and water and kneading the dough while washing it with water to remove the starch. The resulting wet gluten is then dried at a relatively low temperature to prevent degradation of the protein molecules. The product is referred to as vital because it is not denatured during its preparation.

In the first operation, vital gluten is extracted with anhydrous isopropanol to remove the astringent principles. For optimum results, a series of extractions are applied. First the gluten is extracted with anhydrous isopropanol. This initial extraction is followed by another in which 0.05 to 0.1 N foodgrade alkali (e.g., NaOH or KOH) in 5 to 10% aqueous isopropanol is used as the solvent. Following this, the gluten is extracted with 0.05 to 0.1 N foodgrade acid (e.g., hydrochloric or sulfuric) in 5 to 10% aqueous isopropanol to neutralize the gluten.

The extractions may be repeated a number of times to ensure complete removal of astringent principles. Ambient temperatures or lower (e.g., 25°C down to about -15°C) are used to minimize denaturation of the gluten by the applied agents. Following each extraction the solid gluten can be separated from the liquid by conventional means such as decanting, filtering, or centrifuging. Generally, in each extraction a 5 to 20:1 ratio (volume to weight) of extraction solvent to gluten is used. It may be observed that this extraction procedure does not alter the proteinous components of the gluten since both gliadin and glutenin are insoluble in the solvents. Gliadin, for example, is not soluble in anhydrous isopropanol or that containing 10% water; it is only soluble in isopropanol containing at least 20% of water.

Following the removal of astringent principles, the gluten is deamidized by very dilute acid hydrolysis under controlled conditions. In this hydrolysis the primary amide groups ($-CONH_2$) occurring along the protein chains are hydrolyzed to carboxyl groups ($-COOH$) without significant rupturing of the peptide linkages which form the protein chains. It is evident that by changing the conditions of hydrolysis, one may vary the proportion of amide groups which are converted into carboxy groups. For maximum dispersibility in acidic beverages, it is preferred to use the products in which 45 to 70% of the amide groups are converted into carboxy groups. Generally, 0.05 to 1.0 N HCl is used for the deamidation and is applied at a temperature from 25° to 100°C for a period of 20 minutes to 24 hours. After the hydrolysis the deamidized gluten is dried in conventional manner, for example, spray or freeze drying. The dried material is ready for use as a beverage fortifier. Since the solid has a powdery nature, it is excellent for fortification because it allows a specific amount to be metered out for a particular liquid food or beverage. The powder is free-flowing and easy to handle and readily dissolves in either distilled water or water at pH 2 to 7. It is to be particularly observed that deamidized gluten offers considerable advantages, important in the preparation of food products, as compared with unmodified gluten.

Unmodified gluten is sensitive to heat. Thus, if it is used in beverages which are subjected to heat, as for pasteurizing or sterilizing during bottling or canning, deterioration of the gluten occurs which renders it nutritionally useless. Deamidized gluten, on the other hand, is not sensitive to heat and may be used with equal success in cold and hot beverages and in those which are subjected to thermal processes such as pasteurization or sterilization. Unmodified gluten is sensitive to pH and is insoluble or only slightly soluble at pH 2 to 7. This means that the protein cannot be used in beverages having such a pH range. On the other hand, deamidized gluten is soluble over this entire pH range and thus can be used in such products.

The process may be utilized for the nutritional fortification of all kinds of beverages, for example: carbonated beverages such as colas, ginger ale, root beer, cream soda, and the various fruit-flavored sodas such as grape, lemon and lime, orange, cherry, etc.; noncarbonated beverages including fruit-flavored drinks such as orange, lemon, lime, grape, grapefruit, cherry, etc.; fruit juices including orange, apple, pineapple, grapefruit, etc., and simulated juices which contain some proportion of fruit juice plus added sugar, flavorings, citric acid, etc.; vegetable juices such as tomato juice, carrot juice, celery juice and mixtures of tomato juice with other vegetable juices; beverage bases such as the dry products which contain sugar (or artificial sweetener), fruit flavor (real or imitation), citric or other fruit acid, and coloring; coffee and tea, both in liquid and dehydrated forms; wines, beer, ale, and the like.

Deamidized gluten is incorporated in the beverage in a desired proportion. Thus, the amount added is not critical but merely depends on the degree of nutrient enhancement desired. Usually, at least 1% is added preferably at least 3%. If desired, the proportion may be as high as 10 or 15%.

Example 1: Vital gluten (30 grams) was successively extracted with the solvents listed below, applying centrifugation after each extraction to separate the solvent from the treated gluten. The extractions were conducted at -15°C. Anhydrous isopropanol (500 ml); two 500 ml portions of 0.1 N NaOH in 10% aqueous isopropanol; 500 ml of 0.1 N hydrochloric acid in 10% aqueous isopropanol.

The so-treated gluten was then hydrolyzed (deamidized) with 1.4 liters of 0.075 N HCl in a steam-heated mixing bowl for 30 minutes. Temperature of the hydrolysis mixture was about 95°C. Following this, the mixture was lyophilized to remove the liquid components. The resulting dry, nonastringent, deamidized gluten was found to be completely soluble in pH 3 citrate buffer. Its flavor was bland; the astringent taste of the original gluten was totally absent.

Example 2: Dry nonastringent deamidized gluten prepared as described in Example 1 was incorporated in the proportion of 3% in a series of acidic beverages, namely, orange juice, lemon juice, lemonade, grapefruit juice, pineapple juice, pineapple-orange juice, and pineapple-grapefruit juice. It was observed that in each case the modified gluten went into solution, and each of the beverages exhibited the same flavor and texture as the corresponding unfortified one.

SUGAR-ACID SOYBEAN EXTRACT

It has been found that oil-free soybean flake contains a great amount of organic acids which can be effectively utilized in the production of refreshing beverages. Among these organic acids, the principal one is citric acid and others include malic acid, tartaric acid, succinic acid and so forth.

The following details are given by *A. Nagata, H. Yamamoto, S. Terashima, T. Matsumoto, S. Onishi and S. Konishi; U.S. Patent 3,082,093; March 19, 1963; assigned to Ajinomoto KK, Japan* for the preparation of extracts and beverages from oil-free soybean flakes.

One part of oil-free soybean flake is introduced into an extraction vessel and 5 parts of water are added. Then a bag of synthetic textile containing 0.7 part of a cation exchange resin is thrown into the extraction water and extraction is carried out for 1 hour at room temperature under agitation. The bag of resin is removed from the liquid and the solid materials are separated from the extraction liquid by centrifuging. The extraction liquid amounting to 4 parts has a pH of 3.2 and has a refreshing sour taste. The analysis of acids contained in the liquid is as set forth in the following table. (In the table, the composition of the acid content of the juice of a mandarin orange is also described for comparison.)

Acid Composition of Extraction Liquid

	Extraction Liquid, %	Mandarin Orange Juice, %
Citric acid	70	72.5
Succinic acid	trace	15.0
Malic acid	8	9.0
Tartaric acid	5	1.0
Phosphoric acid	14	-

As seen from the above table, the acid composition of the extraction liquid is composed mainly of organic acids such as citric acid, malic acid and tartaric acid and it resembles the acid composition of the juice of a mandarin orange. The sugar content, acid content and the results of an analysis of the other elements in the extraction liquid are set forth in the following table. (In the table, the values of extraction liquid which is concentrated to one-quarter volume are described along with the values of the juice of a Chinese citron and juice of mandarin orange according to the standard of the Ministry of Agriculture and Forest in Japan and the value of a canned orange juice commercially available.)

	Sugar Content (g/dl)	Acid Content (g/dl as citric acid)	Ratio of Sugar Content to Acid Content	Nitrogen (mg/dl)
Extraction liquid	2.5	0.5	5.0	22
Extraction liquid concentrated to ¼	10.0	2.0	5.0	88
Chinese citron	9.0	2.0	4.5	30
Mandarin orange	10.0	0.7	14.3	30
Canned orange juice from market	10.1	0.5	20.2	32

As apparent from the above table, the extraction liquid will have a composition substantially similar to that of a mandarin orange or canned orange juice when it has sugar added and extraction liquid concentrated to one-quarter volume has practically the same composition as that of a Chinese citron.

The extraction liquid obtained by conducting extraction on oil-free soybean flake with water in the presence of a cation exchange resin has a composition which is very similar to that of fruits. However, it contains various elements which may cause a disagreeable or peculiar odor and coloring, and hence it is not suitable for use in a beverage and requires decoloring and deodorizing for such use. However, these elements attributable to the odor and color of soybeans can be easily removed with adsorption by active carbon or decoloring agents such as decoloring resins. In fact, the odor and coloring of soybeans can be removed by simple treatment of the extract liquid with decoloring agents thereby achieving simultaneously decoloring and deodorizing.

The refined liquid obtained by the treatment with decoloring agent is, however, not sweet enough because of insufficient sugar content. It is therefore sweetened by adding sweetening agent such as sucrose, glucose, maltose, honey, sorbitol or synthetic sweetening agent. Further, there may be added, if desired, perfumes, emulsifiers, nutrients, and so forth to obtain any predetermined composition for providing a refreshing beverage. Each step of the process may be modified in various ways as will be explained more fully in the following.

Firstly, the extraction is conducted by maintaining oil-free soybean flake and a cation exchange agent in a common extraction medium such as water. When the extraction is thus conducted, water-soluble substances in oil-free soybean flake, namely carbohydrates, ash contents (mainly potassium or calcium probably present as salts combined with acid) and protein, are dissolved out of the soybeans and they contact the cation exchange resin which adsorbs cations. Consequently the concentration of cations in the liquid is lowered while anions are freed according to the decrease of cations. Thus the value of pH of the extraction liquid is gradually lowered until it reaches a substantially constant value of 3 to 4. As well-known, the isoelectric point of protein is about pH 4, and the solubility of protein is minimum at the isoelectric point.

The extraction is conducted at a point near the isoelectric point of protein in the absence of alkali salt. Hence the solubility of protein is extremely low and there occurs practically no dissolving out of the protein while the carbohydrate and ash content is effectively extracted and the cations are immediately adsorbed by the resin to leave carbohydrates and anions in accumulation in the liquid. The anions are organic acids, the major part being citric acid as already mentioned, while the carbohydrates are composed mainly of sucrose mixed with other sugars such as stachyose and raffinose.

Thus an acid-sugar liquid suitable for a refreshing beverage is obtained. If the extraction step is considered as a step for the recovery of protein from oil-free soybean flake, it is again very useful since it improves greatly the purity of protein of oil-free soybean flake due to the extraction and removal of carbohydrates and ash contents without accompanying loss of protein. Thus this process gives batch refined oil-free soybean material useful in food industries and acid-sugar liquid, and thus attains simultaneously the refining of protein.

The extraction step may be modified as described in the following. Oil-free soybean flake may be extracted solely with water and separated into solids and the extraction liquid. Then the extraction liquid may be treated with a cation exchange resin to decompose salts dissolved in the extraction liquid and to adsorb cations by means of the resin thereby emancipating organic acids and lowering the pH to 3 to 4 thereby coagulating and precipitating protein. When the precipitated protein is separated, an acid-sugar liquid having the same composition as the acid-sugar described before is obtained. The separated protein contains practically no impurities and is a soybean protein of very high purity.

Any resin may be used as long as it is capable of decomposing neutral salt and adsorbing cations, but cation exchange resins which have sulfonic acid radicals as their functional radicals, are particularly suitable. Complex type cation exchange resins having sulfonic acid radicals as their main functional radicals and cation exchange resins having phosphoric acid radicals as their functional radicals can also be used. The amount of cation exchange resin used should preferably be more than 0.3 part per 1 part of oil-free soybean flake and the best result is obtained when 0.6 to 0.8 part of the resin is used.

The extraction medium, that is water, is used at the rate of 4 to 7 parts of water per 1 part of soybeans and 5 parts is preferred. However, the amounts of cation exchange resin and water to be used may be altered as desired and according to requirements as a matter of course. The treatment temperature in the extraction step has scarcely any effect on the result. The cation exchange resin used in the extraction may be regenerated by treatment with dilute mineral acid according to conventional method and reused repeatedly. In the regeneration of the resin, mineral salts, mainly potassium salts, are recovered from effluent.

In the refining step, the disagreeable soybean odor and the coloring materials of the acid-sugar liquid are removed by the treatment with a decoloring agent. This decoloring and deodorizing are important in the preparation of a refreshing beverage which has no commercial value if it is not satisfactory in color, odor, and taste.

For the decoloring agent in the refining step, any active carbon or substantially neutral porous decoloring synthetic resins having intrinsically no ion exchange function may be used. However, it is convenient to use a decoloring resin since this has a strong function of adsorbing materials having a disagreeable odor and can be used repeatedly by regeneration after use. It is also possible to carry out extraction and refining steps simultaneously by having a cation exchange resin and a decoloring agent copresent in the extraction water.

Example 1: One kilogram of oil-free soybean flake (total nitrogen 8.20%, ash 5.8%) is charged in an extraction vessel, and 700 ml of an H-type cation exchange resin are separately charged into a resin vessel. These two vessels are connected to each other with a pipe and 5 liters of water are circulated by means of a pump to carry out extraction. After conducting extraction for about 1 hour at room temperature, solids are separated from the extraction liquid and 705 grams of refined oil-extracted soybeans (total nitrogen 11.3%) are obtained by drying the separated solids.

The extraction liquid amounting to about 4 liters is slightly colored and contains 2.5 g/dl of sugar, 0.5 g/dl of organic acids (as citric acid) and 0.025 g/dl of total nitrogen and exhibits a pH of 3.2. This extraction liquid is sterilized by heating at 80°C for 30 minutes and then is passed through a column filled with 400 ml of decoloring resin. A colorless, transparent, odorless acidic liquid having an agreeable sour taste is obtained. When this acidic liquid has 200 grams of sugar added and 4 grams of sodium cyclosulfamate and coloring agents and perfumes which give the flavor of an orange thoroughly mixed therein, 4 liters of a beverage like orange juice and having an excellent flavor is obtained.

Example 2: Oil-free soybean flake amounting to 1 kg is charged into an extraction vessel and extraction is carried out for 1 hour at room temperature by adding 5 liters of water. By separating solids by centrifuging, 4 liters of the extraction liquid is obtained. 600 ml of an H-type cation exchange resin is placed into a bag of synthetic textile and the liquid is then agitated mildly with the bag introduced into the liquid. Cations are adsorbed by the resin and organic acids are freed to lower the pH thereby coagulating and precipitating protein.

Precipitated protein is separated by centrifuging and settlement and an acidic aqueous solution obtained is subjected to heat treatment and thereafter is decolored and deodorized with active carbon. Then 200 grams of sugar, 4 grams of sodium cyclosulfamate and perfumes and a coloring agent are added. Further added are 5 grams of lysine hydrochloride and 1.6 grams of vitamin C and the flavor is adjusted to obtain 4 liters of a beverage having increased nourishing effect.

STABILIZED VITAMIN C FORTIFICATION

Histidine, Glycine or Methionine Addition

The process of *P.A. Hammes and L.D. Morse; U.S. Patent 3,652,290; March 28, 1972; assigned to Merck & Co., Inc.* relates to beverages which contain vitamin C and particularly to stabilizing this vitamin against deterioration. The process is of especial value in protecting the vitamin C against the destructive action of nutritionally assimilable iron if it is also present.

The beverage may be a natural or manufactured, fruit or vegetable drink containing vitamin C. By a natural drink is meant the whole, concentrated or diluted juice from oranges, lemons, grapefruit, grapes, carrots, prunes, tomatoes and other produce which is squeezed or crushed to supply a beverage which naturally contains an appreciable amount of vitamin C. It may be fortified with added vitamin C and mixtures of different juices may be made. By a manufactured drink is meant one that is wholly or predominately flavored with a synthetic flavoring agent and to which vitamin C has been added to supply all or nearly all of its vitamin C content. Of the several known vitamins, vitamin C is recognized as a most important one as its inadequate consumption produces various human ills one of which is scurvy. As it is very water-soluble it is readily excreted and consequently it is important that it be consumed every day. It is generally accepted that for adults the minimum daily intake is 30 mg and the recommended daily allowance is 60 mg.

Despite the human need for vitamin C many persons are reported to be receiving an inadequate daily intake despite the fact that it is available in substantial amounts in many fruits and vegetables. To help satisfy a widespread need for vitamin C, it has been added to the various beverages mentioned above and particularly to the packaged beverages sold in stores. However, the vitamin C activity deteriorates in the package during storage and even after the package is opened and is awaiting complete consumption. For instance it has been considered advisable to add vitamin C to packaged drinks but it has been found that the other ingredients ordinarily present hasten vitamin C deterioration.

Vitamin C is also known to have an adverse effect on ingredients which sometimes are added to the packaged beverage. An illustration of this is certain dyes which are added to give the beverage a more natural or a more attractive color. The chemical incompatability of the dye in the presence of the vitamin C will cause the dye to fade during storage of the packaged beverage. The process includes as one of its features the discovery that by stabilizing the vitamin C, its deteriorating action on the dye or other labile substance is overcome or minimized.

It is apparent that if beverages and especially packaged beverages can be fortified with iron this would serve as an ideal medium to supply iron to children and adults. It is possible to add iron to both natural and manufactured beverages, which naturally contain vitamin C or which have been fortified with vitamin C. For instance, ferrous sulfate could be added to freshly made orange or tomato juice as well as to their canned or frozen preparations. These preparations may be in their whole form or may be concentrated so that they are to be reconstituted with added water for consumption.

For a manufactured beverage which is packaged it has been found that a preferred addition of vitamin C to the drink in the single strength form, i.e., the dilution in which it is to be drunk, is 0.338 mg/ml but it may be from 0.084 to 0.675 mg. This concentration of vitamin C is, as stated, based on the drink in its single strength form and it will provide a person with a considerable portion of the daily need for the vitamin. The process may be applied equally as well as to concentrates, frozen or unfrozen, such as those which require dilution with water after the package is opened to obtain the single strength form. For instance, if the packaged concentrate is to be diluted with three volumes of water, i.e., the concentrate volume is one-fourth of the intended dilution volume, the percent of vitamin C would be four times higher in the concentrate.

If histidine is selected for stabilizing vitamin C, it preferably amounts to about 4.0% by weight of the vitamin C. If glycine is selected, it preferably amounts to about 1.5% by weight of the vitamin C but from 0.4 to 3.0% may be used. If methionine is selected, it preferably amounts to 3.0% by weight of the vitamin C but from 0.7 to 4.0% may be used. If the HCl salts are used, an additional amount must be added to obtain the equivalent amount of the amino acid base. This concentration of the amino acid in the drink is not noticeable or objectionable to the taste or smell. If more than 0.0675% vitamin C is in the drink it will call for an amount of the amino acid in the lower part of the range so that an objectionable taste does not appear.

Vitamin C and the selected one of the mentioned amino acids are mixed into the beverage before it is packaged. The package may be the conventional capped bottle or sealed metal container. On opening the container for consumption, the vitamin C will assay approximately as high as when it was prepared and this will be true even if there is also present iron or iron compounds or other ingredients which generally tend to deteriorate the vitamin C.

If iron is also to be present it may be added in its elemental form if it is used in an ultra-fine powdered form and if it is kept in suspension preferably with a conventional suspending agent. As elemental iron it would constitute from zero up to three times the weight of the vitamin C. Metabolically available iron compounds are preferable to the metallic iron and these may be the anhydrous or hydrated forms of ferrous lactate, fumerate, gluconate, succinate and sulfate. Ferric compounds which may be used are ferric chloride, phosphate and ammonium citrate. A sufficient weight of the selected iron compound would be used to supply an amount of iron ion within the range mentioned above.

If the whole, fresh juice is fortified with added vitamin C the amount of amino acid added would be based on the resulting total vitamin C. If the fresh juice is concentrated, irrespective of whether it is or is not frozen, the added amino acid would be based on the vitamin C content as mentioned above. If iron is added to these natural juices its amount would constitute the same relationship to vitamin C that is mentioned above, i.e., from zero to three times the weight of vitamin C.

Example: Bottles of the type used to package carbonated beverages (12 oz capacity) were filled with 360 ml of aqueous solution. The solution each contained 0.333 mg vitamin C/ml and 3.25 mg citric acid/milliliter. Their contents varied in respect to ferrous sulfate content and amino acid content as shown in the accompanying tables. All of the bottles had a head space of about 30 ml air above the beverage. The bottles were sealed with traditional crimped caps. Initial determinations of vitamin C were made within 3 hours of bottling. The head space was flushed with nitrogen after sampling; recapped, and stored in the dark at 38°C. Sampling at 1 week, 3 weeks and 7 weeks, the head spaces were flushed with nitrogen each time. Vitamin C determinations are shown in the accompanying tables as percent vitamin C recovered, based upon the original 0.333 mg/ml.

Sample with Histidine –

	$FeSO_4.7H_2O$, mg./ml.	L-Histidine HCl, mg./ml.	Percent vitamin C recovered			
			Initial	1 week	3 weeks	7 weeks
Run No.:						
1	0	0	89.8	75.6	57.9	49.0
2	0	0.01445	89.2	63.3	60.2	56.8
3	0	0.02890	89.2	64.1	58.7	56.1
4	0.0914	0.01445	78.4	67.1	60.2	52.2
5	0.0914	0.02890	96.8	67.9	64.9	62.2

Sample with Glycine in Place of Histidine –

	$FeSO_4.7H_2O$, mg./ml.	Glycine HCl, mg./ml.	Percent vitamin C recovered			
			Initial	1 week	3 weeks	7 weeks
Run No.:						
1	0	0	89.8	65.5	57.9	49.0
2″	0	0.00517	89.2	64.9	60.2	57.6
3″	0	0.01034	87.8	61.8	57.9	49.8
4″	0.0914	0.00517	79.3	66.4	61.0	57.6
5″	0.0914	0.01034	79.3	67.1	61.8	53.0

Sample with Methionine in Place of Histidine –

	$FeSO_4.7H_2O$, mg./ml.	DL-Methionine HCl, mg./ml.	Percent vitamin C recovered			
			Initial	1 week	3 weeks	7 weeks
Run No.:						
1	0	0	89.8	65.6	57.9	49.0
2′	0	0.0102	94.2	70.9	65.6	61.4
3‴	0	0.0204	86.8	64.9	57.1	49.0
4‴	0.0914	0.0102	76.9	64.1	56.8	50.6
5‴	0.0914	0.0204	78.4	61.8	53.3	48.3

Combinations with Iron

In another process by *L.D. Morse and P.A. Hammes; U.S. Patent 3,734,742; May 22, 1973; assigned to Merck & Co., Inc.* a special process is used for the fortification of beverages with vitamin C and iron. Ascorbic acid is stable in the dry condition. In aqueous solution, however, it is rapidly oxidized by oxygen from the air. The initial oxidation product is dehydroascorbic acid and this is further oxidized (with a loss of vitamin C activity) to degradation products. Iron is known, along with copper and other metals, to catalyze oxidation of ascorbic acid so that a liquid product containing ascorbic acid and iron in its conventional ferric ion salt form becomes useless very rapidly.

Consequently, food scientists and nutritionists would like to increase the present day average consumption of iron and ascorbic acid. As soft drink beverages may be consumed several times daily by many people, bottled or canned beverages offer an ideal carrier to supply an increased amount of iron and ascorbic acid. Although canned fruit and vegetable beverages may not be consumed every day, they also offer an excellent means for giving additional ascorbic acid and iron to the consumer.

The addition of ascorbic acid and iron to beverages has not been considered feasible and the reason for this is the catalytic effects of iron on the oxidative deterioration of ascorbic acid. Thus, one does not find bottled or canned soft drinks or fruit or vegetable beverages fortified with both ascorbic acid and iron.

It was found that ferrous ion does not decrease the stability of ascorbic acid in the usual bottled or canned fruit, vegetable and soft drink preparations despite all the flavoring, acidifying and other agents which are ordinarily present. This is in marked contrast to the ferric form. Thus, an anaerobic system, i.e., a canned or bottled system with a small headspace demonstrates the utility of ferrous ion as an additive in the presence of ascorbic acid. Iron can be included in these packaged goods as the ascorbic acid will remain substantially unimpaired during storage. Once opened they are consumed rather quickly and before any substantial nutrient deterioration has occurred.

The critical requirements of the process are accomplished by providing a sealed canned or bottled aqueous beverage containing from 80 to 95% water and having a pH of from 2.5 to 3.4, an ascorbic acid content of from 0.112 to 0.840 mg/ml and a ferrous ion content of from 0.011 to 0.062 mg/ml.

It is preferable to use ferrous sulfate heptahydrate as it is commercially obtainable at a low price but its monohydrate or the anhydrous form may be used as well. It is also possible to use ferrous fumerate, citrate, lactate and the other salt forms which are used medicinally to treat anemia. If ferrous sulfate heptahydrate or other salt is selected, it is weighed out so that its elemental iron content is in the range of 0.008 to 0.150 mg/ml of the bottled or canned preparation. As this is based on the ferrous ion present, it determines the equivalent weight required for whatever ferrous salt is selected.

Vitamin C and ferrous ion are added to the beverage and then the can or bottle is sealed (i.e., exclude outside air) as soon as is practical. Since the oxygen in the air tends to oxidize vitamin C in dilute solutions, the contact of the beverage with air reduces the vitamin C content. Furthermore, in the conventional canning and bottling procedures, a small headspace of air (with oxygen) is left in the bottle or can which also oxidizes the vitamin C. Accordingly, after bottling or canning, the vitamin C continues to be oxidized until the oxygen in the headspace is used up. While an inert gas may be used in the headspace to avoid oxidation, the inert gas is generally not used due to the increased cost of the bottling operation. After the oxygen in the headspace is used up, the beverage becomes quite stable with less than a 20% decrease in vitamin C over a period of 1 month. The sealed canned or bottled beverage of the process has a shelf life of many months.

Example 1: An experimental aqueous solution is prepared to contain 3.25 mg of citric acid per milliliter, 0.338 mg ascorbic acid per milliliter along with ferrous ion as ferrous fumarate, at 0.10 mg Fe^{++}/ml. The control is prepared to contain the citric acid and ascorbic acid only in the same amounts. 200 ml solutions are prepared in 200 ml volumetric flasks. 20 ml of the solutions are then removed from each flask to provide a small air space. The flasks and contents are heated to 90°C in a water bath for 1 minute and then cooled in ice water. The flasks are then stoppered and stored for testing. After 4 hours, 1 week and 3 weeks at room temperature, the results of the analyses for ascorbic acid are shown below. In each case, the water content is 99.6%.

	Ascorbic Acid (mg/ml)		
	4 Hours	1 Week	3 Weeks
Control	0.32800, pH 3.15	0.05430, pH 3.00	0.00121, pH 3.01
Experimental	0.31300, pH 3.20	0.13600, pH 3.10	0.02190, pH 3.10

Example 2: A commercially available lemon still beverage is treated in several ways. One portion (1) is fortified with vitamin C at 0.339 mg/ml beverage. Another portion (2) is fortified with vitamin C at 0.339 mg/ml and with ferrous iron (as ferrous sulfate) at 0.062 mg/ml. Another portion (3) is fortified with vitamin C at 0.339 mg/ml and with ferrous ion at 0.021 mg/ml. The drinks are pasteurized by heating in a coil at 190°F for 20 seconds. 12 oz cans are then filled within 20 seconds at that temperature and the cans are sealed and allowed to cool in an inverted position. In each case, the water content is 88%. Some cans are stored for 2 months at room temperature and some for 44 days at 38°C. The contents of the cans are analyzed for total vitamin C (total ascorbic acid and dehydroascorbic acid, both biologically available as vitamin C) and checked for color change and taste. No color or taste differences are observed. Total vitamin C analyses are as follows.

		Vitamin C (mg/ml)	
Portion	**pH**	**2 Months at Room Temperature**	**44 Days at 38°C**
1	3.0	0.265	0.280
2	2.95	0.350	0.360
3	2.95	0.296	0.286

The above procedure is repeated utilizing vitamin C at 0.339 mg/ml and ferrous ion at 0.062 mg/ml but the cans are not sealed after filling. The cans are allowed to cool at room temperature and are stored at room temperature for 1 month. Total vitamin C analysis indicates the presence of less than 0.1 mg/ml of vitamin C.

EGG ALBUMEN BASE PRODUCTS

Glycine Additive

Egg albumen and glycine are combined to form a water-soluble powder for beverage use in the process of *J.A. Basso, C.V. Fulger and J.J. Thompson; U.S. Patent 3,736,150; May 29, 1973; assigned to Kellogg Company.*

The objectionable egg odor and taste may be effectively inhibited, in fact eliminated, in beverages reconstituted with water from powdered mixes containing acid precipitated albumen as the primary source of protein, by adding relatively small proportions of an amino acid such as glycine to the mix, along with sufficient food acid to cause the pH of the resulting solution to fall within the range of between 1.95 to 4.0. The resulting enriched beverage is substantially indistinguishable organoleptically from a similar beverage with all protein omitted.

The egg albumen is first subjected to acid treatment and converted to powder form by conventional methods. For example, the albumen may be provided in fresh, frozen or other suitable form and may be treated with a food acid such as citric, tartaric, fumaric, or malic to provide a pH of from 3.5 to 4.5. The liquid mixture is stirred thoroughly to dissolve the food acid and cause the denaturation and precipitation of a fraction of the proteins in the liquid egg albumen. The denatured and precipitated fraction of the protein is then removed by suitable means such as centrifugation and the remaining soluble fraction of the protein recovered in a yield of approximately 88 to 90% by weight. This acid stabilized filtrate is then dried under mild conditions such as by freeze drying, vacuum drying or spray drying to prevent further denaturation of the protein, and then pulverized if not already in such condition.

The dry soluble fraction of protein is then mixed with powdered or granulated amino acid and used to make food products such as acidified breakfast juice-type drinks. The treated dried albumen may be agglomerated with sucrose or other suitable polyhydroxy materials to facilitate more rapid dissolution in water. An alternate method of preparing the high protein composition is to dissolve or dispense one or more of the components of amino acid, food acid, or sucrose into the liquid acid stabilized albumen and then to dry the mixture.

It has been found that the ratio between the amounts of albumen and amino acid is critical to the extent that the proportion of amino acid must be above a defined minimum in order to effectively inhibit the egg taste and odor. If the total protein components are to be supplied by albumen and an amino acid such as glycine, the glycine commences to impart its inhibiting characteristic at a level of about 3% by weight of the protein mixture, and at 4.75% achieves satisfactory effectiveness. Proportions of amino acid above about 5% have not been found to be contributing greater organoleptic effectiveness. In the range over 6% the amino acid is used principally to adjust the amount of protein to be contributed by the albumen.

It is theorized that particular amino acids, when present above a given minimum level and within certain ranges of acidity, tend to tie up or otherwise combine with chemical groups in the albumen, such as sulfhydryl groups, thought to be responsible for the characteristic odor and taste of eggs, and thereupon eliminate or take away such odor and taste.

It has also been found that the dry product must be reconstituted within a particular range of acidity in order that the egg taste and odor will be effectively inhibited. The acidity of the reconstituted beverage is preferably maintained within a pH range of 1.95 to 4.0, although some degree of inhibition is achieved throughout the pH range between 1.95 and 7.0. Therefore, a sufficient amount of solid food acid is incorporated into the protein mixture to assure that the water reconstituted product will attain a pH within the specified range. The following is a general formula for a dry mix that may be reconstituted by the addition of water into a breakfast juice-type drink.

Ingredients	Percent by Weight
Protein (17 to 42%)	
Dry acid stabilized albumen	15 - 40
Amino acid	2 - 18
Food acid	1 - 2
Sucrose	30 - 43
Flavor	0.5 - 1.4
Color	0.01 - 0.02
Clouding agent	1 - 2

The above composition is prepared either by mixing the above separate ingredients together or by codrying liquid acid stabilized albumen with one or more of the amino acid, food acid or sucrose and adding the remainder of the ingredients. The resulting mixture may be prepared with or without the addition of suitable wetting and antifoaming agents and dried to the desired moisture level. In order to prepare a beverage for consumption, the dry mix is added to and dissolved in water and diluted such that the mix constitutes, typically, 15 to 30% by weight of the solution. The resultant mix is soluble in water and may be rendered more readily soluble if desired by agglomerating the mixture into porous aggregates. Appropriate additives may include natural dried food pulps, fortifiers such as vitamins and minerals, antifoaming agents.

Substitutes for the amino acid glycine can be others such as tyrosine, alanine, L-leucine, lysine, isoleucine, arginine, methionine, L-cystine, 4-hydroxy-L-proline, valine, D,L-proline, glutathione and various polypeptides. For the purpose of inhibition of egg flavor and odor, it has been found that L-cystine is slightly more effective than glycine and may be added in slightly lower proportions; glycine is preferred, however, because it is less expensive and has a better acceptance in foods. For the purpose of illustration, the following is a specific example of use of this process in the preparation of a breakfast type food drink.

Thirty pounds of fresh liquid egg albumen was stirred thoroughly for 30 minutes with 210 grams of anhydrous citric acid. The product was then centrifuged and the precipitated protein fraction of 12% by weight discarded. The remaining soluble fraction was poured in flat trays, frozen and freeze dried. The treated dried protein was then used to prepare mixes for the preparation of beverages.

Whey Combinations

Whey is also used by *J.A. Basso, N.A. Lemaire and C.V. Fulger; U.S. Patent 3,737,326; June 5, 1973; assigned to Kellogg Company* in combinations with egg albumen to inhibit the taste and odor of egg albumen in a beverage product.

The whey is provided in dry or liquid form and may be separately added in dry form to the dried albumen or dextrin-albumen, or may be incorporated into the mixture by other methods. Although the whey may be used in any convenient form without further processing, it is preferable that the whey be at least partially delactosized and demineralized in order to increase the relative proportion of protein. Thus, whereas natural whey solids may contain 11% protein with the remainder substantially as lactose and ash, the modified whey solids may contain up to 35% or more protein, thereby providing a richer source of protein while decreasing the relative amount and characteristic flavor of lactose.

The whey solids, when mixed directly with the dry albumen or albumen-dextrins, may be added in sufficient quantities to contribute from 30 to 45% of the total protein with the remainder of the protein being provided by the albumen. The whey, when maintained within the above stated ranges, serves to inhibit the egg flavor and odor of the albumen; albumen inhibits the flavor of the whey; and additionally, the prescribed addition of whey optimizes the palatability and mouthfeel of the beverage and minimizes foaming.

If the protein contribution of whey is below 20 or 25%, the resulting beverage will exhibit egg flavor and/or odor and may not be entirely stable. If the protein contribution of whey is above approximately 50% (or in the order of at least about 60% with modified whey), the whey will impart a noticeable chalky flavor to the beverage. An alternative method of incorporating the whey into the beverage mix is to dissolve the liquid or solid whey into the liquid albumen filtrate or albumen-dextrin solution prior to drying, and then drying the resulting solution to solid form. By this method, the solubility of the mix is improved over the method by which the whey is separately added.

After the protein or protein-dextrin or sucrose mixture has been prepared an amount of food acid, such as citric, fumaric or malic, is incorporated into the dry mix in sufficient quantities such that the resulting beverage will exhibit an acid pH in the order of from 2.5 to 4.6. If the pH does not fall within the minimum and maximum levels indicated, the egg flavor and odor of the albumen will not be effectively inhibited. In the case of fruit-type or citrus drinks, the resulting pH is preferably in the order of 3.4 to 4.0 with approximately 3.8 being optimal. In the case of vegetable-type drinks, such as tomato, the resulting pH is in the order of 3.7 to 4.6, with 4.1 being optimal. Normally, at least a majority of the food acid that is required in the final product is added to the liquid prior to drying. The resulting solution is then dried, preferably by spray methods, to yield a composite acid-whey-albumen mixture.

If required, sucrose may also be added to the protein base or mix in order to attain the desired sweetness level in the resulting beverage. Suitable flavors, such as orange, grapefruit, pineapple, tomato, raspberry or others, are also added, together with a suitable coloring agent. Other appropriate additives may include natural dried fruit pulps, fortifiers, such as vitamins and minerals, antifoaming agents. The resulting composition is then mixed, and suitable wetting agents may be added concurrently with mixing. The resulting mix is then ready for use, although a further treatment step may be instantizing by agglomerating the mixture. Also, the food acid, as well as sugar, may be added to the other materials which have previously been agglomerated and dried.

The resulting mix is a free-flowing powder or agglomerate that is completely storage stable at room temperatures and for extended periods of time. In order to prepare a beverage for consumption, about 15 to 30% by weight of the mix is dissolved in water. The mix is readily dissolved in cold water and the resulting solution remains stable for relatively long periods of time. As an alternative, all necessary ingredients may be combined in the form of a concentrated solution and stored in containers as a frozen concentrate.

Example 1: The following formula was used in the preparation of an orange flavored drink:

Ingredients	Amount (grams)
Dry egg albumen acid precipitated	16
Dry whey solids (11% protein)	16
Sucrose	24
Citric acid	1.1
Orange color	0.013
Orange flavor	0.6

The above ingredients were mixed together, and then dissolved in 235 ml of cold water to yield a beverage with a pleasant taste and aroma. The beverage contained about 2 grams of egg albumen per fluid ounce.

Example 2: About 100 lb of fumaric acid were mixed with and dissolved in 5,500 lb of liquid egg white containing about 11% protein. About 650 lb of precipitate were removed by filtration. About 400 lb of dextrin, 1,220 lb sugar and 1,320 lb of whey containing about 33% or about 440 lb of protein were then dissolved in the liquid albumen filtrate, and the resulting solution was spray dried to yield about 3,465 lb of a composite product containing about 915 lb of protein, 2,490 lb of carbohydrate and 60 lb of fumaric acid. The resulting product was used in a beverage mix where the whey protein constituted about 48% of the total protein, and the egg albumen contributed the remainder.

Example 3: About 3 lb color, 1,370 lb sugar, 1,610 lb modified whey containing 410 lb protein, 2,222 lb dextrin and 175 lb fumaric acid were dissolved in acid precipitated egg albumen, and the resulting solution was spray dried to provide a basis for a beverage mix. Of the total protein, the whey constituted about 25.5% and the albumen constituted the remainder.

SOYBEAN MILK

Full-Fat Oilseed Method

Traditionally, soybean milk has been made by the water extraction method, the original procedure used in China for several centuries. Protein and oil are extracted in about the same ratio as they occur in the beans and are in the form of an emulsion. The oligosaccharide carbohydrate fraction of soybeans which has been reported to be responsible for flatulence remains in the soybean milk by this process. To circumvent this problem, alternate methods have used water-soluble protein isolates prepared from defatted soy products by isoelectric precipitation. These methods start with protein isolate powders and make an emulsion by adding emulsifiers, fats, minerals and sugars. Since these nonfat protein isolates are obtained in rather low yields (25 to 50% of the total protein) they are relatively more expensive than other soy fractions. However, the isolate route permits better control of the composition and concentration of the milk; also, the milk has a bland flavor as compared to the whole bean extract.

In the method of *G.C. Mustakas; U.S. Patent 3,809,771; May 7, 1974; assigned to the U.S. Secretary of Agriculture* the following steps are used to produce full-fat oilseed-protein beverages:

(a) suspending full-fat oilseed flour in from 8 to 10 times its own weight of water, which has been adjusted to a temperature of from 150° to 212°F and a pH of from 3.5 to 4.5, for a time sufficient to inactivate lipoxygenase enzyme present in the full-fat oilseed flour;

(b) cooling the suspension to and holding at a temperature of from 50° to 75°F for a time sufficient to precipitate lipid-protein present in the full-fat oilseed flour;

(c) separating the precipitated lipid-protein from the resulting liquid whey;

(d) resuspending the precipitated lipid-protein in water at a pH of about 9, the amount of the water equaling from 7 to 10 times the weight of the oilseed flour of step (a);

(e) heating the suspension of step (d) at from 205° to 212°F for from 1 to 5 minutes;

(f) cooling the suspension of step (e) to 70° to 75°F and adjusting the pH to about 7; and

(g) clarifying the cooled and neutralized suspension to form a full-fat oilseed-protein beverage.

Full-fat oilseed flours useful as starting materials include any whole, dehulled, full-fat oilseed product, either raw or cooked, and having nitrogen solubility index (NSI) values in water ranging from 15 to 98%. This includes products from soybean, peanut, cotton, rape, sesame, sunflower, and safflower seeds. These starting materials are ground to flours having mesh sizes such that 95 to 100% will pass a 100-mesh screen. However, oilseed products having greater mesh sizes, including flakes, will result in acceptable milks with somewhat reduced lipid-protein yields. Soybeans are considered representative of the above-mentioned oilseeds and are used for the purpose of the descriptions.

To obtain a soybean product having optimum nutritional value and consumer acceptability, it is preferred that the starting material be prepared in the following manner: crack and dehull whole raw soybeans, lightly cook the dehulled cracked beans in dry heat at 150°F for 5 minutes, flake and pin mill to a mesh size such that 95 to 100% of the product will pass a 100-mesh screen.

The object of step (a) of the process is to impart to the final product a superior flavor and shelf stability by deactivating the lipoxygenase enzymes that catalyze oxidation of fats, thereby causing rancidity. Enzymes were inactivated in less than 1 minute when soy flours were slurried in boiling water at a pH of 7. Complete deactivation also occurs within about 10 minutes when the starting materials were slurried in water which had been previously adjusted to pH 3.5 to 4.5 and temperatures of 150° to 212°F. It is best that the amount of water be from 8 to 10 times the amount of soy flour (by weight), that the temperature be about 205° to 210°F, and that the pH be about 3.5 before adding the soy flour, because these parameters are somewhat more compatible to those of the second step and yields of precipitated protein are somewhat larger than those obtained using other parameters. Products obtained using these preferred parameters also have a more bland flavor.

Lipid-protein precipitations were carried out by modifications of the acid washing and the isolated defatted soy protein processes. A comparison of two methods was made. In the acid wash technique, soy flour was stirred into water at a pH of about 4.5 and the mixture allowed to stand for 20 minutes. The pH rose to about 6. The mixture was then centrifuged and washed twice with water at pH 4.5 to separate the lipid-protein curd from the lipid whey fraction. Lipid-protein curd was resuspended in water, homogenized, cooked at 160°F for 5 minutes, and cooled.

The isolated soy protein process was accomplished by suspending full-fat soy flakes in water at a pH of 4.5 and allowed to stand for 20 minutes while the pH was maintained at pH 4.5 with additions of H_3PO_4. Lipid-protein curd was separated from the whey by centrifugation, washed twice with water at pH 4.5, and resuspended in water. The suspension was adjusted to pH 7, and homogenized. The milk fractions resulting from the two methods above were analyzed for recoveries, and distribution of solids, protein, and fat, are shown in the following table.

	Distribution between milk and whey fractions					
	Solids, percent in—		Protein, percent in—		Fat, percent in—	
Process	Milk	Whey	Milk	Whey	Milk	Whey
Acid wash	66	34	75	25	73	27
Isolated soy protein	77	23	90	10	91	9

Since the distributions of protein and fat in the isolate process were considerably better than in the acid wash process, the isolate procedure is preferred. It is preferred that lipoxygenase enzyme deactivation and lipid-protein precipitation should be carried out in one step by dispersing soy flour in water at a temperature of 206° to 212°F at a pH which has been adjusted and maintained at 3.5 by dilute sulfuric acid. Any mineral acid is operative for this part of the procedure, but others may require slightly different pH values to accomplish maximum precipitation.

However, at pH values of 4 or over, there is a tendency to form soluble proteinates that are lost in the whey fraction. A 5 to 10 minute holding period at the high temperature is sufficient to inactivate the enzyme. The preferred time and temperature parameters also have some effect on the protein and fat such that increased yields are obtained. At the end of the holding time, the mixture is cooled to 50° to 75°F and stirred for 10 to 20 minutes to assure maximum precipitation of the lipid-protein curd. This acid water cook accomplishes another very important object, that is, the reduction of flatulent properties by solubilizing carbohydrates which are removed in the whey fraction.

Separation of lipid-protein curd from the liquid whey fraction can be accomplished by any suitable method such as filtration or centrifugation. Centrifugation is preferred because of its efficiency and shorter operation times. It is also more compatible with continuous or semicontinuous operations. Using a pilot-plant size perforated basket centrifuge covered with duck cloth, preferred speed was between 3,600 and 3,800 rpm. The centrifuge cake (i.e., the lipid-protein curd) was washed two or more times with tap water or water at the same pH at which the curd was precipitated. In an alternate method the cake was resuspended in 50% w/w aqueous ethyl or isopropyl alcohol followed by centrifugation and washing from two to three times with aqueous alcohol and/or water. Washing the cake with aqueous alcohols resulted in products having more bland flavors. The washed cake was resuspended in tap water in amounts of from 7 to 10 times the original amount of starting material.

When defatted protein isolates were the desired products, the amounts of resuspension water were from three to five times the original amount of starting material. The next step is for the purpose of inactivating trypsin inhibitor which, when present in the final product, reduces the nutritional value of the protein. Inactivation of this enzyme proceeds very slowly under acidic conditions. Therefore, the acid cook in the first step (lipoxygenase deactivation) did not accomplish the inactivation of trypsin inhibitor.

It was discovered that under alkaline conditions, trypsin inhibitor is deactivated very rapidly. In the example, 100% deactivation was obtained by adjusting the pH of the suspension to 9 with NH_4OH and heating quickly with sparge steam to a temperature of about 205° to 212°F and holding for 1 minute. The suspension is quickly cooled to about room temperature. High temperature short time (HTST) plate exchangers can be used to further reduce heating times. The advantages of short cooking times in this step are: (1) reduced viscosity of the soy milk product (i.e., as low as 6.5 cp; cow's milk has a viscosity of 4.2 cp); (2) less heat-polymerized curd clinging to the walls of the vessel, thereby reducing protein yield; (3) a more stable emulsion; and (4) reduced losses of curd in the final clarification step. The pH of the suspension should be maintained at at least about pH 9 because at pH 8 an inactivation of only about 60% was obtained.

After this alkaline cook step, the suspension is subjected to colloid milling and homogenization, which were accomplished in a colloid mill having maximum plate clearances of about 0.001", and in a high-pressure homogenizer under pressure of 8,000 to 10,000 psi. Low-pressure homogenizers such as the Manton-Gaulin two-stage homogenizer are also suitable. The use of these optional steps decreased the amount of solids removed from the soy milk in the clarification step.

When milling and homogenization were not used, the products had poorer mouthfeel and developed some sedimentation upon standing. After the HTST alkaline cooking step, or after colloid milling-homogenization when they are used, the pH of the suspension is reduced to a pH of about 7 with a mineral acid. This completely removes the alkaline flavor present at the higher pH.

The final step is a clarification process best carried out in a centrifuge as described for separation of the lipid-protein curd. This removes any agglomerated particles that have resisted complete dispersion, leaving a highly stable, bland, nonflatulent soy milk product having a mouthfeel of essentially the same quality as cow's milk.

In the production of defatted protein isolates it is preferred that the suspension not be subjected to colloid milling, homogenization, or the final centrifugation step. After the HTST cooked suspension is cooled to a temperature of from 70° to 100°F and the pH is adjusted to about 7, the fat is removed by liquid-liquid extraction using a suitable water-immiscible solvent (e.g., pentane, hexane), and the defatted product dried. However, alternatively the water is first removed by means such as evaporation, spray drying, or freeze drying and the fat removed from the dried product by solvent extraction. In this manner defatted protein isolate products were obtained in total yields of up to 95% of the original protein. By using a gas-liquid chromatographic (GLC) procedure, it was found that the hexane defatted isolates contained small amounts of bound lipids (fats). Three commercial soy protein isolates analyzed in the same manner as the products from the process contained 1 to 2% bound lipids, while the beverage products contained from 4 to 5%. However, products containing as high as about 90% protein can be obtained essentially fat-free by extracting the full-fat products with polar solvents such as aqueous alcohols.

The final beverage products are suitable for further processing such as spray drying to obtain dried reconstitutable soy milk powders as food supplements. Additives such as fats, vitamins, minerals, and flavorings can also be used to make beverages suitable to any taste and nutritional need. When extra fat is added, it is preferred that it be added to the suspension as it is being fed to the colloid mill.

Example: (A) Preparation of Raw Preheated Full-Fat Soy Flour – Field dried soybeans were subjected to a single break (one pass) cracking on corrugated rolls (10 corrugations per inch, 0.040" clearance). The cracked meats were then passed over a two-deck screen shaker (1/8" round hole top screen and a No. 14 bottom screen) with dehulling aspiration on the meats as they entered the top screen and on the No. 14 screen tailings. The dehulled meats were then dry heat-treated by passing through a screw paddle conveyor maintained at 150°F by jacketed steam and having an overall retention time of about 5 minutes. The hot meats were passed through flaking rolls, cooled, and ground (99% through 100-mesh screen) in a wide-chamber air-swept pin or stud mill. The resulting flour had an NSI value of 97.

(B) Preparation of Soy Milk – Soy flour (500 grams), prepared as described in (A) above (39.5% protein, 22.0% fat, 4.8% moisture, all by weight), was slurried in 4,000 cc of tap water that had previously been adjusted to pH 3.5 with 10% sulfuric acid, heated with sparge steam to 206° to 210°F, held at that temperature and stirred for 10 minutes, cooled to room temperature (70° to 75°F) and stirred for 15 minutes. The slurry was maintained at pH 3.5 with additions of 10% sulfuric acid. The slurry was then fed over a 5 to 10 minute period to the bowl of a perforated basket centrifuge running at 3,800 rpm (1,500 *g*). The perforated basket was covered with a cotton duck cloth. The resulting cake (lipid-protein curd) was washed two times with 500 ml portions of tap water, pH 7, spun dry for 5 minutes.

The liquid whey was recycled through the centrifuge and the recovered curd combined with the cake. The wet curd (i.e., cake) was reslurried in 4,000 cc of tap water and the suspension adjusted to pH 9 with 10% NH_4OH solution. The alkaline suspension was quickly (2 to 3 minutes) heated to about 205°F using open sparge steam, held at 205°F for 1 minute, and quickly (2 to 3 minutes) cooled to room temperature (70° to 75°F). The alkaline-cooked suspension was fed to a Manton-Gaulin colloid mill having a plate clearance of 0.001", then to a Manton-Gaulin single-stage homogenizer at 8,000 to 10,000 psig. The milk was adjusted to pH 7.2 with 10% H_2SO_4 and centrifuged. The small amounts of cake obtained at this stage can be recycled to the soy flour starting material at least five times with no detrimental effect on the finished product.

Dispersible Powder Base

Processes have been developed that produce protein-rich, highly stable soy flours that have a mild, acceptable flavor, good nutritional properties, and a long shelf life. However, heat treatments used in production of soy flours reduce the solid suspension properties and yield sand-like, hard particles that give very poor mouthfeel when suspended in water. Good mouthfeel is used in this context to mean having a smoothness similar to cow's milk.

It is desirable to produce a protein food which can be easily mixed with water to give a beverage. The process of *G.C. Mustakas, W.J. Albrecht and G.N. Bookwalter; U.S. Patent 3,639,129; February 1, 1972; assigned to the U.S. Secretary of Agriculture* involves the following steps for conversion of vegetable protein flours into dried powders that form stable emulsions when mixed with water.

The preferred starting material is full-fat soy flour. Defatted soy flour gives an equivalent product with respect to dispersibility and mouthfeel but requires the addition of greater amounts of vegetable fat in the formulation of milk substitutes. Any vegetable protein flour or grits, i.e., flour or grits from oilseeds or other difficultly dispersible protein product, can be processed by this method to produce a highly water-dispersible powder similar to that produced from soy flour.

Enough water is added to the starting material to give a 10 to 20% by weight flour-water slurry. This ratio of flour to water is specified because thicker slurries, greater than 20% flour, clogged the colloid mill and homogenizer used in the examples. However, another kind of mill could be used which might operate with a thicker slurry. Theoretically, the percentage of flour in the slurry could become almost zero since the particles going through the rollers of the mill would still be broken down. However, practical limitations (e.g., removal of the water, volume handled) keep the percentage from going lower than 11% by weight flour in water. A reduction in flour particle size is achieved by wet milling (i.e., milling flour in a hydrated condition). The flour-water slurry was run through a mill which can be any one of the many different types used industrially; hammer mills, roller mills, single and double disc mills, or disintegrators such as the Rietz type. The clearance of the mechanism through which the slurry will pass should be 0.001" or less to insure a material ground fine enough for the next step. The above dimension is the minimum clearance for the equipment used in the examples. One of the major criterion is the small particle size of the final powder. The smaller the particle size, the more dispersible it will be and the better mouthfeel the final beverage product will have.

Particle size of the final product of Example 1 and of full-fat flour were compared by scanning electron microscopy. Measurements of particles magnified 500 times showed that the full-fat flour had particles that appeared as aggregate clusters approximately 75 microns in diameter. The product of Example 2, however, consisted of particles having diameters of between 5 and 40 microns.

Homogenizing and emulsifying can be done simultaneously in apparatus such as ultrasonic vibrators, impulse rotary hydromills, high-pressure piston-type, or gear-type homogenizers. The equipment used in the examples was a Manton-Gaulin two-stage homogenizer, which was run at 3,500 psi. The maximum pounds per square inch of apparatus is about 5,000 which is also operative for this purpose. At lower pounds per square inch, about 2,500, the emulsifying action is much less efficient.

Spray drying was effected in a Bowen-type apparatus at an inlet temperature of 250° to 275°F. At 275°F brown specks appeared in the product which make it unusable for a beverage base. The lower limit of spray drying temperature depends on air flow. A greater air velocity permits the use of lower temperatures. The preferred temperature is between 250° and 265°F. However, because of the flexibility of this type of drying system, the process should not be limited to this temperature range. The dried product was dry blended with flavorings, sweeteners, vitamins, and minerals, and then stored in closed containers for at least 6 months at elevated temperatures with no deterioration (shown in the following table).

Storage		Flavor Score*	
Days	°F	Initial	Final
56	0	8.8	8.8
	120	8.8	8.3
182	0	8.8	7.9
	77	8.8	7.7
	100	8.8	7.5

*Score based on a scale of 0-10.

One part by weight of the dry formulation was mixed with 9 parts of water to make the beverage that was tested by the taste panel. The dispersibility, as measured by the nitrogen solubility index (NSI) of the final product, is improved as much as 300%. The high NSI for final spray dried product shows that no elaborate mixing means are necessary when water is added to make a beverage.

Example 1: Six pounds of full-fat soy flour were slurried with 48 lb of water and the slurry of about 11% by weight flour in water processed through a colloid mill having a clearance of 0.001". The Manton-Gaulin two-stage homogenizer emulsified the milled slurry at 3,500 psi, after which the emulsion was spray dried in a Bowen-type spray dryer at 250° to 265°F inlet air temperature and 150° to 160°F outlet air. The soy protein powdered product had a NSI of 57.2 and a particle size of 5 to 40 microns as measured by scanning electron microscopy.

Example 2: Five pounds of defatted soy flour in 40 lb of water (about 11% by weight flour in water) were processed in the same manner as Example 1. The final product had a NSI of 49.7.

Example 3: Five pounds of full-fat flour were processed in the same manner as in Example 1 with the exception that 1.0 lb of hydrogenated soybean oil, MP 100° to 105°F, and 0.055 lb of polyoxyethylene sorbitan monostearate emulsifier were added to the slurry before colloid milling. Three pounds of the final spray dried product were blended into the following formulation:

		Percent
(a)	Spray dried product (3 lb):	
	Protein	32.1
	Fat (total)	31.1
	Carbohydrates	17.7
	Fiber, ash	8.1
	Emulsifier	1.0
(b)	Sucrose	7.3
(c)	Sodium chloride	0.7
(d)	Vitamin premix, with antioxidant, Formula No. 2 for use in blended food products for U.S. Government A.I.D. programs	0.1
(e)	Mineral premix, CSM Mallinckrodt Chemical Works,	1.8

	Percent by Weight
Mineral premix:	
Calcium carbonate, precipitated FCC grade	46.1
Zinc sulfate, hydrated, FCC ($ZnSO_4 \cdot 7H_2O$)	0.3
Ferrous fumarate, purified food grade	1.8
Iodized salt (0.007% I_2) food grade	51.8
Vitamin premix:	
Thiamine mononitrate	1.88
Riboflavin	2.63
Pyridoxine hydrochloride	1.13
Niacin	33.80
Ca D-pantothenate	18.78
Folacin	0.22
Vitamin B_{12}	0.02
Vitamin A (stabilized retinyl palmitate), 15.0 million USP units; Vitamin D (stabilized), 1.8 million USP units; Alpha-tocopherol acetate, 14,000.0 IU	11.54
Butylated hydroxyanisole	15.02
Butylated hydroxytoluene	15.02

Soy-Whey Combinations

The method of *A.C.-Y. Peng; U.S. Patent 3,798,339; March 19, 1974; assigned to Swift & Company* involves the mixing of a full-fat soybean material in an aqueous solution of milk material or milk by-products including whole milk, skim milk, neutralized buttermilk (the liquid remaining following butter churning of sour cream previously neutralized to about pH 6.6), lactose, and particularly certain cheese wheys. Whey is the serum which remains after the coagulation of the casein and fat when cheese is manufactured and comprises principally water and 6 to 7% solids such as lactose, lactalbumin, and most of the ash of the original milk.

Cheese whey usually is slightly acid due to the ripening process in cheese manufacture. The degree of acidity is dependent upon the extent ripening is continued, but it is believed that most fresh cheese wheys, excepting cottage cheese whey, immediately upon removal from the processing vat are pH 6.0 or higher. (Cottage cheese whey is usually pH 5.0 or lower.) However, the whey will continue to become more acid, if bacterial action is not halted, until pH declines to about 4.5, whereupon most bacterial action will cease. Pasteurization of whey will also halt bacterial action and fix the pH value. Similarly spray drying whey to form a powder will arrest the pH value and dried food-grade wheys are usually found to be about pH 6.3.

It is necessary that the mixture of milk material and soybean material be no more than slightly acid, displaying about pH 6 or higher. Subsequently, the mixture is heated to boiling (which is about 214°F for cheese whey) and held for a period of 10 to 30 minutes. This cooked product may be used without further processing as a nutritional and relatively flavorless and odorless milk-substitute type product. However, the mixture may be dried to remove moisture and produce a powdered milk-substitute type beverage base product. This powdered product can easily be rehydrated, either immediately or after storage in dry form, into a bland, nutritive beverage of the milk-substitute type. The product may also be used as a milk solid substitute in the baking industry and the like.

The method requires soaking a soybean material such as dehulled whole soybeans, cracked soybeans, soy meal, soy flakes or soy flour, in a solution of certain milk materials. The more finely divided material requires less soak time, with soy flour requiring no soaking and whole beans requiring 7 to 8 hours. Although the temperature at which the beans are soaked in the solution is not critical, where long soaking is necessary it is preferred that the temperature be fairly low (e.g., in a range of from 32° to 40°F) to prevent deterioration of the soybean protein during the soaking period.

The soybean material may be either toasted or untoasted and may be either defatted or full fat. The greatest reduction in beany flavor probably is obtained by processing soy flour. Highly improved taste can also be obtained with full-fat material to produce a more nutritious product. Also full-fat soy material will result in a product of superior appearance and texture more closely resembling cow's milk.

The milk material solution may be either fresh (i.e., obtained in liquid form from milk or food processes utilizing milk), for example, fresh cheese whey; or a solution prepared from a dried milk product, such as whey powder or skim milk powder, etc. The ratio of soybean material to milk material should be in the range of from 1:4 to 2:1 (solids basis).

Acid mixtures below about pH 6 are thought to be inoperative in that insufficient flavor improvement is obtained. It appears that milk materials having acidity of pH 5 and lower will not result in suitable mixtures. In selecting operative ingredients it must be considered that the soybean material in water will usually be about pH 6.5 and will act as a buffer when acid materials are added. Similarly certain milk materials, such as whey, have a buffering effect. Thus, the relative amounts of some ingredients will affect their operativeness. It was found that lactic acid solutions per se and most cottage cheese whey are too acid and will not provide good results. However, most Italian and cheddar cheese wheys, skim milk, neutralized buttermilk, lactose and whole milk are satisfactory.

A ratio of soybeans to cheddar cheese whey of from 1:2 to 2:1 (solids basis) is used with a most preferred ratio in a range of from 1:2 to 1:1 (solids basis). It has been found that if the milk material content of the mixture is greater than about 4 parts to each part of soybean used, certain unpleasant milk flavors may predominate. For instance, a mixture having greater than 4 parts whey solids per part of soybean material will exhibit a distinct whey flavor and aroma which will ordinarily render the product undesirable for consumer acceptance. Also in the case of whole milk or skim milk, certain undesired boiled flavors become detectable.

On the other hand, when greater than about 2 parts of soybean material is used per part of milk material, the final product will tend to exhibit a beany flavor which will also render the product generally undesirable. The pH of the soybean-whey mixture may be adjusted to insure a liquid product of pH comparable to cow's milk (which is in the range of pH 6.3 to 7.3); however, this step is not necessary, other than pH 6 is a minimum requirement for the mixture. The mixture displays a buffering action and if the initial pH is adjusted between 6 and 8 the end product will be in the range of 6.3 to 7.3. The higher pH is believed to promote slightly superior taste.

After the soybean material has been allowed to soak in the milk material solution for a sufficient period of time to render the material soft enough to be milled or ground, the mixture is cooked to boiling. For best results, it has been determined that boiling should be continued in an open or exhausted vessel for a period of about 15 minutes. This boiling step has been found to achieve several desirable results. It has been found that boiling the mixture eliminates the beany or grassy flavor and aroma, and produces a product of a bland unobjectionable taste and odor.

The mixture of soy material and milk material should be milled or ground (particularly where whole, cracked or flaked beans are used), either before or after cooking of the mixture, in order to reduce the particle size in solution so that the final product will have a uniform and fine texture. The cooked mixture, which has been milled, may be utilized without further processing as a whole milk-substitute type beverage. However, additional processing steps such as evaporation and homogenization may be used subsequently if desired, and particularly if a dried product is to be prepared. Vacuum evaporization is believed to impart a further slight improvement in taste. Sugars and desired flavoring materials may also be added.

If it is desired to prepare a dry milk-substitute type beverage base, the cooked mixture can be dried by any convenient method such as jet-drying, spray-drying and freeze-drying. Homogenization prior to drying is highly beneficial for developing stability in the reconstituted product. The dried, powdered product prepared is easily rehydrated into a stable, nutritive beverage very similar in composition and physical characteristics to cow's milk.

Whole milk having strong, undesirable or offensive flavors and/or odors may be benefited and upgraded by the process. Such whole milk is obtained in certain seasons from bovines grazing on certain plants; and from any other animals, such as goats, which milk is considered by many persons as being unpalatable. The milk of some species displays an inadequate protein to fat balance from the standpoint of human nutrition, and may also be benefited and improved by this process. For example, whole milk from the Indian buffalo (Bubalus Bubalus) is very rich in fat content but relatively low in protein. Such milk can be upgraded to higher protein content and effectively extended to supply the nutritional needs of greater numbers of humans.

Example 1: A 1:1 ratio (solids basis) of clean, dehulled cracked soybeans (7 lb) and fresh Italian cheese whey (100 lb with 7% whey solids) of pH 6.3 were soaked for about 8 hours. Then the mixture of pH 6.5 was heated to boiling in a 45-gallon round processor and held at boiling for 15 minutes. The whole mixture was then ground and filtered through a wire screen. The resulting slurry was then jet dried. The dry product was a slightly yellowish white powder, and had a sweet taste and a bland flavor. The dried product was readily rehydrated into a milk-substitute type liquid product of neutral pH around 6.8.

Example 2: Twenty-seven pounds of clean, dehulled, full-fat cracked soybeans were soaked in 400 lb of fresh cheese whey of pH 5.6 for about 8 hours at 32° to 40°F. The mixture was boiled in a 45-gallon round processor for 15 minutes, evaporated under vacuum of 27" for 15 minutes, and cooled down to 100°F. The mixture was then processed by the method of Example 1 and the final dry product was slightly yellowish white, had a sweet taste and a bland flavor, and dispersed readily.

Example 3: Five pounds of clean, dehulled, cracked, full-fat soybeans and 8 lb of 62% solids of concentrated cheese whey (pH 5.6) were diluted with 87 lb of water and mixed thoroughly. The whole mixture was heated to boiling in a stainless steel steam-jacketed vessel with constant agitation, and held at boiling for 4 minutes under a careful control of steam to avoid foaming. The boiled product was ground, filtered through a wire screen and jet dried. The final product had a sweet taste and bland flavor and was a light yellow color.

Example 4: Seventy grams of full-fat, cracked soybeans and 140 grams of Italian cheese whey powder were mixed with 1,000 grams tap water, producing a mixture of 1:2 solids basis ratio and pH 6.4. This mixture was soaked only during the time required to heat to boiling and was boiled for 15 minutes and then blended for 2 minutes in a Waring blender operated at medium speed. The mixture was filtered through four layers of cheesecloth and the filtrate cooled to room temperature. A taste panel of eight trained members sampled the liquid product and rated it according to a scale of 1 to 6, representing a range of no beany flavor to very much bean flavor, respectively, on average at 2.91. The test was repeated using a 2:1 solids basis ratio by mixing 140 grams cracked soybeans and 70 grams whey powder in 1,000 grams water. Taste panel rating was 3.16. These figures compare favorably with a reference taste rating of 3.94 assigned to a mixture of 70 grams beans ground and boiled in 1,000 grams water.

Low Isoelectric Point Process

In the process of *W.E. Koski, D.E. Smith and A.R. Touba; U.S. Patent 3,653,912; April 4, 1972; assigned to General Mills, Inc.* the starting material may be a soy milk or other bland soy material containing 40 to 60% protein. Such materials may be obtained by removing at least a part of the nonprotein constituents of defatted soy flour, meal or flakes by any of various means. Such materials may be obtained by alkaline extraction and acid precipitation using conventional techniques. For example, they can be prepared by the extraction of defatted soybean flakes or similar source material with dilute aqueous solutions of sodium hydroxide followed by precipitation of the extracted protein with an acid at the isoelectric point. Typical acids used in such processing include hydrochloric acid, sulfuric acid, and sulfurous acid.

Any of a variety of water-soluble alkaline materials can be used to raise the pH of the starting soy protein material. Preferred alkaline materials are the inorganic bases and salts such as sodium hydroxide, ammonia, ammonium hydroxide and trisodium phosphate. The amount used will depend somewhat upon the particular alkaline material. It is preferable to carry out the alkaline treatment for from 2 to 15 minutes. Any of a variety of acids may be used in lowering the pH to neutral or near neutral. Typical acids would include citric acid and phosphoric acid.

However, other acids such as hydrochloric and lactic acid may be used. During the addition of the acid it is preferable that the soy protein solution be stirred in order to disperse any beads of precipitated protein that might form as the acid is being added. The pH of the solution should be reduced to about 7.0 to 8.0. The modified soy protein, at or near the neutral point has a substantially improved solubility and therefore beverages of very high protein content can be produced. For example, if reconstituted with water to form an aqueous solution, the resulting beverage will usually contain at least as much protein as cow's milk and, if reconstituted with cow's milk may contain 1½ to 2 times as much protein as normal cow's milk.

The modified soy protein may be used as the resulting aqueous solution or it may be dried and reconstituted by the addition of the dry modified soy material to an aqueous liquid prior to use. The drying is carried out using spray drying techniques. However, other drying techniques may be used with less satisfactory results. Other materials may be added prior to or subsequent to the drying step. For example, one may add flavoring such as chocolate or fruit juices to the modified soy material.

The beverage can be prepared by adding a nutritionally significant amount, for example, up to 20 parts of the dry modified soy material to 100 parts of water, by weight. If desired, flavoring may be added. The beverage, alternatively may be prepared by adding a nutritionally significant amount, for example, up to 15 parts of the dry modified soy material to 100 parts of cow's milk by weight.

Example 1: A beverage was prepared using Textrol. Textrol contains about 58.9% soy protein, 5.7% ash and 7.1% moisture. Prior to modification, about 38% of the protein was soluble. The Textrol was modified to improve flavor and dispersibility. Approximately 150 lb of the Textrol was placed in a 200 gallon tank containing 1,134 lb of water and mixed for 15 minutes using two Lightnin Mixers. Eighteen pounds of a 50% aqueous sodium hydroxide solution was added raising the pH to 12 and mixing was continued for 10 minutes. Twenty-six pounds of a 50% aqueous citric acid solution was added at the rate of approximately ½ lb/min. The mixing was continued during the addition of the acid and the final pH was 8.0. The product was then stored in a room having an ambient temperature of 0°F for about 6½ hours and then in a room having an ambient temperature of 40°F for about 8 hours. The product at this point may be used as a beverage.

The product, however, in this instance was spray dried. The spray drying conditions were: inlet temperature, 480°F; outlet temperature, 200°F; feed pressure, 5,000 psig. The spray dried product contained 50.5% protein, 85% of which was soluble. The product further contained 7.8% moisture and 11.5% ash. A tasty beverage was prepared by mixing 12 grams of the dry modified soy material, 20 grams of Hershey's Instant Chocolate and 8 ounces of cow's milk. The resulting beverage had 6% protein content which is substantially above that of the original cow's milk.

Example 2: A beverage was prepared using Soy Milk A. The Soy Milk A prior to modification contained 43.6% soy protein, 6.0% ash and 9.1% moisture. 78% of the protein was in the soluble form. The Soy Milk A was treated in a manner similar to that described with respect to Textrol in Example 1. Approximately 154 lb of the Soy Milk A was added to 1,140 lb of water and mixed for 1 hour in a 200 gal tank equipped with two Lightnin Mixers. Then 13¼ lb of 50% aqueous solution of sodium hydroxide was added to the dispersion of Soy Milk A and mixing was continued for 10 min. The pH at that point was 12.2. 24 lb of a 50% aqueous solution of citric acid was added slowly over a period of 31 min thereby reducing the pH to 8.0. The product was then stored at 40°F for about 16 hours.

The product was then spray dried as described in Example 1; however, operating conditions of the spray dryer were: inlet temperature, 415°F; outlet temperature, 215°F; and feed pressure, 4,000 psig. The dry product contained 39% protein, 10.5% ash and 7.4% moisture. The protein was 95% soluble. A beverage was prepared by dissolving 15 grams of this product together with 20 grams of Hershey's Instant Chocolate in 8 ounces of milk using a Waring Blender. The treated Soy Milk A remained dissolved or suspended in the milk. The beverage thus prepared was tasty and provided approximately twice as much protein as ordinary cow's milk.

Example 3: A product was prepared by dry blending 7.0 lb of the modified, spray dried product of Example 1; 9.5 lb of nonfat milk solids; 3.5 lb of granulated sugar and 1.1334 pounds of a flavoring and coloring premix (the premix was prepared by mixing 0.1150 lb of imitation strawberry flavor, 0.130 lb of red food coloring, 0.0054 lb of yellow food coloring and 1.00 lb of granulated sugar in a Hobart mixer at speed No. 2 for 10 minutes).

The ingredients were first mixed in a ribbon-type blender for 5 minutes and subsequently passed through a Fitz mill. The blended material was agglomerated in an agglomerating tube substantially like that described in U.S. Patent 3,288,115. The agglomerating tube had an internal diameter of 4" and was 20' in length. The air velocity in the tube was 5,000 ft/min. The air temperature was 280°F. The relative humidity in the tube was about 42%. The residence time of the material in the tube was about one-half second. The material was then dried at 240°F for between 1 and 2½ minutes and cooled for 30 seconds. The product was passed through a No. 8 mesh screen.

A tasty beverage was prepared by adding 42 grams of the agglomerated powder to 156 ml of fresh whole milk and shaking for 15 seconds. The beverage had a consistency and appearance much like a milk shake. The beverage had a protein content of about 9.6%. Another beverage was prepared by adding 42 grams of the agglomerated powder to 156 ml of water and shaking for about 30 seconds. The powder dissolved immediately. The resulting beverage was tasty and had a consistency slightly thicker than milk but not as thick as a milk shake. This beverage contained about 6.2% protein.

Carrageenin Stabilizer

The process of *K.-S. Lo; U.S. Patent 3,563,762; February 16, 1971; assigned to Hong Kong Soya Bean Products Co., Ltd., Japan* is directed primarily to the development of a soybean liquid which can be conveniently bottled and stored without the danger of spoilage, nor the need for continuous refrigeration. The product is a refreshing beverage and can be marketed as a soft drink in the nature of Coca-Cola or 7-Up. This product has the basic nutritional qualities which are found for example in animal milk.

Liquids have been developed from soybeans in the past. In one typical method of obtaining such liquids, the beans are soaked in water to soften them. They are then mashed and extracted with a large quantity of water. This type of process, where the nutritive ingredients of the soybeans are extracted by soaking and using the resulting fluid, is quite common. This process is distinctive in that it involves the convenient production of a soybean beverage from a full-fat soybean powder. This powder is suspended in water to provide a beverage having substantially the same consistency as cow milk. The resultant beverage is not subject to settling out and need not be refrigerated in order to avoid spoilage.

To produce the full-fat bottling powder, the process illustrated in Figure 5.1a is used. The whole soybeans are first cleaned **10** to remove dirt, splits, and other foreign matter. After cleaning, the beans are heated or toasted **11** in a horizontal toaster, which may be of the nature where the beans are placed upon a wire screen conveyor and transported through a hot air flow for a period of time approaching 5 minutes. The hot air may be maintained at a temperature of approximately 300°C and is used in order to remove excessive moisture content from the beans prior to dehulling.

After heating, the beans are immediately transferred by means of the wire screen conveyor into a cooling chamber where the beans are cooled **12** by currents of cold air. The abrupt drop in temperature is such that the hull shrinks away from the meat and thus, dehulling is facilitated. After cooling, the soybeans are transferred into a conventional dressing and blending machine where the impact action of paddles removes the hulls from the meat **13**.

The next step is to separate the beans from the hulls **14** and this is done in conventional scouring apparatus where the soybeans are agitated to permit aspiration of the hulls from the top of the unit while the heavier meat falls to the bottom of the unit. The dehulled soybeans are next fed into a hammer mill in order to compress them into the form of flakes **15**. Once flaked, the soybeans are then cooked in a Wenger expansion cooker. Within the cooker the flakes are first preconditioned with steam at a temperature of 212°F to establish a moisture content of 18 to 21%. In this condition, they are extruded through ¼" die openings and come out as pellets. The cooking is performed at high pressure in order to prevent boiling.

In effect, the extrusion output renders the unit a continuous pressure cooker and the product of this type of process has the growth inhibiters of the soybean removed. In addition, it provides a flour that does not have the objectionable bitterness that is inherent in soybeans. Rather, the powder is bland in taste and entirely suitable for the addition of flavoring materials in order to produce any kind of beverage that will suit local markets. The removal of growth inhibiters from the powder renders the subsequently produced beverage more digestible. This is an important factor inasmuch as the beverage is particularly valuable for its nutritional value and as a protein supplement.

The pellets that are provided at the output of the Wenger cooker are again toasted **17** in order to reduce the moisture content to a point between 3½ and 4%. Following toasting, the beans are again cooled **18** rapidly by bringing the temperature to a level of approximately 90°F. The dry pellets are ground **19** into a fine powder, e.g., in an Alpin pin mill, and classified **20** in order to obtain a final full-fat soybean powder of approximately 270 to 300 mesh. The resulting powder has substantially the following composition:

	Percent
Protein	45
Fat	20
Ash	5
Fiber	3
Other carbohydrates	23
Moisture	4

Three types of soybean beverages may be prepared from the bottling powder produced. The first of these beverages will contain all of the insoluble carbohydrates. The second of these beverages will contain partially insoluble carbohydrates, and the third beverage will be entirely free from insoluble carbohydrates.

FIGURE 5.1: PROCESS FOR PREPARING A SOYBEAN BEVERAGE

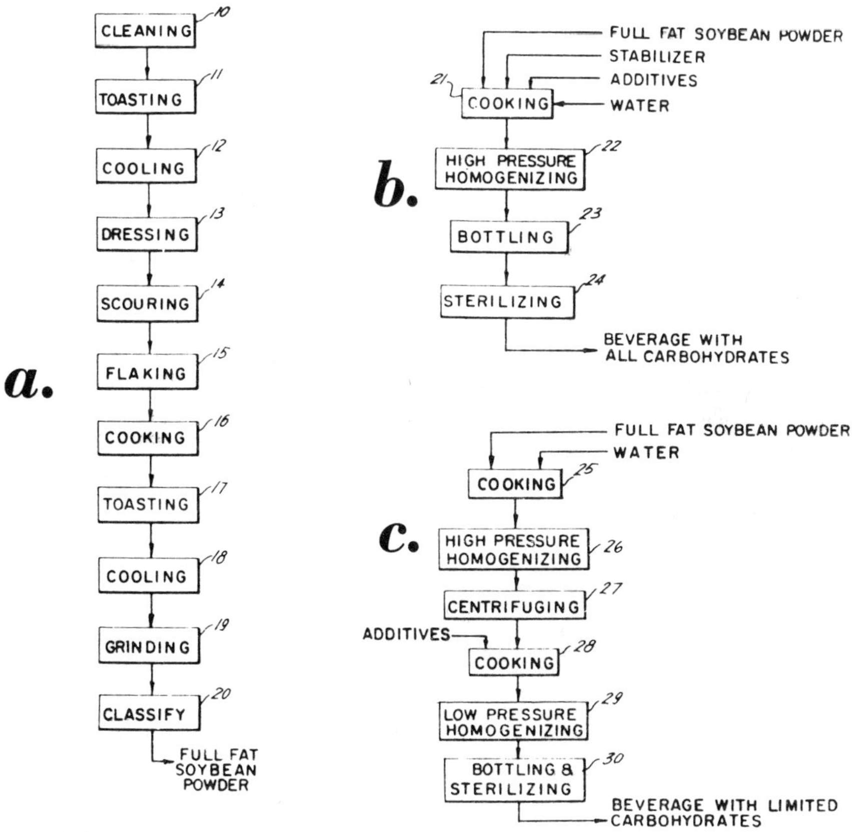

Source: K.-S. Lo; U.S. Patent 3,563,762; February 16, 1971

Carbohydrate refers to the carbohydrate components within the soybean itself and not to any other form such as may be added in the form of sucrose, starch, etc. A flow diagram for the production of a soybean beverage containing all of the insoluble carbohydrates is illustrated in Figure 5.1b.

The full-fat soybean powder produced as described in connection with Figure 5.1a is placed in a cooker along with water, a stabilizer, and additives for flavor and vitamin supplementing. During this cooking **21**, the mixture is heated and agitated to assure adequate mixing of all components. After cooking, the mixture is subjected to high pressure homogenization **22**. It has been found that a Manton-Gaulin high pressure homogenizer is particularly suited to this process. A pressure of approximately 8,000 psi is maintained during homogenization in order to develop a soybean particle size ranging from 2 to 10 microns.

Particularly satisfactory results have been obtained by the use of carrageenin as the stabilizer. This is a dried plant of seaweed *Chondrus crispus* which is edible and useful for its gelatinizing properties. The combined effect of high pressure homogenization and the addition of a stabilizer provides a permanent suspension in the soybean beverage. After homogenization, the beverage may be bottled **23** or canned for commercial distribution. Sterilization **24** should be performed, and it may be conveniently performed after bottling.

In order to prepare soybean beverages partially containing insoluble carbohydrates, the soybean solution is first reduced within a homogenizer, as previously described, and centrifuged in order to remove the coarser particles within the range of 8 to 10 microns. These latter particles represent approximately 10 to 15% of the original dry weight powder added. A smaller amount of carrageenin is added to the solution in the event of this second method in order to yield a permanent suspension. The beverage with only partial insoluble carbohydrates is of lighter consistency than that of the first method and may be more suitable for certain uses.

In order to prepare a soybean beverage that is free from any insoluble carbohydrates, the initial steps are again utilized, as illustrated in Figure 5.1c. Thus, the full-fat soybean flour is homogenized **26** under high pressure in order to reduce the particle size. Thereafter, the product is centrifuged **27** in order to eliminate all of the coarse particles and again cooked **28** and homogenized **29** at lower pressure to yield a stable soybean beverage which is substantially free of insoluble carbohydrates. Here, too, the beverage may be conveniently bottled and sterilized **30**. The bottling or canning of this beverage may be done with any conventional equipment and is particularly simple because refrigeration is not required.

Specific examples of processes for developing several typical soybean beverages will now be described. In order to produce a soybean beverage containing all of the insoluble carbohydrates, the following ingredients were placed in a cooker equipped with a steam injection heater and an agitator.

Water	50 gal
Bottling powder	67 lb
Sucrose	50 lb
Carrageenin	0.55 lb
Refined soybean oil	20 lb
Tablet salt	0.63 lb
Sodium bicarbonate	0.32 lb
Vitamin A (1.0 m IU/g)	5.81 g
Vitamin B_1	1.31 g
Vitamin B_2	1.56 g
Niacinamide	14.0 g

The carrageenin, which functions as the stabilizer, is dispersed in small amount of water prior to adding it to the mixture. Following the initial mixing, water was added to a final volume of 100 Imperial gallons. The mixture was then heated by direct steam injection and maintained at 180° to 190°F for one-half hour while constantly stirring. The heating period is important, because it assists in dissolving the carrageenin into the solution and it also softens the particles of the soybean powder. The mixture was next high-pressure homogenized with a Manton-Gaulin homogenizer at a pressure of 8,000 psi. It was then stored in a holding tank at a temperature of approximately 180°F where it was constantly stirred until bottled. The functions of the Manton-Gaulin homogenizer at this stage of the process not only give a homogeneous oil and solid dispersion, but also reduce the powder particles to a size within the approximate range of 2 to 10 microns.

The finished beverage was then placed in regular beverage bottles by means of an automatic filler, the bottles being sealed with a metal cap. Finally, the bottled beverage was sterilized at about 250°F for 15 minutes. After sterilization, the bottled beverage was retained in the sterilizer to cool down to approximately 100° to 110°F with constant shaking. Shaking of the contents at temperatures ranging from 120° to 140°F is essential, because it encourages the carrageenin to start gel formation, a reaction responsible for holding up the particles in suspension. The soybean beverage obtained from this method contained an average composition of the following:

	Percent
Protein	3.0
Fat	3.1
Sucrose	5.0
Other carbohydrates	1.7
Ash	0.4
Water	86.8

Every 6½ fluid ounces of this beverage contains ⅓ RDA (recommended daily dietary allowance by the Food and Nutrition Board, National Research Council, U.S.A.) of vitamins A, B_1, B_2 and niacinamide, as follows: vitamin A, 2,400 IU; vitamin B_1, 0.55 mg; vitamin B_2, 0.67 mg; and niacinamide, 6.00 mg.

In order to produce a soybean beverage partially containing insoluble carbohydrates, 78 lb of bottling powder was first dispersed in approximately 50 gallons of water in a cooker. The mixture was heated to 180° to 190°F and held at that temperature for one-half hour with constant stirring. It was then high pressure homogenized in a Manton-Gaulin homogenizer at a pressure of 8,000 psi. After homogenization, it was centrifuged with an in-line auto-centrifuge which was adjusted so that approximately 15% of the coarse particles of the full-fat soybean powder was centrifuged out. After centrifuging, the beverage (now approximately 85% of its original volume) was pumped into another cooker and mixed with the rest of the ingredients as listed in the above table.

The ratio of the ingredients remained the same as in the preceding methods, except for the carrageenin, of which only 0.25 of a pound was added. Water was added to a final volume of 100 Imperial gallons and then the mixture was heated to 180°F and thoroughly mixed. It was thereafter low pressure homogenized, this time with an ordinary dairy homogenizer at a pressure of 2,500 psi. After homogenization, the beverage was filled in regular beverage bottles which were then sealed and sterilized.

In this latter method, the quantity of bottling powder used for producing 100 Imperial gallons of soybean beverage was more than the first method, because part of the powder must be centrifuged out. The quantity of carrageenin was less. The quantity of the other ingredients remained the same as in the first method. The soybean beverage obtained from the second method contained approximately the same composition and vitamin content as that of the first.

In order to produce a soybean beverage containing no insoluble carbohydrates, a similar procedure was followed, except: (a) the amount of full-fat soybean bottling powder used was 80 lb; (b) the centrifuge was adjusted to centrifuge approximately 30% of the coarse particles out; and (c) no carrageenin was added. The soybean beverage obtained from this method contained the following average composition:

	Percent
Protein	2.4
Fat	2.7
Sucrose	5.0
Other carbohydrates	1.0
Ash	0.3
Water	88.6

The pH of beverage is 6.8.

For every 6½ fluid ounces it contains ⅓ RDA of the following vitamins: vitamin A, 2,400 IU; vitamin B_1, 0.55 mg; vitamin B_2, 0.67 mg; and niacinamide, 6.00 mg. By using high pressure homogenization, there is obtained an unusually effective oil and solid dispersion within the beverage. At the same time, the powder particles are reduced to sizes within the range of 2 to 10 microns and are kept in suspension within the liquid.

In addition, a stabilizer is introduced to further assist in preventing settling-out of the solids from the liquid. Finally, the results of centrifuging have been added to remove the heavier powder particles and thereby also assist in preventing settling-out and to improve the palatability of the beverage.

LOW CHOLESTEROL EGG PRODUCTS

LOW CHOLESTEROL EGGS

Fat Solvent Extraction

Eggs are a very popular and desirable component of the human diet, both in terms of organoleptic appeal as well as the high nutritional quality of the protein component. Hence, the modification or alteration of egg yolk solids to significantly decrease the cholesterol content and the fat of low ratio of polyunsaturates to saturates while retaining the quality of the egg protein would constitute an important innovation.

Several methods of improving eggs for inclusion in diets for the control of serum cholesterol have been suggested. One recent approach involved decreasing the saturated fat content and increasing the polyunsaturated fat content of egg yolks through changing the diet of the hens. Although the fatty acid composition of the yolk was markedly improved by this means, the cholesterol content of the yolk was found to be essentially independent of the type of fat used in the diet of the hens. In fact, there was an indication that the cholesterol level of the egg yolk increased as the degree of unsaturation of the dietary fat increased. Such eggs proved to be no better than conventional eggs in the control of the serum cholesterol level.

Another approach in providing the consumer with an egg-like product, free of cholesterol, has been through the inclusion in the formulation of an imitation yolk which contains no natural yolk material at all, but is fabricated from vegetable sources. This product, after hydration and frying, is so different from scrambled eggs in odor, texture and taste as to be unacceptable to most users. As indicated in the following table, the yolk portion of the egg from hens on conventional feeds contains low levels of polyunsaturated fatty acids, moderately high levels of saturates, and an unusually high amount of cholesterol.

Composition of Egg Yolk Lipids

Fatty acid composition:	
Saturated fatty acids, (S), percent	34.0 - 36.0
Monounsaturated fatty acids, percent	51.0 - 52.0
Polyunsaturated fatty acids, (P) percent	7.0 - 14.0
P/S ratio	0.2 - 0.8
Cholesterol content, mg percent	*3,700 - 7,700

*This corresponds to 2,200-4,600 mg percent cholesterol in yolk solids, or 200-400 mg cholesterol in an average egg of 50 grams.

D. Melnick; U.S. 3,563,765; February 16, 1971; assigned to CPC International Inc. gives a method for producing an edible egg yolk product of reduced fat content by solvent-extracting from egg yolk solids at least a substantial portion of the free fat content of the solids, while leaving in the solids a portion or all of the bound fat. A solution of extracted fat is obtained.

The residual, extracted egg yolk solids are then separated from the fat solution. The solution of extracted fat is fractionated into a fat by-product and solvent, the latter being recovered for reuse. The residual, extracted yolk solids are of reduced fat content as well as reduced cholesterol content.

The extraction process is carried out by mixing dry egg yolk solids with a fat solvent in proportions such that the solvent is at least twice in weight in relation to the weight of the dried yolk solids. The mixture is heated with constant agitation to a temperature not in excess of 120°F, for a period of time sufficient to permit extraction of a major portion of fat and of the cholesterol.

After extraction, the solids are separated from the solvent, as by filtration or by allowing the solids to settle and decanting the extracted fat in solvent solution. The solids are then rinsed with additional solvent in proportions such that the weight of added solvent is at least equal to the weight of the initial egg yolk solids. The yolk solids, freed of excessive solvent by filtration, are heated to temperatures high enough to evaporate the solvent, under 120°F to prevent denaturation of the egg yolk protein. The egg-fat is recovered from solvent by distilling off the latter for reuse in the system.

The extracted, edible egg yolk solids constitute a product from which at least 50% but not more than about 90% of the fat and at least 50% of the cholesterol have been removed. Such a product will contain at least 5 parts of the native fat based on 100 parts of original yolk solids.

The partially defatted yolk solids, must possess the following attributes:

(1) The partially defatted yolk solids must be readily soluble and/or dispersible in water, with or without salt in the aqueous phase.

(2) The partially defatted yolk solids on rehydration must exhibit the functional characteristics of unaltered egg yolk protein.

(3) The partially defatted yolk solids must be free from objectionable flavors to a degree at least equivalent to that noted with conventionally dried egg yolk solids and preferably to a degree equivalent to that noted with fresh egg yolk.

It is preferred to use nonpolar fat solvents in making the defatted yolk. Egg yolk contains high-density lipoprotein, apparently consisting of protein chains almost completely covered with cholesterol and phospholipids. The proposed structure of the low-density lipoprotein in egg yolk appears to be an outside layer of protein, phospholipid, and cholesterol, surrounding a lipid core, thus resembling a micelle or microemulsion. Nonpolar solvents extract mainly neutral lipids from the low-density lipoprotein microemulsion, and hardly any of the bound lipid from the high-density liproprotein. The large yield of cholesterol and cholesterol esters in the solvent extract, and the relatively lower yield of phospholipid in the nonpolar solvent extracts, indicate that the coalesced (easily extractable) lipids and cholesterol but not phospholipids, may be removed from these entities without disrupting their fundamental integrity.

When extracting egg yolk solids with a nonpolar solvent only, such as n-hexane, there appears to be little, if any, damage to the functional properties of the remaining protein. The treatment of egg yolk solids with a nonpolar solvent extracts readily the coalesced or easily extractable fat. The percentage of theoretical or coalesced fat, that can easily be removed by a nonpolar solvent, is in excess of 70%. Fortunately, much of the cholesterol in egg yolk solids is contained in the low-density lipoprotein fraction and is removed with the coalesced fat fraction. Consequently, extraction of yolk solids by a nonpolar solvent, when removing about 70% to 90% of the total fat, also removes 70 to 98% of the total cholesterol on a dry weight basis.

In yolk solids that have been extracted with a nonpolar solvent, it is no simple matter to replace the neutral, extracted egg fat with a vegetable (polyunsaturated) oil. If severe extraction conditions have been used, or if a polar solvent has been used, the replacement is even more difficult, if it is desired to restore the functional properties of the yolk, since it is difficult to reconstruct the natural lipid-protein relationship which exists between the protein and the natural egg fat. Hence, it is preferable to use nonpolar solvents such as hexane, cyclohexane, heptane, or tetrachloroethylene.

Analytically, more efficient or complete removal of the lipid constituents from protein material consists of disrupting the lipid-protein bonds by a polar solvent such as alcohol, thus permitting the subsequent extraction of the lipids by a nonpolar solvent such as ethyl ether. These two processes can be combined through the use of solvent mixtures such as ethanol-ether. Although a polar-nonpolar solvent extraction system would be desirable from the extraction efficiency standpoint, the use of a polar solvent for extraction may impair the functional characteristics of the remaining protein.

The amount of information available on egg yolk protein and lipids and their complex lipid-protein structure is very meager in comparison to that available on egg white protein. It is only in the past few years that investigators have been interested in the yolk portion of the egg and have attempted to separate the lipid from the protein fraction. Conventional solvent extraction processes yield a defatted product unsuitable for reconstitution.

The defatted egg yolk has many uses in the formulation of food products. However, it is preferred to utilize the unique physical, chemical, and organoleptic properties of the defatted egg yolk product, in making low-cholesterol products in which the egg oil of low P/S ratio removed by the process, is replaced with vegetable oil of high P/S ratio.

Example 1: In one demonstration, dry egg yolk solids having a moisture content less than 5% were slurried in hexane in proportions such that the solvent to solids weight ratio was 3:1. The temperature of the slurry was raised to about 80°F for a period of about 30 minutes. The solids were then separated from the hexane solution of fat by filtration. The filter cake was then rinsed with an amount of hexane equal in weight to the original weight of the dried egg yolks, filtered, and the rinse solution was then added to the initial solution of fat and hexane. The solids were then air dried in a tray dryer at 100°F for 25 hours, to remove the solvent.

The solution of fat was distilled, to separate the hexane and egg oil. The hexane was condensed for subsequent use in further extractions. The egg product is a valuable component for use in feed preparations and the like.

As shown in the following table, the extracted egg yolk solids contained 19% of the original free fat and 18% of the original cholesterol content. Because the free fat and cholesterol made up 57.6% of the egg yolk, and inasmuch as 81% of the free fat was removed, the egg yolk solids product weighed exactly half its original weight, prior to the extraction. The extracted yolk solids, after drying, were a free-flowing, finely divided, particulate mass, of good color and when rehydrated, had a natural, clean, egg flavor.

The protein solubility index is indicative of the extent of denaturation of the protein in the egg yolk due to heat treatment during extraction. The test for determining the protein solubility index is basically described in the article by Bishov and Mitchell, *Food Research,* vol. 19, pages 367-372 (1954).

The solubility index for conventionally dried egg yolk solids is 2.3. Any lessening of the solubility index represents denaturation of the protein to a degree. The defatted egg yolk solids of Example 1 exhibit a protein solubility index of 1.8. Since the reproducibility of the method is ±0.1 units, this difference would appear to indicate that some denaturation had occurred during the defatting operation. However, a maximum protein solubility index of 1.8 was found for egg yolk solids defatted in the laboratory under extremely mild conditions; hence, a protein solubility index of 1.8 represents no significant protein denaturation for these purposes.

	Conventional, Stabilized Egg Yolk Solids	Defatted Egg Yolk Solids*
Moisture, percent	4.8	1.0
Total fat, percent	57.6	11.1
Protein, percent	33.4	33.4
Cholesterol, percent	2.98	0.53
Protein solubility index	2.3	1.8

*Results are calculated on the basis of the original product weight.

The defatted, low-chloesterol egg yolk solids may be used in formulations as a replacement for conventional egg yolk solids where a low-fat, low-cholesterol, low-calorie egg product is desired. Because the nonlipid components of the egg yolk have been concentrated by defatting, a smaller quantity of the defatted egg yolk solids can be employed in many food formulations to achieve equivalent results.

It is preferred, however, to replace the fat extracted from the egg yolk solids with an equivalent amount of a vegetable oil that is free of cholesterol and that has a high P/S ratio. This product would find use in foods to the same extent as conventionally dried egg yolk solids.

After prolonged storage (one year) of the dry, extracted yolk solids at room temperatures, no deterioration of the functional properties or flavor of the yolk solids could be detected.

Example 2: The hexane-egg oil solution from the extraction of Example 1 was further treated to separate the egg oil from the hexane. The hexane was separated by distillation, using mild heat, and was subsequently condensed for reuse as a solvent in the system. As shown in the following table, a comparison of the composition of the hexane extracted egg oil from Example 1 with the actual composition of the lipid portion of egg yolk indicates close agreement except for the phospholipid content. It is apparent that a substantial portion of nonextractable fat is in the form of phospholipids.

Composition of the Lipid Portion of Egg Yolk and the Hexane Extract of Egg Yolk Solids

	Egg Yolk Lipids Theoretical	Hexane Extract of Egg Yolk Solids
Total lipids, percent	100	99.6
Phospholipids, percent	27 - 30	10.8
Triglycerides, percent	65 - 66	84.5
Cholesterol, percent	5.2	4.3
Polyunsaturated/saturated ratio	0.2 - 0.8	0.45

Yolk Reduction Process

The process of *R.E. Jones; U.S. Patent 3,475,180; October 28, 1969; assigned to Tillie Lewis Foods, Inc.* involves removing a specific amount of the yolks from a desired quantity of liquid whole eggs, replacing the protein content of the withdrawn yolks by adding milk and soy proteins, and supplementing functional performance by adding carboxymethylcellulose, then pasteurizing, homogenizing and spray-drying the resulting mixture.

Rather than remove part of the yolks from a batch of whole eggs, in practice it is more desirable to blend liquid whole eggs with liquid egg white (i.e., eggs from which all the yolks have been removed), thereby obtaining a mixture equivalent to liquid whole eggs from which part of the yolks have been removed. To this end it is possible to blend three parts by weight of liquid egg white with one part of liquid whole egg to obtain four parts of a mix that is equivalent to that obtained by removing about two parts by weight of yolk from six parts of liquid whole egg. Whole fresh eggs are composed of 60% white and 40% yolk.

The use of only a small amount of the yolks present in whole egg is an important feature of the process. The elimination of most of the yolk is necessary to reduce the fat, the cholesterol, and caloric content of the product. When more fat and cholesterol are desired, more yolk (whole egg) is used. Thus, the fat and cholesterol level can be controlled.

On the other hand, the small amount of yolk used is required to maintain real egg flavor and thereby provide a fully satisfactory product. The amount of egg yolk used may vary between about one-fifth to one-half of that present in liquid whole egg.

To replace the value of the protein content (16%) of the yolk removed, or omitted from, the whole eggs a mixture of dry nonfat milk solids, dry sweet whey, and isolated soy protein is added. Thus, when two parts egg yolk are removed from six parts of liquid whole egg to obtain four parts of a preferred liquid mix, the protein content may be reduced from about 13% protein in the whole egg to about 11% in the mixture. Then by adding about 5 parts by weight of nonfat dried milk solids (36% protein), 7 parts by weight of sweet whey (13% protein) and about one part of isolated soy protein (75% protein) to 200 parts of the liquid egg white-whole egg mixture, the protein content of the resulting composite will be restored to about 13%.

This product contains by weight, approximately: 150 parts of liquid egg white; 30 to 100 parts of liquid whole egg; 8 to 16 parts of solids chosen from the group consisting of nonfat milk solids, sweet whey solids, and a mixture of the nonfat milk and sweet whey solids with each other; and ½ to 4 parts of isolated vegetable protein solids, 0.1 to 2 parts of hydrophilic thixotropic material chosen from the group consisting of sodium carboxymethylcellulose, pregelled starch, methylcellulose, guar gum, gum arabic, gum tragacanth, agar-agar, carrageen, and alginic acid intimately mixed and adjusted to a pH of at least 9.

This product is pasteurized, after mixing, to 100% Salmonella negative, and then dehydrated to a powder. The protein solids may contain between 0.1 and 15 parts of soy lecithin, and the 8 to 16 parts may be equal portions of nonfat milk solids, and sweet whey solids. Also, the product is homogenized, before pasteurization.

Nonfat milk solids are an excellent source of protein for replacing the protein lost in the reduction of the yolk content. The use of too much nonfat milk tends to impart an unnatural sweet characteristic to the product. Liquid nonfat milk or condensed or evaporated nonfat milk may be used by employing the solids value of each and arriving at the desired amount.

Sweet whey comprises the noncoagulable solids of milk, having a protein value of 12.9%, fat of 1.1%, and a carbohydrate or lactose content of 73.5%. Sweet whey is used in preference to high acid whey, because sweet whey does not cause a lowering of the pH in the final product that would otherwise affect the blandness associated with fresh whole egg. Since sweet whey is noncoagulable with excellent solubility, it helps to retain stability of the product during pasteurization, acting as a kind of temperature buffer. Too little sweet whey reduces the ability of the other proteins in the liquid egg to withstand exposure to the pasteurization temperature. Too much may make the taste too sweet. The sweet whey may be used dry or in liquid form, as may the milk solids noted above.

The egg product, when reconstituted with water has a total solids content of 21.5% in comparison to a 25.5% solids content of whole egg. The sweet whey content may be increased in this product to gain a 25.5% solids content, without affecting palatability. Beyond this point it would not be duplicating whole egg as to sugar content. The additional solids may be divided between the nonfat milk solids and the sweet whey solids. The additional solids would increase the dryer yield and lower the cost of dryer operation materially. The two milk by-products may supplement each other for some uses, such as baking, where the taste in relation to that of fresh egg is not too critical and may be masked. In fact, the dry sweet whey content could be increased to replace all the dry nonfat milk content (or vice versa) without any palatability problems occurring in an association with fresh whole egg.

Isolated soy protein is another source of protein, one of exceptionally high protein value. The protein content is 74.9%, fat is 0.1%, and carbohydate is 15.1%. The solubility is good and it is noncoagulable when in solution at temperatures far in excess of the pasteurization temperature. The incorporation of more soy protein than recommended presents flavor and odor problems, a slightly foreign taste as well as mealiness, and a pronounced toughness in a scrambled egg.

To improve the consistency of the product, particularly when reconstituted with water, a small amount of a thixotropic, hydrophilic colloid, such as carboxymethylcellulose is used, preferably sodium carboxymethylcellulose, such as about 0.05 to 1% of the liquid mixture or about 0.2 to 4% of the finished dry product.

A suitable grade of sodium carboxymethylcellulose is that designated as 7LP which has a relatively low viscosity readily adapted to the processing steps hereof. The low-viscosity sodium carboxymethylcellulose is preferred over the grades creating a higher viscosity, as the flow through the processing system is more fluid. It has not only exceptional emulsification and colloidal properties but also a desirable protecting action by coating the coagulable proteins. It was found that one gram of sodium carboxymethylcellulose is the equal of 7/10 gram of dried whole egg in creating a suspension in reliquefying the dried egg product. It acts as the binder with the other solids.

To introduce sodium carboxymethylcellulose alone into the liquid egg material prior to or after pasteurization does not seem possible. Excessive foaming is created and small undissolved particles are noted throughout the liquid egg material. When sodium carboxymethylcellulose is used as an ingredient in conjunction with nonfat milk solids, sweet whey, and soy protein, which act as carrier and wetting agents, and when the liquid mixture is passed through a shear pump, there is no evidence of carboxymethylcellulose particles in the buildup of solids in the liquid egg. The application of heat as in pasteurization causes the suspended sodium carboxymethylcellulose to become more fluid and helps to prevent cook-on during the exposure to heat. The preferred amount is ideal in relation to the 1 to 5 reliquefying ability of the egg product. Any change in the reliquefying ratio of egg product could necessitate a change in the carboxymethylcellulose content, much less being used when less water is used.

About 0.05 to 0.1% of potassium sorbate, a water-soluble form of sorbic acid, is used to retard spoilage. This material not only acts to protect the finished product against contamination due to improper handling by the customer during use, but also is an excellent wetting agent which aids in introducing the milk and soy solids into the liquid egg mixture. It may also act as a pH regulator or buffer and it is more effective at the pH used than would be sodium benzoate, for example.

Often in bakeries it is noted that the container for the liquefied egg product is left open, thus exposing the egg product to airborne bacterial contamination, yeast and molds. Many times, a reliquefied egg product is held at room temperature, then placed in a refrigerator for further use. Potassium sorbate helps to safeguard the product against customer abuse and the resulting consequences.

The solids are introduced into the liquid egg material to enable copasteurization followed by codrying. The effects of copasteurization and codrying are quite dramatic. The egg product produced in this manner is far superior in every respect to one made from the same ingredients by determining the solids present in liquid whites and in the liquid whole egg, allowing and compensating for the moisture content in each, then taking the dried solids of the whites and solids of the whole egg, blending them with other solids, and arriving at a simple mechanical mixture. The process is conducted in the following manner.

The nonfat milk solids, sweet whey, soy protein, sodium carboxymethylcellulose or other colloid and any other dry ingredients are mixed together in proper proportions in a solids mixing zone. The mixed solids pass to a liquids-solids mixing zone, where they are added to the liquid whole egg and liquid egg white, which may have been mixed together and are preheated to 135° to 140°F. To aid in mixing and preheating, a recirculation line is provided. The preheating helps to prevent coagulation by keeping all the ingredients in a fluid state. Blending them at this stage enables them to cooperate together to protect the egg materials when the temperature is raised.

The mixed and preheated mixture is conducted to a homogenization zone, where homogenization may be achieved at 1,500 psi. While there is nothing new in the homogenization of liquid egg, there is this of interest in the homogenization of this low-fat egg product: the flavor of an egg is largely contained within the yolk, and in this product, the fine flavor is apparently associated with the distribution of the remaining fat throughout the egg product. Homogenization multiplies the fat globules approximately 10 times by dividing them into smaller globules. This improves the flavor characteristics of the liquid egg product much in the same manner that homogenization does with the fat content of milk.

From the homogenization zone, the homogenized mixture is conducted to a heating zone, where the temperature is raised to between 148.5° and 155°F. Then the material may be taken to a high heat zone where it is pasteurized by flashing under vacuum at 161° to 165°F. This may be done by steam infusion or injection.

From the heating zone, the material passes to a vacuum tank cooling zone. This zone may be a vacuum tank where there is sufficient vacuum to cool it to 150° or even down to 140°F. From the vacuum tank or from the flashing zone the product flows to and through a cooling zone, where the liquid egg product is brought down to a temperature of about 38° to 40°F and then sent either to refrigerated storage or to a dryer at a dehydration zone.

When the product has completed the pasteurization cycle and has been homogenized, it may be held in storage at 34° to 38°F for storage and/or drying purpose. At this temperature the liquid egg product may be packaged for use, frozen in cartons or in cans, or it may be dried by spray-drying.

Example 1: 9,000 pounds of liquid egg white and 3,000 pounds of liquid whole egg are charged into a 2,000 gallon open-top tank equipped with a recirculating overshot system operated by a shear pump capable of intimate mixing of its throughput. Meanwhile, in a horizontal type ribbon mixer 320 pounds of nonfat dry milk solids, 455 pounds of dry sweet whey and 50 pounds of isolated soy protein are thoroughly and intimately mixed with 91 pounds sodium carboxymethylcellulose, 7 pounds of potassium sorbate and 5 oz of certified edible color. The completely mixed solids are then added to the liquid egg mixture by introduction into the suction of the shear pump while circulating the egg mixture from the bottom to the top of the tank, but under the liquid surface.

After introduction of all the solids, circulation is continued until the solids are completely dispersed into and throughout the liquid egg. After the mixture is complete, the temperature of the tank contents is raised to 135° to 140°F by circulation through a heating tube. The product is then charged through a conventional homogenizer operating at 1,500 psi. The temperature of the homogenized mixture is then raised, and the liquid egg mixture at a temperature of about 155°F is charged through a steam injection conduit where the temperature is raised to 161° to 165°F for 2 to 5 seconds (flashing) after which the heated material is charged into a vacuum tower, where the temperature is immediately dropped to about 150°F and then through a cooler, where the temperature is reduced to 38° to 40°F. The cooled liquid is then dehydrated, in a conventional vacuum dehydrator, to a dry powder.

Due to the specific order of processing conditions and proportions used, no foaming, no coagulation of solids, and no cook-on is encountered during the operation. The powder may be reconstituted by adding five parts by weight of water to one part by weight of the powder.

Example 2: Scrambled Eggs — One pound of the product from Example 1 is mixed with five pints of water. The mixture is whipped into a batter, permitted to stand five minutes, and again whipped until all particles are dispersed. The mixture has the same consistency and functionability as a batter similarly prepared from one pound of dried whole egg and 2½ to 3 pints water. The mixture may be cooked into an omelet or scrambled eggs on a greased griddle, stirring as necessary. In place of water, whole milk may be used. The cooked omelet (or scrambled eggs) is found to be tasty and to have a satisfactory consistency and texture.

Example 3: Custard — In place of two large fresh eggs weighing about 114 grams having 30 grams of solids content, 20.5 grams of the egg product from Example 1 is used. At the same time, the ordinary requirement of 3 cups of milk may be reduced to 2½ cups of milk. With this is blended ⅓ cup of sugar and ¼ teaspoon of salt. Vanilla or nutmeg or both are added to flavor. In place of 2½ cups of milk, 2 cups of milk and ½ cup of water may be used. The product gives an excellent custard, especially as a filler for custard pies; and saves considerably in egg solid content, 9.5 grams thereof. Many other foods using eggs are similarly made with satisfactory results.

Extraction with Oil

In the process of *J.A. Fioriti, H.D. Stahl, R.J. Sims and C.H. Spotholz; U.S. Patent 3,717,474; February 20, 1973; assigned to General Foods Corporation* wet egg yolk is combined with an edible oil using a high shear mixer. During the mixing, the cholesterol is extracted from the yolk into the oil. Simultaneously, the polyunsaturated fat content of the yolk is increased, raising the P/S ratio, if polyunsaturated oil is used. The yolk is then separated gravitationally from the oil and thus recovered. The yolk is still wet, having its water content intact.

This process makes use of the known fact that cholesterol is soluble in fats and oils. However, this does not necessarily predict success in using oil to extract cholesterol from egg yolk, since, it is possible that a water barrier exists within the yolk which surrounds the yolk fat globules which would prohibit contact between those globules and the solvent oil droplets.

If this is correct, this barrier would have to be broken down to achieve substantial removal of cholesterol from the yolk. This would require a mixing device which could impart the needed energy to disrupt the yolk particles and thus hopefully the water barrier, and obtain intimate contact between the yolk fat globules and the solvent oil droplets. Therefore, a Waring Blendor, a kitchen type liquid blender which can supply a vigorous shearing action by high speed rotation of its blades, was used with considerable success as shown in some of the examples below.

Wet egg yolk is placed in a high shear mixer capable of imparting the needed energy to disrupt the water barrier in the yolk, with an amount of an edible oil sufficient to keep the yolk in a free-flowing state within the oil. The material is mixed at high energy for a time short enough to keep the rising temperature of the mixture from cooking the egg yolk.

After blending, the oil yolk mixture is placed in a centrifugal separator to recover the wet yolk. Oil and egg yolk analyses show that at least half the cholesterol is removed from the yolk. When the edible oil is polyunsaturated, the P/S ratio increases from less than one-half to at least one.

Example 1: The yolks of two eggs averaging 20.8 grams in weight and 283 mg of cholesterol, were placed in a Waring Blendor with 180 ml of safflower oil. The mixture was blended at high speed for 5 minutes, the temperature of the mixture increasing from room temperature to 131°F. After blending, the oil yolk mixture was placed in a batch type centrifuge at 1,000 RCF x g for 30 minutes to separate the yolk from the oil. Oil and egg yolk analysis by gas liquid chromatography showed that 221 mg of the cholesterol in each yolk had been extracted, or in other words 78% extraction had been achieved. Furthermore, fatty acid analysis disclosed that, per yolk, the fat content comprised 21.8% saturated fats, 15.4% monounsaturated fats and 62.8% polyunsaturated fat. The P/S ratio thus increased from less than ½ to 2.85.

The recovered yolks were than cooled to 35°F and molded into yolk-like shapes. They were recombined with liquid egg white and the composite was cooked. A fried egg, acceptable in appearance taste and texture resulted.

Example 2: Four wet egg yolks averaging 16 grams in weight and 202 mg cholesterol content were placed in a container with 360 ml of safflower oil. The probe of a Model W 185 D Sonifier Cell Disruptor was immersed in this mixture and run at maximum output (150 watts) for 5 minutes. The well emulsified slurry was placed in a centrifuge to recover the yolk. Analysis indicated that 101.6 mg per yolk of the cholesterol had been extracted or 50.3%.

The process may be extended to obtain any amount of cholesterol extraction that is desired. Semicontinuous or continuous extraction may be used; for example, the yolk may be blended successively with batches of fresh oil with recovery being accomplished after each blend as before. In a single extraction, as the amount of cholesterol dissolved in the oil increases, the ability of the oil to extract cholesterol decreases. Using multiple extractions with fresh oil eliminates this problem.

Example 3: 41 pounds (5.4 gallons) of safflower oil were placed in a ten gallon jacketed kettle, equipped with a turbine type self-baffled mixer and heated to 125°F with hot water. Ten pounds of wet egg yolk were added to the oil with mild agitation. The mixture was then blended at full speed (7,000 rpm) for 15 minutes with the temperature being held between 125° and 135°F. The resulting mixture was separated in a desludging centrifuge to recover the yolk. The recovered yolk was then added to 5.4 gallons of fresh preheated safflower oil and blended for an additional 15 minutes at full speed. The yolk was again separated in the desludging centrifuge. The third extraction was performed with 5.4 gallons of fresh preheated safflower oil under the same conditions. The yolk was again separated and saved as the final product.

The initial cholesterol in the yolk totaled 0.15 pound. After the first extraction analysis revealed that 0.069 pound of the cholesterol had been removed, or in other words 46%. During the second extraction 0.037 lb were removed. During the third extraction 0.02 lb were removed. The total cholesterol removed for all three extractions equalled 0.126 lb, or in other words 84.3% of the cholesterol had been extracted.

Safflower oil is the preferred solvent since it is by weight 77% polyunsaturated fat. However, for purposes of cholesterol extraction from egg yolk, any edible oil would suffice, since all these oils act as solvents for cholesterol. But the types of fats in edible oils can vary widely with each particular oil. To achieve a low cholesterol wet egg yolk with an increased P/S ratio, it is necessary to use an oil which, like safflower oil, is high in polyunsaturates. Thus, corn oil, soybean oil, cottonseed oil and sunflower or seed oil would be appropriate.

The amount of oil needed per yolk isn't crucial so long as the amount used is great enough to support the blended yolk in a free flowing state within the oil. Best results in cholesterol extraction are obtained when the yolk is fresh and has not been frozen for storage. Freezing produces centers of coagulation in the yolk, thus reducing the shearing efficiency of the extraction process.

The amount of cholesterol extracted varies directly with the increase in temperature due to the high energy blending. When the temperature rises much above 149°F although almost all the cholesterol can be extracted, cooking of the yolk commences which will result in a poor egg product upon reconstitution. When the blending time is kept to about five minutes, adequate temperature is maintained.

Gravitational separation by batch centrifuge is used for maximum separation of the oil and yolk after blending. On a larger scale, a continuous desludging centrifuge, such as a Model SAAOH-205 Westphalia Separator can be used.

It has been seen that other mixers besides a Waring Blendor can be used. The sonic oscillator of Example 2 can be used to provide the shearing energy to break up the yolk particles and achieve sufficiently intimate contact with the oil molecules to produce cholesterol extraction. On a larger scale, high-shear mixers such as the Beloit-Jones Model C-020 Hi-Intensity Mixer, which was used in Example 3, the Eppenbach Homogenizer, and the Cowles Dissolver, a colloid mixer, can be used.

The yolks obtained from this process can be made into a variety of egg products. They can be recombined with liquid egg white, packaged, and frozen. The contents of the frozen package when thawed and cooked, or when cooked from the frozen state, will result in an acceptable fried egg with good taste and texture. This is the form illustrated in Figures 6.1a and 6.1b.

FIGURE 6.1: EXTRACTION WITH OIL

a.

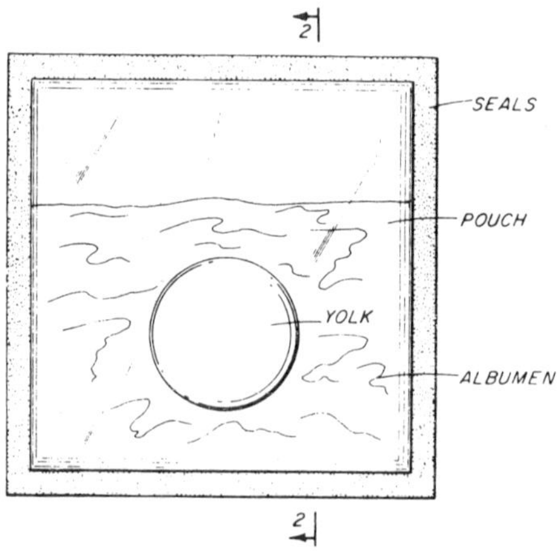

b.

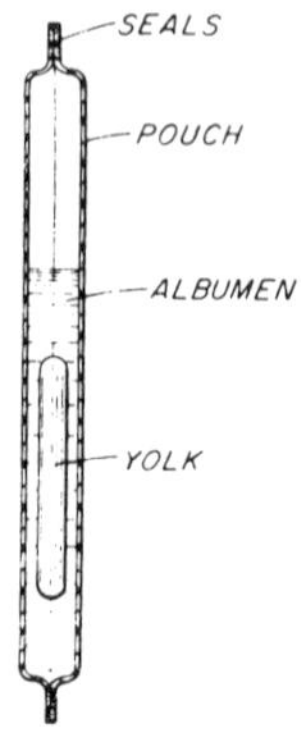

Source: J.A. Fiorite, H.D. Stahl, R.J. Slims and C.H. Spotholz; U.S. Patent 3,717,474; February 20, 1973

When in the liquid state, the yolk and white may be scrambled while in the package by kneading the contents and an acceptable scrambled egg will result. Also the sealed package may be immersed in boiling water to produce a poached egg. Another advantage of this process is that when a low cholesterol yolk is frozen and subsequently thawed, the functionality of the yolk is basically the same as it was before freezing. When natural egg yolk is frozen, gelation takes place, and when the yolk is thawed it is seen to have centers of coagulation and to possess a pasty jell-like texture. This does not result with the low cholesterol yolk of this process.

Aside from the product just described, a frozen egg roll has been proposed, comprising a cylindrical core of the low cholesterol egg yolk surrounded by a wider cylindrical shell of egg white, the whole being frozen. An egg product resembling the shape of a fried egg can be obtained by cutting off a cross-sectional piece of the egg roll to be cooked as is.

When the egg yolk is to be used for baking purposes, it has been found that adding up to two grams of lecithin to the yolk after recovery as an emulsifier promotes better consistency in the baked product. It prevents falling apart of the baked goods and surface cracking which otherwise seems to occur. Lecithin need not be used when the yolk is for direct consumption.

NUTRITIOUS EGG WHITE POWDER

A dry egg white powder low in calorie content and cholesterol level which includes glycerides predominant in unsaturated fats is described by *W.A. Gorman, C.K. Stearns and S.M. Weisberg; U.S. Patent 3,207,609; September 21, 1965; assigned to National Dairy Products Corporation.*

The dry egg white product contains dried egg albumen (dry egg white), a water-binding carbohydrate, agglomerated milk solids, and an edible oil. The edible oil is not intimately distributed within the indicated dry constituents, but is blended therewith so that it is present primarily on the surface of the dry constituents.

The dried egg albumen is a spray dried product. Preferably, in order to provide a product having optimum color and flavor, the albumen is subjected to treatment to effect depletion of its contained glucose prior to spray-drying.

The water-binding carbohydrate of the product is a cereal or root starch or flour which is capable of absorbing and binding water by virtue of its polymeric structure. Examples of such water-binding carbohydrates includes corn, tapioca, amioca, rice, potato starches and flours. Preferably, the water-binding carbohydrate is capable of retaining water at elevated temperature conditions so that weeping does not occur during frying. One carbohydrate which is particularly suitable is rice flour, which provides substantial advantages over other water-binding carbohydrates in preventing weeping under frying conditions.

The agglomerated milk solids contain nonfat milk solids which have been treated to improve the water dispersibility. At present, such treatment involves subjecting dry milk powder to water vapor, and effecting controlled adherence of the dry particles into larger agglomerates.

The edible oil in the product is a vegetable oil which is normally liquid at room temperatures. Desirably, the edible oil should be a high percentage of unsaturated glycerides, such as safflower oil, corn oil, and soybean oil.

The oil is incorporated into the product by mixing the oil with the remaining constituents in dry form. By such a dry mixing operation, the edible oil is not intimately distributed throughout the dry constituents, but instead is present primarily on the surface of the dry constituents. The method of incorporation of the edible oil into the product is important and is believed to be functionally related to the substantially improved dispersibllity of the product.

In this connection, a mixture of dried egg albumen and water-binding carbohydrate, without oil, is difficultly dispersible in water. While a mixture of these two constituents can be dispersed in water or batters and doughs and cooked to provide an acceptable food product, the time required for dispersion renders such a mixture of questionable value.

Also, in this connection, a mixture of dried egg albumen, water-binding carbohydrate, and agglomerated milk solids, without oil, while providing somewhat improved dispersibility in water, nevertheless requires substantial mixing and agitation to effect dispersion.

The dry egg white product is rapidly and efficiently dispersible in water, milk, or batters and doughs. It is believed that the mechanism through which such improved dispersibility is obtained involves the provision of a partially hydrophobic surface on the dry constituents by the edible oil. The term partially hydrophobic denotes the property of repelling water therefrom. It is thought that a partially hydrophobic surface tends to slow wetting of the dry constituents, and thus inhibits clumping, which clumping is known as a cause of poor dispersibility of dry powders.

The product may further include minor proportions of sugar, salt, spices, and coloring. In order to obtain maximum benefits from the added coloring, it is incorporated in the liquid egg white prior to drying, rather than incorporating the coloring into the dry ingredients.

The body and texture of the products can be desirably improved by providing a buffering salt. The buffering salt is added at a level such that the product has a pH of about 8.5 to 9.5. Examples of salts which may be utilized are trisodium phosphate and tetrasodium pyrophosphate. Other condensed phosphate compounds may also be used as long as the indicated pH is obtained.

Example: A quantity of spray dried egg albumen was prepared by desugaring liquid egg white by fermentation of contained glucose, and adding to the liquid egg white minor proportions of FD&C Yellow #5 and Yellow #6. The colored egg white was spray dried to provide a colored dried egg albumen. The resultant dried egg albumen was blended with the remaining dry constituents in the amounts set forth.

Constituents	Weight, grams
Dried egg albumen	20
Agglomerated nonfat milk solids	10
Rice flour	10
Trisodium phosphate	2
Dextrose	0.4
Salt, spices	1.3

After the dry constituents were thoroughly blended, 5.34 grams of refined safflower oil was added to the dry constituents while the dry constituents were tumbled until satisfactory distribution of the oil was effected. The resultant product had the following composition in weight percentages:

Constituents	Percent by Weight
Dried egg albumen	40.8
Agglomerated nonfat milk solids	20.4
Rice flour	20.4
Safflower seed oil	10.9
Trisodium phosphate	4.1
Dextrose	0.8
Salt, spices, color	2.6
	100.0

The calorie content of the product was 3.5 calories per gram. In contrast, the calorie content of whole egg solids is about 6.2 calories per gram. The cholesterol content of the product was about 0.01 mg per gram, whereas the cholesterol content of whole egg solids is about 17.6 mg per gram.

The ratio of polyunsaturated fat to saturated fat in the product was about 7.1:1. In contrast, the ratio of polyunsaturated to saturated fat in whole egg is about 0.8:1. Thus, the ratio of polyunsaturates to saturates in the product was nine times greater in whole egg.

The dry product of the example may readily be dispersed in sufficient water or milk to provide a mixture having about the same solids level as liquid whole eggs (24 to 28% solids). The mixture may be used in place of liquid whole eggs, in a variety of dishes to provide a dish substantially equivalent thereto.

For example, the product was used to prepare a dish substantially equivalent to scrambled eggs, ordinarily prepared from whole eggs, 19.5 grams of the product was dispersed in about 78 grams of water to provide a batter. The powder dispersed rapidly and effectively in the water upon gentle stirring, to provide a smooth batter. The following table sets forth the nutritional composition of the resultant batter and of conventional liquid whole egg:

	Batter, Percent by Weight	Liquid Whole Egg, Percent by Weight
Fat	2.4	11.5
Protein	9.6	12.8
Water	78.7	72 - 78
Solids	22.3	22 - 28

The calorie content of the batter was about 0.8 calorie per gram, whereas the calorie content of liquid whole egg is about 1.6 calories per gram.

The foregoing batter was fried to provide a scrambled egg dish. There was substantially no weeping of water during the frying operation, and the dish had desirably firm body and texture, and its flavor was excellent and fully equivalent to whole-egg scrambled. Ham, mushrooms, or cheese may be added to the batter to provide an omelet-type dish substantially equivalent to whole-egg omelet in flavor, texture, and other characterisitcs.

Other foods which may be prepared utilizing the product of this process in place of whole eggs includes waffles, eggnog, biscuits, french toast, frying batters, custards, puddings, and the like.

LOW CHOLESTEROL SYNTHETIC EGG

The removal of fats and cholesterol from egg products is difficult, and therefore *M.A. Perret; U.S. Patent 3,806,608; April 23, 1974; assigned to Rhodia, Inc.* investigated the possibility of a synthetic egg composition.

It is known that liquid whole chicken eggs are composed of about 64% egg white, and about 36% egg yolk. The composition of each of these components is quite complex, and even now is not fully understood. Egg white contains approximately 12% matter which is proteinaceous and is referred to as albumen (as opposed to albumin, which is the class of pure protein of this type, but is derived from any other sources, as well as eggs). The egg white or albumen is in the form of an aqueous solution of the phosphoproteinaceous material, containing small amounts of minerals and sugars, and only a trace of fat. The egg yolk contains virtually all of the fat and cholesterol of the egg, and imparts the characteristic egg-like taste and flavor properties.

The yolk is composed of about 50% solid matter, with both bound and unbound lipids, as well as cholesterol and proteins of a complex nature, probably in some obscure relationship with the lipid material as lipoprotein complexes. A part of the lipids, the free lipids, is easily extractable with fat solvents, but the remainder, the bound lipids, is not.

The taste of the egg yolk is the part of the egg that is the most difficult to duplicate in an egg substitute, probably because of the complexity of the materials present in the yolk. It is the failure to duplicate the egg-like taste and flavor properties of the egg yolk that has led to the failure of most egg substitutes, and this in turn has forced those seeking to develop egg substitutes to rely at least in part upon the flavor imparted to the egg substitute material by the addition of naturally-occurring egg yolk-derived components.

The synthetic egg composition contains either a water-soluble albumin or an edible fatty oil, or both, at least one amino acid, and an inorganic sulfide, and has an alkaline pH within the range from about 7 to 9.5 in the presence of water. The composition can be formulated as an aqueous solution or dispersion, a paste, or a dried powder, and can be made to approximate an egg, including egg yolk and total egg (white + yolk), in flavor and texture. It is believed that the combination of amino acid, inorganic sulfide and alkaline pH of the composition is primarily responsible for the egg-like flavor of the product. The water-soluble albumin and fatty oil contribute body and texture to the composition, and usually both are used to approximate a whole egg. For an egg yolk, only one need be used, but both can be used, if desired.

One amino acid can be used, but for better flavor it is best to use a mixture of amino acids. The amino acid can be selected from the group consisting of DL-isoleucine, DL-leucine, DL-valine, DL-methionine, threonine, DL-tryptophane, L-cysteine, DL-phenylalanine, L-tryosine, L-lysine, L-histidine, L-arginine, β-alanine, glutamic acid, cystine, glycine, aspargine, aspartic acid, serine, gentamine, proline, hydroxyproline, thyroxine, thiozine, and hydroxylysine. Two or more of these amino acids can be used, in any combinations. A preferred amino acid mix is indicated below.

	------ Parts by Weight ----	
Amino Acid	Preferred	Range
DL-isoleucine	2	0.5 to 5
DL-leucine	3	0.5 to 5
DL-valine	2	0.5 to 5
DL-methionine	1	0.2 to 3
L-cysteine hydrochloride	0.5	0.1 to 3
DL-phenylalanine	1.5	0.2 to 3
L-tyrosine	1.5	0.2 to 3
Threonine	1.5 to 0	0 to 3
DL-tryptophane	0.5 to 0	0 to 3
L-lysine monohydrochloride	2.0	0.5 to 5
L-histidine	0.5	0.1 to 3
L-arginine hydrochloride	2.0	0.5 to 5

The amino acid can also be in the form of a suitable protein hydrolysate such as hydrolyzed vegetable proteins derived from peanuts or soybeans, hydrolyzed animal proteins such as hydrolyzed deflavored cod flesh, keratin protein, as well as the hydrolyzed plant proteins occurring in cereal grain. The protein hydrolysates generally contain the constituent amino acids in peptide nonultimate amino acid form, representative of the source protein in the complex mixture in which they occur naturally, and in this respect differ significantly from individual mixtures of amino acids prepared synthetically. The amount of amino acid is within the range from about 0.01 to 3% by weight of the solids of the composition.

An inorganic sulfide is added also, in a very small amount. Free hydrogen sulfide can be used as the inorganic sulfide. Any alkali metal or alkaline earth metal sulfide or sulfhydride, for example ammonium sulfide, sodium sulfide or hydrosulfide, potassium hydrosulfide, and calcium hydrosulfide, which may generate hydrogen sulfide, can also be used as the inorganic sulfide. A trace is adequate, for instance, one drop of a 20% aqueous solution of ammonium sulfide in 350 grams of composition. It is preferred to incorporate the inorganic sulfide into the mix by diffusion in a smooth but indirect manner, such as by impregnating the inorganic sulfide into a porous surface, which is then brought into contact with the composition.

If neither the amino acid nor inorganic sulfide is present, the composition is wholly devoid of the flavor of eggs. Both must be present, for good egg flavor. Body and texture, when desired, can be provided to the synthetic egg product by a water-soluble albumin or a fatty oil, or both.

The albumins are a group of simple proteins that yield mostly amino acids upon cleavage by enzymes or acids, and are found in nearly all living body tissues. Examples of albumins are grain albumin, soybean albumin, ovalbumin (albumin from egg white), serum albumin (albumin from blood), and lactalbumin (albumin from milk). The albumins are soluble in water.

Alternatively, an edible fatty oil can be included to improve body and texture, as well as impart a fatty flavor, in order to approximate the flavor and texture of whole egg. An example is a spray-dried vegetable oil base Vee-Kreme, the constitution of which is partially hydrogenated vegetable oil, lactose, sodium caseinate, dipotassium phosphate, and butylated hydroxyanisole.

If the egg product is used in a preparation where eggs are one of several other ingredients, the egg body and texture are less important, and either the fatty oil or the albumin can be omitted.

The pH is brought to above 7, within the range from about 7.1 to 7.5 by addition of an alkaline compound. The amount of the alkaline compound is within the range from about 0.1 to 3%, to bring the pH to within the desired range.

In addition to the above indicated essential ingredients, substances such as mineral salts can also be added, such as sodium chloride, especially phosphates as, for instance, potassium phosphate, sodium phosphate, ammonium phosphate, calcium phosphate; lecithin; sugars such as sucrose, dextrose, glucose, fructose, galactose, xylose, mannose, ribose, arabinose, maltose, and lactose; milk products such as dried milk, dried whey and casein; surfactants such as the mono and di fatty glycerides or polyols; sodium inosinate; sodium guanylate; sodium and calcium alginate compositions; and coloring agents. Thus, the synthetic egg compositions have the following formulation:

Ingredient	Amount (%) Overall Range	Amount (%) Preferred Range
Amino acid	0.001 to 5	0.1 to 3
Inorganic sulfide	0.00001 to 0.1	0.0001 to 0.01
Albumin	0 to 90	15 to 85
Fatty oil	0 to 90	0 to 75
Lecithin	0 to 25	2 to 15
Milk products	0 to 20	0 to 10
Sugar	0 to 10	0 to 5
Miscellaneous (emulsifiers, palate	0 to 10	1 to 5
sensitizers, coloring agents, mineral salts)	0 to 10	1 to 5
pH	7 to 9.5	7.1 to 7.5

For a whole egg, taken as an egg, such as a scrambled egg dish, approximately equal amounts of albumin and fatty oil best approximate the texture and mouthfeel of a fresh egg. For other uses, where texture and mouth feel are not important, the proportion of albumin and fatty oil can vary widely within the stated ranges.

The synthetic egg product can be in the form of a liquid, paste, or powder, depending mainly on the type of oil or fat employed, a liquid oil leading to a liquid or semisolid material, and a solid fat leading to a powder. If the product is to be used as a liquid, an amount of water is added to obtain a consistency approximating a fresh liquid egg. In cooking and other applications the liquid, pasty, or dried egg-like concentrate can be used as is, i.e., as obtained after formulation.

The product can be prepared by simple mixing or blending of the ingredients. The solids can be mixed first, and then mixed into the liquids. If water is present, the water-soluble ingredients can be dissolved in water, and then the other ingredients blended in. A homogeneous solution or dispersion is desired in the case of liquids. A uniform solid mix can be obtained by first dispersing or dissolving the ingredients in a volatile liquid, to recover the composition as a solid residue, paste, or powder.

Example 1: An amino acid blend (18 grams) was prepared, having the following composition:

Grams	Amino Acid
2.0	DL-isoleucine
3.0	DL-leucine
2.0	DL-valine
1.0	DL-methionine

(continued)

Grams	Amino Acid
0.5	L-cysteine hydrochloride
1.5	DL-phenylalanine
1.5	L-tyrosine
1.5	Threonine
0.5	DL-tryptophane
2.0	L-lysine monohydrochloride
0.5	L-histidine
2.0	L-arginine hydrochloride

The synthetic egg composition was prepared to the formulation:

Grams	Ingredient
325.0	Water
50.0	Powdered egg albumen
0.25	Amino acid blend
2.5	Dextrose
1.5	Sodium chloride
45.0	Crisco shortening (a partially hydrogenated vegetable oil)
5.0	Lecithin
2.0	Sodium bicarbonate, to adjust pH between 7.2 and 7.5
1 drop	Ammonium sulfide (20% aqueous solution)
Trace	Yellow food color
Trace	β-carotene to give egg color

This synthetic egg composition was blended with 1% ham flavor and cooked in a drying pan in the manner of a scrambled egg. The resulting composition was excellent in flavor and had a texture very similar to real scrambled eggs and ham.

Example 2: An amino acid blend (18 grams) was prepared, having the following composition:

Grams	Amino Acid
2.0	DL-isoleucine
3.0	DL-leucine
2.0	DL-valine
1.0	DL-methionine
0.5	L-cysteine hydrochloride
1.5	DL-phenylalanine
1.5	L-tyrosine
1.5	Threonine
0.5	DL-trypotophane
2.0	L-lysine monohydrochloride
0.5	L-histidine
2.0	L-arginine hydrochloride

An egg composition was prepared to the formulation:

Grams	Ingredient
50.0	Egg albumen (dried powder)
1.5	Amino acid blend
3.0	Dextrose
2.0	Sodium chloride
10.0	Lecithin (dry)
2 drops	Ammonium sulfide (20% aqueous solution)
2.0	Sodium bicarbonate

Example 3: An amino acid blend (18 grams) was prepared, having the following composition:

Grams	Amino Acid
2.0	DL-isoleucine
3.0	DL-leucine
2.0	DL-valine
1.0	DL-methionine
0.5	L-cysteine hydrochloride
1.5	DL-phenylalanine
1.5	L-tyrosine
1.5	Threonine
0.5	DL-tryptophane
2.0	L-lysine monohydrochloride
0.5	L-histidine
2.0	L-arginine hydrochloride

A synthetic egg composition was prepared to the formulation:

Grams	Ingredient
250.0	Powdered egg albumen
1.3	Amino acid blend
13.0	Dextrose
8.0	Sodium chloride
250.0	Shortening (Crisco)
25.0	Lecithin
0.3	β-carotene
20 drops	Yellow food color
5 drops	Ammonium sulfide (20% aqueous solution)
28.0	Dry milk
5.0	Sodium bicarbonate

Example 4: An amino acid blend (18 grams) was prepared, having the following composition:

Grams	Amino Acid
2.0	DL-isoleucine
3.0	DL-leucine
2.0	DL-valine
1.0	DL-methionine
0.5	L-cysteine hydrochloride
1.5	DL-phenylalanine
1.5	L-tyrosine
1.5	Threonine
0.5	DL-tryptophane
2.0	L-lysine monohydrochloride
0.5	L-histidine
2.0	L-arginine hydrochloride

A synthetic egg composition was prepared, to the following formulation:

Grams	Ingredient
125.0	Serum albumin (dried powder)
170.0	Vee-Kreme (spray dried vegetable oil)
10.0	Lecithin
5.0	Sodium bicarbonate
0.3	Amino acid blend (as above)
2 drops	β-carotene (30% in vegetable oil)
5.0	Kelset
20 drops	Tween 60
20 drops	Atmos 300
0.3	Monosodium glutamate
0.025	Ribotide
1 drop	Ammonium sulfide (20% aqueous solution)

One drop of ammonium sulfide was put on filter paper, and placed in a jar of the composition. The solid materials were mixed together. The emulsifiers and color were blended, and added slowly to the dry mix while mixing on a Sunbeam mixer. When thoroughly mixed, the mixture was passed through a 30 mesh screen. The dried synthetic egg composition was excellent in all respects.

NO-CHOLESTEROL MAYONNAISE

Salad dressings in order to be properly labelled mayonnaise, must contain a minimum amount of egg yolk. The composition has been set in the Federal Standards which requires the presence of at least 4% of egg yolk.

The presence of egg yolk while desirable for its flavor and its emulsifying effect, nevertheless possesses two serious drawbacks which has limited both the wider use of mayonnaise as a salad dressing and its storage capability between the time of manufacture and the ultimate use by the consumer.

First, egg yolk is composed of cholesterol which has been recognized by various medical authorities as being an undesirable constitutent, above certain prescribed levels, in the blood of humans, particularly adult males. The presence of egg yolk in mayonnaise produces an undesirable level of cholesterol in all mayonnaise products and renders them unacceptable in the diet of many who are knowledgeable and concerned over the presence of cholesterol in foods that are part of a common diet.

The other inherent characteristic of egg yolks which makes them an undesirable, although legally necessary, constituent of mayonnaise is the fact that the shelf life of mayonnaise is at best within a range of three to four months. Eggs, and particularly egg yolks, are a principal factor in the reduced storage capability due to the presence of cholesterol in foods that are part of an inherent susceptibility of egg to microbiological spoilage. The egg yolk may initially contain various microorganisms prior to the incorporation of the egg yolk into mayonnaise and in time, even with sanitary manufacturing conditions, the growth of the microorganisms in the egg yolk spreads to ultimately contaminate and spoil the entire mayonnaise product. The growth of these microorganisms is further enhanced by the opening of the mayonnaise container by the consumer.

According to *J.J. Katz; U.S. Patent 3,764,347; October 9, 1973; assigned to United Food Industries, Inc.* it is possible to prepare mayonnaise without egg yolk. In place of egg yolk it has been found that egg white in an amount between 3 to 9 parts by weight will produce a stable suspension between oil and starch which are the other essential ingredients.

The oil that is to be used should be of vegetable origin and can be any one of a number of such oils such as soybean, corn, cottonseed, peanut, palm, cocoanut and safflower oil. The amount of the vegetable oil may range from 35 to 50 parts by weight.

The starch constituent of the salad dressing is a conventional ingredient and may be derived from corn or tapioca or other vegetable. The starch is formed into a paste with sufficient water to gelatinize the starch through heating. The conventional spices may include between 2 to 6 parts by weight of the total composition and may be incorporated into the starch paste or added subsequently.

Example:

	Parts by Weight
Starch paste	47.2
Egg white	5.64
Spices	3.19
Soybean oil	43.04
Vinegar	1.00

In order to mix the ingredients to obtain satisfactory and stable suspension, the egg white is to be added to the starch paste in the mixture while mixing, then the soybean oil and vinegar are added. To enhance the coloring, if desired, oleo resin of paprika solution may be added with the soybean oil. The mixing is usually complete in about five minutes, and then, the premixed ingredients are pumped through a colloid mill in about six minutes in order to obtain the stable suspension.

While this composition produces a satisfactory product, it is possible to substantially enhance the quality of the taste, texture and appearance as well as to improve the stability of the suspension by incorporation of optional ingredients such as an inert, nontoxic emulsifier and chelating agent.

The emulsifier is any one of the natural plant hydrocolloids or gums such as tragacanth, carrageenan, karaya, ghatti, arabic, agar, locust bean, pectin, pectinates, guar, algin and the alginates. The alginates may be any of the alkaline metal salts of alginic acid such as sodium, calcium or the glycerol esters of the algins or algenic acid. Also carboxymethylcellulose may be used.

Among others found useful are the polyoxyethylene sorbitan fatty acid esters such as monooleate, stearate, palmitate, laurate, etc., generally known as a Tween. Still other nonionic emulsifiers that are useful include the fatty acid esters of sorbitan and other polyglycols and their partial anhydrides. The term fatty acids shall include all organic, monocarboxylic acids, saturated and unsaturated having 5 to 18 carbon atoms. Within this group are those commonly known as Span, a sorbitan ester of oleic, stearic, palmitic, lauric, etc., fatty acids. The above emulsifiers may be present in amounts between, preferably 0.07 to 0.15 part by weight.

The purpose of the emulsifier is to effect even greater stability than that produced by the egg white. The emulsifiers, however, will not be adequate to produce the desired results and obtain the shelf life stability without the presence of egg white. The egg white alone, though, will provide sufficient stability to make the use of the emulsifier optional.

It is also necessary to incorporate a chelating agent such as the sodium salt of ethylenediaminetetraacetic acid (EDTA). The EDTA is incorporated in amounts ranging from 0.005 to 0.01 part by weight. A typical example of the salad dressing of the mayonnaise type having no cholesterol present is shown below.

	Parts by Weight
Starch paste	47.02
Egg white	5.64
Algin derivative (sodium alginate)	0.1
EDTA	0.01
Spice blend	3.19
Safflower oil	43.04
Vinegar	1

The procedure in combining the ingredients which should be followed in order to produce the most stable suspension may be as follows: (A) Blend the spices such as the salt, sugar, mustard, onion powder and soluble pepper. (B) Dissolve the EDTA in about one pint of water; mix for about one minute. (C) Dissolve the sodium alginate in the safflower oil in amounts between ½ to 1 gallon of oil to approximately 1 pound of the alginate; mix for about two minutes. (D) Combine the ingredients in the premixer in the following order. Starch paste (previously made as stated above); egg white; alginate and oil mixture in (C) above; EDTA-water from (B) above; spice [from (A) above]; remainder of the safflower oil and vinegar which may be 1 to 5 parts by weight although it is optional; and a minor amount of oleo resin of paprika solution in amount to give the desired color. Agitate for about five minutes while adding the ingredients. (E) After having been premixed, the ingredients should be passed through a colloid mill in about six minutes. The product is found to have a shelf life between 6 and 9 months and contains no cholesterol. Its appearance, taste and texture is that of mayonnaise commonly available in the commercial market.

HIGH POLYUNSATURATE MARGARINE

In recent years, nutritionists and medical practitioners have become increasingly aware of the dietary importance of polyunsaturated fats. The polyunsaturated fatty acids are, according to authorities, a dietary requirement for protection against fat-deficiency symptoms. Linoleic, linolenic, and arachidonic acids have been recognized as the chief polyunsaturated fatty acids which possess an appreciable biopotency in counteracting fat deficiency.

Polyunsaturated acids of the type described above are believed to be required for growth and for the maintenance of normal skin conditions by a wide variety of animals, including man. These fatty acids mediate not only growth but also certain diverse functions such as protection from x-irradiation injury, maintenance of capillary resistance in the skin capillaries, and the normal transport and metabolism of cholesterol.

Many medical experts believe that there is a relationship between the intake of polyunsaturated fatty acids and the level of plasma cholesterol. It has been reported that high levels of blood cholesterol can be reduced when large amounts of vegetable fats are included in the diet, and that the level will be increased when large amounts of animal fats are taken.

The present-day trend toward the use of hydrogenated oils in food products leads to a lessened intake of essential fatty acids, resulting in the esterification of body cholesterol with saturated fatty acids. These esters are not readily disposed of by the body, nor are phospholipids containing saturated fatty acids, but are retained to a great degree in the plasma. They may cause atheroma, a disease characterized by fatty degeneration of the inner coat of the arteries, and may increase the coagulability of the blood, and contribute to coronary and cerebral thrombosis. *J.P. McNaught; U.S. Patent 3,746,551; July 17, 1973; assigned to Lever Brothers Company* discusses the use of polyunsaturated fatty acids to produce a highly nutritive margarine.

Arachidonic acid occurs to some extent in marine oils, although these oils are not considered suitable for use in margarine without hydrogenation. The acid occurs only in very small amounts in animal fats, and for all practical purposes not at all in vegetable oils. Thus it is impractical to attempt to introduce a high arachidonic acid content into a food product by the incorporation of a naturally occurring fat. This acid however is synthesized in the body by conversion from linoleic acid. Linoleic and linolenic acids are not synthesized in the body and must be taken in the diet.

Linoleic acid, usually along with small amounts of linolenic acid occurs at a relatively high level combined as glycerides in a great number of well-known edible vegetable oils, for example safflower, tobaccoseed, sunflower, corn, soy, cottonseed, peanut, sesame, etc.

Within this group of oils, those having 65% or more polyunsaturated fatty acids are used. Normally these will be safflower and tobaccoseed oils, although any triglyceride oils of which special high-linoleic acid varieties may be grown are suitable provided a margarine having a ratio of polyunsaturated acid to saturated acid of at least about 3.5 to 1.

Sunflower oil is an example of an oil having one or more varieties with a higher linoleic acid content than does the conventional variety. Varieties are known having 65% or more linoleic acid.

Safflower oil and other polyunsaturated fats are currently marketed as salad oils, and persons desiring to enrich their intake of polyunsaturated fats can avail themselves of these products. It is evident that there is a need for other edible products containing high proportions of polyunsaturated fatty acids to enlarge the scope of availability of such products to persons desiring to increase their intake of polyunsaturated fatty acids. Margarine contains at least 80% oleaginous matter (imitation margarines contain substantially less), and in view of its widespread use, is a particularly suitable product for the incorporation of an oil having at least 65% polyunsaturated fatty acids.

In order to introduce as much polyunsaturated fatty acids, relative to the saturated fatty acids, as possible into the diet through the use of margarine, the liquid oil fraction of the margarine will be nonhydrogenated and will advantageously be an oil that contains the highest proportion of polyunsaturated acids combined as triglycerides.

Safflower oil is available in commercial quantities and has the highest linoleic acid content of the vegetable oils available in large volume, and for this reason is the preferred oil for use to provide the highest possible polyunsaturated acid content and the lowest saturated fatty acid content.

However when safflower oil or other oil having at least 65% polyunsaturated acids is used, the margarine has a serious defect in that when it is stored under deep-freeze conditions, e.g., about -10° to -15°F, it is physically unstable, that is, the margarine texture breaks down and develops a granular, lumpy appearance, accompanied by loss of body, making the product unsightly and unsalable.

The problem of instability of margarine having a high content of polyunsaturated fatty acids at deep-freeze temperatures can be alleviated by using oil that has been randomly interesterified.

Randomization of natural safflower oil changes its properties in margarine with respect to texture stability at deep-freeze temperatures. This change in property, i.e., texture stability at deep-freeze temperatures, is a property on which this product is predicated, and since the margarine is unstable when natural safflower oil (or other oil having at least 65% polyunsaturated fatty acids) is used, but texture-stable when the oil is subjected to a randomization process, it becomes clear that the randomizing process causes a change in structure.

The liquid oil component may be a single oil having a polyunsaturated fatty acid content of at least about 65%, or a mixture of such oils. The liquid oil component may also contain a minor proportion of oils having a polyunsaturated content slightly lower than 65%, for example from about 50 to 65%, provided that the margarine having both liquid and hard stock components, has a polyunsaturated-to-saturated fatty acid ratio within the range of about 3.5 to 1 to about 5 to 1.

To correct the texture problem under deep-freeze conditions the higher polyunsaturated content oil or mixture may be randomized alone or may be mixed with the lower polyunsaturated content oil before randomizing provided that at least 25% of the higher polyunsaturated content oil is randomized.

By random interesterification is meant the interchange of fatty acid radicals on the glyceryl radicals in random fashion. The interchange continues dynamically when the oil is in the liquid state under the influence of a catalyst, until the mixture of molecular configurations reaches an equilibrium in accordance with the laws of probability. The term molecular configuration refers to the identity of the three fatty acid, or acyl radicals combined with one glyceryl radical.

Randomization includes random positioning of the fatty acid radicals within a triglyceride molecule with respect to alpha-beta positioning, as well as random distribution of the fatty acids among the glycerol molecules.

In the randomizing process, any of the recognized alkaline catalysts, such as a sodium alkoxide having from 1 to 4 carbon atoms, sodium metal, sodium-potassium alloy, sodium hydroxide, etc. may be used. The process may be carried out at temperatures as high as about 150°C and as low as about 15°C, so long as the oil remains liquid. Usually it is not necessary to use a temperature outside of the range of about 50° to 100°C, and within this range a temperature of about 70°C is commonly used. Processing time is usually ½ to 1 hour.

It is theorized that the stability obtained through the use of randomized safflower oil is due to the slight increase in the content of trisaturated and disaturated molecules, along with possibly some alpha-beta shifting, to produce crystal matrices that are less subject to phase changes with slight temperature fluctuations under deep-freeze conditions.

It has been observed, as shown below in Example 1, that when the entire amount of safflower oil used is randomized, the margarine is stable at minus 10°F, but exhibits slight emulsion instability when stored at room temperature for several days. However when the same amount of nonrandomized safflower oil is used, the margarine is texturally unstable at minus 10°F, but the emulsion is considerably more stable at room temperature.

The emulsion instability can be demonstrated by disturbing the surface of the margarine such as by cutting or scraping, whereupon some liquid oil collects at the disturbed area.

In the normal channels of commerce, the margarine is desirably stored under deep-freeze conditions for relatively long periods of time, while once in the hands of the consumer, the time during which the margarine would be exposed to room temperature conditions is relatively short, consequently protection of the margarine at deep-freeze temperatures is of prime importance.

The margarine is the type that has become increasingly popular in the past few years, referred to as soft margarine. This margarine is only slightly softer than the conventional print margarine, but too soft to withstand commercial handling if packaged in print form. Consequently the soft type margarine is packaged in tubs, or small containers, usually made of plastic, and which are sometimes decorated for a more attractive appearance.

The solid fat index (SFI) of the margarine is somewhat lower than conventional print margarines due to the high content of liquid oleaginous ingredients. The margarine of the process has the following range of SFI values: about 2.6 to 9 at 32°F; about 1.8 to 6 at 70°F; about 0.4 to 3.5 at 92°F; and about 0 to 1.0 at 104°F.

Example 1: Safflower oil is randomized by the following procedure. 800 to 1,000 pounds of safflower oil is pumped into a tank fitted with a jacket, a stirrer, an evacuating system, and a draw-off valve. The oil is refined by alkali treatment, and then dried for one hour at 150°C. The oil is cooled with stirring to about 70°C and 0.2% sodium methoxide powder added, and the oil stirred for an additional hour at 70°C, under a nitrogen blanket, whereupon the oil becomes interesterified in random configuration. This process is conducted under a nitrogen blanket to inhibit oxidation.

After randomizing, the oil is washed with water to inactivate the catalyst and to remove alkaline substances, e.g., sodium hydroxide and soaps, derived from the sodium methoxide catalyst, then refined, bleached, and deodorized.

For the preparation of a margarine, 90 parts by weight of the safflower oil treated as above and 10 parts by weight of 42.3 I.V. 50/50 blend of peanut oil and cottonseed winter oil stearin having a saturated fatty acid content of 53% are weighed into a churn at about 55°C and a monoglyceride made from 82 I.V. cottonseed oil added, along with lecithin, vitamin A, and carotene.

Into a separate tank is placed nonfat dry skim milk which is reconstituted with 10 parts water, pasteurized, and cooled to 40°F. About one-tenth of this milk product is cultured and returned to the tank. Salt, potassium sorbate, and flavor are then added. Nineteen and eight-tenths parts by weight of this aqueous mixture are blended with 80.2 parts by weight of the oil to form a water-in-oil emulsion. The temperature of the emulsion is adjusted to 40° to 42°C and fed at the rate of 220 pounds per hour through a scraped-surface heat exchanger after dissolving sufficient nitrogen gas (12 to 14% by volume) under pressure to provide the desired specific gravity to the margarine. The margarine is run in a semifluid state into eight ounce containers at a temperature of about 14° to 15°C and held at about 7°C until firm. The composition of the margarine is as follows:

	Parts by Weight
Oleaginous mixture:	
Glycerides	79.794
Mono/diglyceride emulsifier	0.18
Lecithin	0.22
Vitamin A and beta-carotene	0.006
Aqueous phase:	
Reconstituted skim milk (9.1% solids)	17.924
Salt	1.75
Potassium sorbate	0.10
Flavor	0.026
	100.000

The polyunsaturated content is 68.1% and the saturated acid content is 14.5%, based on the total of safflower oil and hard stock. The penetration is 235 at 45°F and 402 and 70°F. The ratio of polyunsaturated acids to saturated acids is 4.7.

The margarine is stored at -10°F for 26 weeks. At the end of this time the texture is smooth. When stored at room temperature it is observed that the emulsion exhibited some instability as evidenced by the appearance of some free oil in a cavity formed by removing some of the margarine from the surface. This indicates a limited storage stability at room temperature.

Example 2: For economic reasons, it will be advisable to mix as much nonrandomized safflower oil with the randomized oil as the system will tolerate. Experiments performed to determine the stability of margarine in which the ratios of randomized to nonrandomized safflower oil are 50/50, 35/65, and 25/75, in addition to ratios of 100/0 and 0/100 provide stability data, after storage at -10°F, in terms of the development of a lumpy, granular texture. A preferred product is one in which about 50% of the liquid oil is randomized. At this blend, the texture and body after storage at -10°F are satisfactory, and the free oil formation at room temperature is minimal.

Example 3: A margarine is made as described in Example 1 except that the glycerides are composed of 44.625% randomized safflower oil, 44.625% nonrandomized safflower oil, and 10.75% of a mixture of equal parts of cottonseed winter oil stearin and peanut oil having an iodine value of 41.6, and SFI values of 71.5 at 21.1°C, 62.6 at 33.3°C, and 41.7 at 40°C, and a saturated fatty acid content of 52.1%. The texture of the margarine does not change after storing more than 4 weeks at -10°F. The penetration at 45°F is 230 and at 70°F is 343. The polyunsaturated to saturated fatty acid ratio is 4.6.

SALT SUBSTITUTION

POTASSIUM CHLORIDE IMPROVEMENTS

Fumaric Acid Additive

Many attempts have been made to provide saline-tasting compositions as a substitute for table salt in trying to overcome the general lack of palatability of the salt-free diets. As examples of such substitutes there may be mentioned ammonium chloride, potassium chloride, and mixtures thereof and/or with other substances. The use of these chlorides, and particularly combinations of them leaves much to be desired as to taste. Neither of them tastes like sodium chloride. Each alone has a disagreeable taste, and so also do both taken together, with the resulting tendency to discourage continued voluntary use.

Potassium chloride per se leaves an aftertaste which is usually characterized as bitter by most people. Ammonium chloride also has a bitter aftertaste, and in addition when ingested disturbs the acid-base balance in the blood, produces acidosis and stimulates the bronchial mucosa causing excess mucous secretion and coughing. It must be eliminated by the body as urea thus putting a strain on the liver and the kidneys. At cooking and baking temperatures, ammonium chloride decomposes readily thereby changing the pH value, taste and other properties of the food containing it.

To mask or overcome the bitter taste of potassium chloride a variety of diluents, additives or adjuncts have been proposed. Among the many additives which have been employed are potassium and calcium formate, magnesium citrate, dipotassium succinate and a variety of citrates, tartrates, gluconates, ascorbates, cyclamates, glutamates and ion exchange resins. For a salt substitute to be commercially acceptable, it must be free-flowing even after long periods of storage. Another is that it must be noncorrosive.

Noncorrosiveness is needed because of the convenience package, a one-portion container which has become widely used in hospitals, institutions and restaurants throughout the country. These packages usually have an aluminum foil lining and for this reason cannot be used with corrosive substances. The various additives that have been suggested and used as adjuncts to potassium chloride have not been a commercial success mainly because of their characteristic deliquescence and corrosiveness. Two salt substitutes, one containing glutamic acid and potassium chloride and the other containing citric acid and potassium chloride were field-tested over a period of many months and both were found to be unsatisfactory. The deliquescence of the citric acid potassium chloride combination caused caking, and the glutamic acid potassium chloride combination was so corrosive that it actually ate holes in the aluminum lining of its packages.

A sodium-free dietary substitute for table salt which is free-flowing was developed by *H.C. Miller; U.S. Patent 3,505,082; April 7, 1970; assigned to Morton International, Inc.* The substitute contains on a relative proportion basis from about 95 to 99% by weight of potassium chloride and from about 1 to 5% by weight of fumaric acid. This corresponds to from about 0.01 to 0.05 part of fumaric acid per part of potassium chloride. The composition also includes a minor proportion of an anticaking agent selected from the group consisting of sodium silicoaluminate, colloidal silica, and tricalcium phosphate.

In an optimum form the fumaric acid is present in amounts of from 0.01 to 0.02 part per part of potassium chloride and anticaking agent from 0.005 to 0.02 part per part of potassium chloride. This corresponds to from about 96 to 98.5% potassium chloride, 1 to 2% fumaric acid and 0.5 to 2% anticaking agent. The particle size of the potassium chloride and fumaric acid should fall within the size range of 150 to 600 microns (U.S. mesh 30 to 100).

Fumaric acid used in the specified concentrations overcomes the bitter taste without the distinct acid taste appearing. Since fumaric acid has a relatively low solubility, large amounts can be used without producing the acid taste. The following tests and specific examples are given to illustrate the properties of this salt substitute.

Salt Substitute Preference Test: The purpose of this test was to determine, in a blind test, the taste preference among leading salt substitutes. In the test the process salt substitute was compared with two commercial salt substitutes designated items A and B. Items A and B contained 96% potassium chloride, 2% glutamic acid and 2% anticaking agents and other additives such as potassium iodide, tricalcium phosphate and potassium glutamate. The process composition contained 96% potassium chloride, 2% fumaric acid and 2% tricalcium phosphate (anticaking agent).

The taste comparison used cantaloupe cubes. Panelists were instructed to take a sip of cold tea, salt a cantaloupe cube and eat it, and then repeat the procedure. After tasting both salt substitutes, the panelist wrote the code letter of the preferred salt substitute on the form provided. Then each panelist proceeded to the second table and repeated the test, using the remaining two salt substitute samples. The salt substitute compares favorably with leading commercial brands.

Deliquescence Test: The following ingredients were mixed thoroughly until a homogeneous composition was obtained: potassium chloride, 96% by weight; fumaric acid, 2% by weight; and tricalcium phosphate, 2% by weight. In a laboratory test this composition was maintained over a saturated NaCl brine at room temperature and at 120°F and no deliquescence was observed.

Field Test: Under field test conditions a 2% tricalcium phosphate, 2% fumaric acid and 96% potassium chloride salt substitute composition was packaged in a small blister packet made of paper, plastic, and aluminum foil. Again there was no evidence of deliquescence or corrosion.

Free-Flow Test: Polyethylene plastic bags were filled with midget containers of regular table salt and midget containers of a 2% fumaric acid, 2% tricalcium phosphate and 96% potassium chloride salt substitute. These were exposed under two different temperature and humidity conditions. The salt substitute is relatively as free-flowing and nonhygroscopic as ordinary table salt.

Corrosion Test: A sealed jar of midget packages containing 2% fumaric acid, 2% tricalcium phosphate and 96% potassium chloride salt substitute was stored at 100°F for 4 weeks. A microscopic examination of material in midget packages showed no acid or chemical attack.

Nonsodium Gluconate Additive

Many attempts have been made to provide saline-tasting compositions as a substitute for table salt in trying to overcome the general lack of palatability of the salt-free diets. Perhaps more of these substitutes use ammonium chloride as the primary ingredient than any other substance, with those using primarily potassium chloride possibly coming next, and then followed by those using primarily both of these substances. Each of these two chlorides, and particularly their joint use, leaves much to be desired as to taste. Neither of them tastes like sodium chloride. Each alone gives a disagreeable taste, and so also do both taken together, with resulting tendency to discourage continued voluntary use. Potassium chloride usually leaves a somewhat bitter aftertaste.

Ammonium chloride has a bitter taste, and disturbs the acid-base balance in the blood, and produces acidosis. It stimulates the bronchial mucosa, causes mucous secretion and promotes coughing. Ammonium chloride decomposes with liberation of ammonia at cooking and baking temperatures, with accompanying change in taste and with the released hydrogen chloride changing the acidity and other properties of the foods. Other ingredients considered from time to time have not overcome the bitter aftertaste of potassium chloride, nor produced a salt substitute with a suitable saline taste, and have manifested unpleasant side effects and other physiological drawbacks as to restrict their usability in cooking as well as on foods at the time they are going to be eaten.

The salt substitute of *J.C. Munch; U.S. Patent 2,968,566; January 17, 1961; assigned to Pharmacodynamics, Inc.* has potassium chloride as its highest content constituent and a nonsodium gluconate as the next higher, together with smaller amounts of potassium citrate and of ascorbic acid and/or calcium ascorbate, and a very small amount of a synthetic sweetening agent such as calcium cyclamate and/or saccharin. This salt substitute may also include glutamic acid and/or calcium glutamate usually to less than about the minimum for the total gluconate content.

The sodium-free table salt substitutes are prepared by commonly used methods of granulation merely from mixtures of the several ingredients, preferably initially conveniently smaller than 40 mesh. Such starting mixtures are wetted, granulated, and their granulations dried as usual and screened to give a product about the usual size of common table salt, for example, by taking the fraction passing through the 30-mesh screen and retained on the 40-mesh screen. The product is nonhygroscopic, free-running and elegantly similar to sodium chloride in taste and flavor.

Example 1:

	Parts by Weight
Potassium chloride	70.73
Potassium gluconate	7.5
Calcium gluconate	7.5
Potassium citrate	6.5
Calcium ascorbate	7.5
Thiamin hydrochloride	0.25
Calcium cyclamate	0.02

Example 2:

Potassium chloride	70.8
Potassium gluconate	7.5
Calcium gluconate	7.5
Potassium citrate	6.5
Calcium ascorbate	7.5
Saccharin	0.2

Example 3:

Potassium chloride	70
Potassium gluconate	7
Calcium gluconate	5.98
Potassium citrate	6
Calcium glutamate	5

(continued)

	Parts by Weight
Calcium ascorbate	5.75
Thiamin nitrate	0.25
Calcium cyclamate	0.02

In any of the Examples 1, 2 and 3, the calcium gluconate can be replaced in part or as a whole by the equivalent amount of gluconic acid, and the calcium ascorbate similarly can be replaced by its equivalent amount of ascorbic acid. The latter replacement is preferred in Example 3 where the resulting reduction in calcium content is compensated by the quantity of it added by the calcium glutamate. The saccharin in Example 2 can be replaced in part or as a whole by its equivalent in sweetening capacity of saccharin soluble and/or sodium, potassium, magnesium or calcium cyclamate. Similarly, the latter in Examples 1 and 3 can be replaced by its equivalent of either saccharin in sweetening capacity, or even by the corresponding equivalent of magnesium or sodium or potassium cyclamate.

The very small amount of sodium that the user would consume if sodium saccharin soluble were used as the sweetening agent would not be prohibitive. The potassium cyclamate is prepared by replacing the sodium salts used for making the sodium cyclamate by the corresponding potassium salt. Also, the magnesium cyclamate is prepared by replacing the calcium salts used for making calcium cyclamate by the corresponding magnesium salt.

The gluconates and other ingredients of these salt substitute compositions contribute to mask the bitter aftertaste that ordinarily occurs with potassium chloride, and enhance the effectiveness as well as the taste. The calcium gluconate, glutamate and ascorbate overcome the undesirable effects of calcium deficiency so often experienced by persons on a salt-free diet, and provide a better physiological balance between potassium and calcium. The calcium cyclamate and the saccharin, in spite of their sweetness, enhance the saline taste of these compositions.

POTASSIUM-SODIUM CHLORIDE COMBINATIONS

R.L. Frank and O. Mickelsen; U.S. Patent 3,514,296; May 26, 1970; assigned to Morton International, Inc. discuss the need for low-sodium diets and the use of potassium chloride as a table salt substitute. There is a body of literature which suggests that diets high in sodium actually cause the condition known as hypertension. The dietary intake of potassium ion is important from a nutritional and physiological point of view. Indeed, the ratio of sodium to potassium in the diet and in the body fluids appears to be at least as important as the intake of sodium alone. For example, a high intake of sodium chloride, which may be harmful in a disease such as hypertension, appears to be not harmful at all if a potassium salt is also included in the diet.

Some in the medical field have found that there is a physiological relation between sodium and potassium, the one ion, in many bodily reactions, counteracting the other. Potassium deficiencies in the diet are commonplace and are caused in part by the refining of foods, in part by overcooking and in part by the exclusion from the diet of leafy green vegetables, a particularly rich source of potassium. Some clinical conditions which have their genesis in potassium deficiencies are overall muscle weakness, hypotension, weak extremities, poor intestinal tone with intestinal distension, cardiac weakness, weakness of respiratory muscles, even to the point of ultimate failure.

To combat these conditions, a variety of enteric-coated tablets of potassium chloride and various potassium supplements have been made available by the pharmaceutical industry. Restriction of sodium intake and increase of potassium intake have therefore been prescribed as therapeutic measures for the treatment of the above-described clinical conditions. Where restriction of sodium in the diet is the primary objective, a dietary regimen in which sodium chloride is entirely excluded is usually prescribed. This approach is generally not very satisfactory over a prolonged period of time because of the lack of acceptability by the patient of a bland, unpalatable diet. This has led to the development of a variety of salt substitutes which contain no sodium chloride and still have varying degrees of salinity to the taste buds. Among the various salt substitutes which have been used for this purpose, most contain potassium chloride and ammonium chloride. Potassium chloride, in the form of enteric-coated pills, has been prescribed for many patients to prevent the disturbing symptoms associated with a deficiency of potassium ion. This concentrated form of potassium chloride has recently been shown to produce intestinal ulceration. All studies indicate that this condition does not occur when potassium chloride is consumed in smaller doses with meals by sprinkling on food.

To mask or overcome the bitter taste of potassium chloride, an assortment of diluents, additives or adjuncts have been proposed. Among the many additives which have been employed are potassium and calcium formate, magnesium citrate, dipotassium succinate, and a variety of citrates, tartrates, gluconates, ascorbates, cyclamates, glutamates and ion exchange resins. One drawback inherent in the use of these additives is that the true salty or zesty flavor associated with salt is lost or impaired. Another drawback is that many of these additives stratify or segregate due to the dissimilarity in crystal structure and density between the additive and the potassium or ammonium chloride.

Critical and synergistic combinations of potassium and sodium chloride can be used to minimize the bitter taste of potassium chloride and at the same time provide nutritionally balanced amounts of potassium and sodium.

The compositions of this product are prepared by thoroughly and efficiently mixing the components in the indicated proportions until a homogeneous product is attained. A homogeneous nonsegregating composition is provided since sodium chloride and potassium chloride are similar in crystal structure, both belonging to the cubic system; they have similar specific gravities, so that mixtures do not lose their homogeneity; the refractive indices of the crystals are similar, so that individual crystals of sodium and potassium chlorides look alike; and the water solubilities of sodium chloride and potassium chloride are not greatly different.

Combinations of potassium chloride and sodium chloride containing from about 20 to about 80 weight percent of potassium chloride, and from about 80 to about 20 weight percent of sodium chloride, actually represent from about 10.5 to 42.0 weight percent of potassium ions and correspondingly from about 31.4 to 7.8% of sodium ion. To arrive at a quantitative determination of that proportion of the population which detects a decreased bitterness and enhanced saltiness phenomenon exhibited by the above combinations of potassium chloride and sodium chloride, statistically sound taste panel evaluations were carried out:

(1) on sodium chloride alone to establish the minimum concentration of sodium chloride which can be detected as salty by a majority of the population;

(2) on potassium chloride alone to establish the minimum concentration of potassium chloride which can be detected as bitter by a majority of the population;

(3) on mixtures of sodium chloride and potassium chloride containing a relative percent of potassium chloride and sodium chloride ranging from 20 to 80 weight percent of potassium chloride and from 80 to 20 weight percent of sodium chloride to establish the proportion of the population classifying such mixtures as bitter and as having enhanced saltiness when compared to sodium chloride alone.

Example 1: Procedure for Determining Saltiness Taste Threshold for Sodium Chloride – Aqueous solutions of pure sodium chloride were prepared containing 0.25% NaCl, 0.05% NaCl, 0.10% NaCl and 0.20% NaCl, respectively. A 5 ml portion of each of these solutions was given repetitively to each member of a six-member taste-test panel for his or her taste evaluation. The solutions were coded and not otherwise identified to the panelists. Each panel member rinsed his mouth with water thoroughly after each taste test. The saltiness of each solution was noted and compared with the preceding solution and rated in increasing order of saltiness. The results are set forth in the table below.

Concentration of Sodium Chloride (percent)	Percent Panelists Rating Solution Saltier than Water
0.025	54
0.05	50
0.10	92
0.20	92

These results demonstrate that the taste threshold for sodium chloride for the vast majority of the population, i.e., more than 90%, is at a sodium chloride concentration of 0.10%. This concentration of sodium chloride was therefore the minimum selected for combination with potassium chloride in subsequent taste evaluations.

Example 2: Procedure for Determining Taste Threshold for Potassium Chloride – As stated, potassium chloride is acknowledged to have a bitter taste to a majority of the population. To evaluate this on a quantitative basis, solutions of pure potassium chloride were prepared containing 0.025, 0.064, 0.075, 0.15, and 0.256% potassium chloride. The taste-testing procedure followed was the same as that set forth for sodium chloride. The results obtained are set forth in the table below.

Concentration of Potassium Chloride (percent)	Percent Panelists Rating Solution Bitter
0.025	25
0.064	67
0.075	68
0.15	74
0.256	83

It is thus evident that a concentration of KCl in excess of about 0.025% is required before a significant proportion of the population can detect its presence. It is further evident that a substantial majority of the population classifies as bitter the taste of potassium chloride when it is present in excess of about 0.025%.

Example 3: Determination of Bitterness of Combinations of Potassium Chloride and Sodium Chloride — In accordance with the lower taste threshold determinations for both sodium chloride and potassium chloride, respectively, solutions were prepared containing mixtures of 0.10% sodium chloride with 0.025% KCl, 0.06% KCl, 0.15% KCl, and 0.40% KCl, respectively. Each of these solutions was evaluated for palatability with respect to bitterness by the taste-testing procedure described earlier for pure sodium chloride. There is observed a marked decrease in bitterness of the mixture of potassium chloride and sodium chloride over potassium chloride per se.

The mixture affords an enhanced or synergistic decrease in bitterness over that which would be expected from the individual bitterness contributions of the separate components. Thus, for example, 0% of the population classifies 0.10% NaCl alone as bitter and 67% of the population classifies 0.064% KCl alone as bitter. Yet when a solution is prepared which contains both 0.10% NaCl and 0.064% of KCl in combination, only 11% of the population classifies this combination as bitter.

In Figure 7.1a, the broken line graph represents a plot of the percent of the population detecting a bitter taste versus the concentration of potassium chloride; the solid line represents a plot of the percent of the population detecting a bitter taste versus the indicated concentrations of potassium chloride in mixture with 0.10% sodium chloride, respectively. As is evident from the graph, the combinations of potassium chloride and sodium chloride provide compositions which taste bitter to a significantly smaller segment of the population than potassium chloride alone in comparable proportions.

FIGURE 7.1: TASTE COMPARISON OF SALT SUBSTITUTES

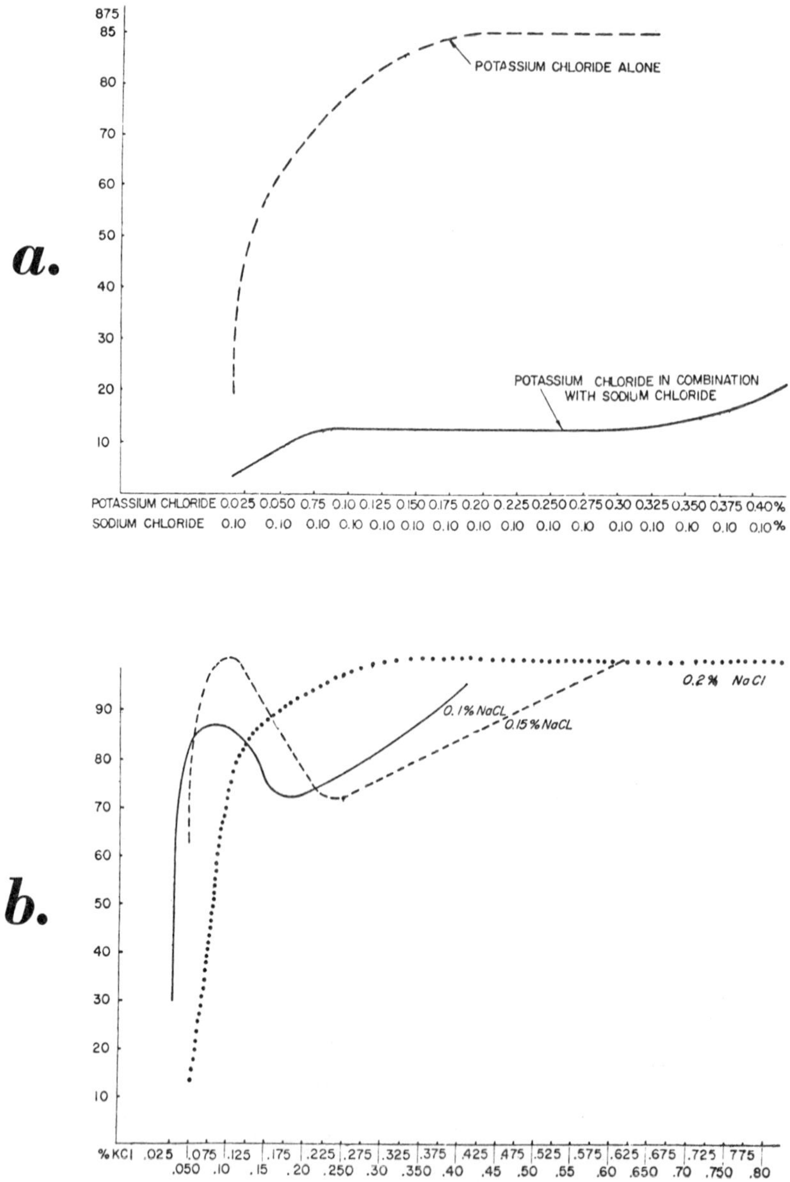

Source: R.L. Frank and O. Mickelsen; U.S. Patent 3,514,296; May 26, 1970

Nutritious Beverage

protein fortified

(above pH5)

casein, soy protein are insoluble at the isoelectric point of the protein (PH 4-5)

As previously stated, another unexpected attribute of these compositions is that of enhanced saltiness. The solid line graph in Figure 7.1b represents a plot of the percent of the population detecting enhanced saltiness versus the indicated concentration of potassium chloride in combination with 0.10% sodium chloride. The dashed line graph represents a plot of the percent of the population detecting enhanced saltiness versus the indicated concentration of potassium chloride, in combination with 0.15% sodium chloride. The dotted line graph represents a plot of the percent of the population detecting enhanced saltiness versus the indicated concentrations of potassium chloride in combination with 0.20% sodium chloride.

POTASSIUM-ENRICHED SALT

Pure sodium chloride, as a salt for food, is deficient in other mineral nutrients, and also has the tendency to agglomerate, especially when exposed to humid weather. It is known that the use of tricalcium phosphate, referred to as TCP or $Ca_3(PO_4)_2 \cdot Ca(OH)_2$, in salt as a conditioning agent does render it free-flowing even when exposed to humid weather conditions, and although it does give it the added nutritive elements of calcium, phosphorus, hydrogen and oxygen, still such a salt is not nutritively balanced. In addition to all these elements in the conditioning agent, there is a further requirement for potassium in order to achieve nutrition balance.

H.N. Norsen; U.S. Patent 3,306,753; February 28, 1967 provides a proper combination, by selected proportions, for producing potassium-enriched conditioning agent which, when added to salt, makes it much more tasty and beneficial. While the addition of a 1% quantity of TCP alone to sodium chloride makes the salt free-flowing so that it will not agglomerate, such a treated salt soon forms a cloudy film or coating of TCP on the inside of a glass salt shaker.

On the other hand, if pure sodium chloride did not have the tendency to agglomerate, it would keep the inside of a salt shaker crystal clear. It should be noted, however, that when the potassium-enriched conditioning agent is combined with salt, it coats the inside of a glass salt shaker with such a thin transparent film of potassium-enriched conditioning agent as to be hardly detectable, and yet the salt so treated is free-flowing and will not agglomerate.

The process is applied to the use of potassium chloride and monopotassium phosphate, with or without sodium chloride added in proper proportions to the correct amount of TCP. The mixture is dry blended, and then, to insure homogeneousness of composition, is thoroughly wet with water to form a thin smooth paste. It is uniformly stirred and dried by suitable means, such as using trays for batch or a rotary kiln dryer for continuous operations, thus evaporating all the water that had been added. The dried material is then ground to the fineness of the original TCP to yield the potassium-enriched, nutritively balanced, coditioning agent, which, when added to the salt to the extent of even less than 1%, makes the salt free-flowing.

Also, unlike other foods, with their inorganic-organic chemical combinations that act as excellent buffering agents, sodium chloride, as a salt, stands alone and apart without this buffering ability, and thus can be eaten only in small quantities without immediately encountering distress. The potassium-enriched conditioning agent for salt, is especially formulated; the combination of potassium chloride and monopotassium phosphate in a molar ratio of 1 to 6 is contained in a mixture with tricalcium phosphate, such that the mixture exhibits a calcium-to-phosphorous ratio of 1:15.

When a combination of these three chemicals is added and thoroughly dry-blended or mixed in such a way as to become a part of sodium chloride to the extent of 0.70%, the resulting conditioned salt has the optimum amount of inorganic buffering ability to make it nutritively balanced. The inclusion of the potassium-enriched conditioning agent for salt in the amount of 1.00% or over has been found to give a mixture richer than necessary for daily usage, and therefore, it is not recommended. Likewise, using quantities of the potassium-enriched conditioning agent for salt in the amounts of less than 0.40% are not rich enough to effect the desirable characteristics. The optimum range of potassium chloride to monopotassium phosphate is expressed by a molar ratio greater than 1 to 5, but less than 1 to 7. Also, the preferred calcium-to-phosphorus ratio is within the range of 1.05 to 1.27.

DIETETIC SALT FORMULATION

J.F. Lontz and C.A. D'Alonzo; U.S. Patent 3,782,974; January 1, 1974 discuss the problem of formulating dietetic salt substitutes and propose new formulations. Many of the past formulations for use as dietetic salt substitutes are based on the replacement of the sodium cation with potassium, ammonium, lithium, and calcium on the one hand, and by anions such as the phosphates and glutamates, all of which introduce other excesses into the diet that may lead to other health dangers.

Some of these formulations hardly fulfill the broad range of salt-like sensory reaction, because the biophysical mechanisms of taste as a neutral response is quite complex, involving not only the initial sensation of the taste, but also its intensity, persistency, and aftertaste that cannot be as yet accurately defined or measured in some acceptable psychophysical scale. The substitution of the sodium cation by the potassium or ammonium is frequently cautioned against by the physician, as these cations can also lead to excesses that have other undesirable effects.

In the case of potassium, prolonged utilization will lead to complications in the gastrointestinal tract, in the heart, or in musculature. In the case of ammonium, the activity of the cation is somewhat so specific to the mucous areas of the nose and throat that the user abandons the use of the substitute entirely, and often returns to sodium chloride usage.

It has been found that the salt-like taste can be made acceptable to a tasting group of individuals by the use of a multicomponent mixture of the anions of organic acids, provided that the mixture of selected anions are formulated in groups of well-defined chemical structures. This discovery came about by the realization that the replacement to some degree of the sodium cation by the hydrogen cation, such as does exist with organic acids, can be augmented by specific anions of organic acids. Furthermore, this replacement requires a highly restricted proportion, as an excess of any one throws the salt-like taste into another taste that can be sour, bland, or excessively tart.

An improved dietetic salt formulation has been attained by the discovery of a three-component mixture of organic acids preferred by a panel of tasters applying the mixtures of organic acids containing: (A) monobasic or monodissociative organic acid; those eliminated from the series include acetic, propionic, butyric, aminoacetic, aminopropionic (alanine), glyoxalic, gluconic, levulinic, and ascorbic acid; (B) dibasic or dissociative organic acid selected from the elimination of malonic, succinic, glutaric, adipic, malic, tartaric, maleic, fumaric acids; and (C) tribasic or tridissociative organic acid, notably citric acid.

Following a preliminary rating for taste response described as sour, acidulant, tart, and bland, a taste panel of from three to six individuals narrowed the three-component formulation from a starter 1:1:1 mixture of (A), (B), and (C), which proved excessively acidulant with a prolonged aftertaste as being by a majority as undesirable and even unpalatable by some, to a mixture of (A) ascorbic acid, (B) fumaric, and (C) citric acid. Following universal recognition and aversion to excess of the last-named citric acid, the taste panel, from statistical assessment of nearest similarity or approximation to common table salt, selected the following tabulated range of the three-component admixture of the crystalline salt substitute.

Component	Acid	Proportion range
(A)... Monocarboxylic	Ascorbic	30 to 70.
(B)... Dicarboxylic	Fumaric	70 to 30.
(C)... Tricarboxylic	Citric	1 to 2.

This synthetic formulation was found to be unique in several respects. Firstly, no known proportions of the above type have been found among the published analyses of natural plant foods, extracts, or prepared components thereof. Secondly, the ratio of the ascorbic acid to the fumaric acid was readily detected as the proportion of the fumaric acid exceeded the 70 to 30 proportionated range indicated in the above table. For instance, a proportion of ascorbic acid:fumaric acid of 20:80 was rejected by several panels as being excessively tart or acidulant. On the other hand, proportions of the ascorbic acid:fumaric in excess of 80:20 was immediately rejected as lacking any of salt-sour flavor, but rather one that was bland and even tended to be sweet. Further preferences were conducted as illustrated in the following examples.

Example 1: The following series of crystalline or crystalline-like mixtures were made available to a tasting panel to rate the salt-like flavor sprinkled on single sections of celery stalks with results indicated in the following tabulation. The rating by the panel was 1 to 5, with 5 being the best and 1 the lowest.

Seasoning	Panel Rating
(a) Common salt	5
(b) Proprietary monosodium glutamate	1
(c) Formulation D-1: 70 parts ascorbic acid, 30 parts fumaric acid, 1 part citric acid	3
(d) Proprietary formulation with potassium chloride, ammonium chloride, potassium glutamate, and other ingredients	2
(e) Formulation D-4: 44.5 parts ascorbic acid, 44.5 parts fumaric acid, 1.0 part citric acid	4

While none of the above formulations approached common table salt universally among the panelists, although some were not able to distinguish between (a) and (c), the formulations (c) and (e) above were rated as acceptable salt substitutes that would not tire, so to speak, the user from continued dependence. Formulation (d) was advised against by the referring physician as having an excessive level of potassium along with an organic activator of the choline type not recommended for general usage, but only upon advice of the physician. The formulation based on sodium monoglutamate was least desirable because of a peculiar bland-to-sour or meaty aftertaste with the celery.

This example indicates that a universal dietetic salt substitute is available, provided the critical range of the three-component ingredient mixture is applied, a feature that would be substantiated by the fact that all of the organic acid ingredients are of the chemical structure found in many edible plants, vegetables, and fruits, but not replicated so restrictively as specified in this process.

Example 2: The same series of crystalline-like mixtures D-1 and D-4 were rated by the panel for a salad consisting of a quarter slice of lettuce and two tablespoonfuls of a 1:1 mixture of olive oil and vinegar, allowing the panel members to sprinkle each of the salt substitutes as generously as they were accustomed to. The preference ratings are based on 5-4-3-2-1 accumulated for each formulation and summed as shown below.

Common salt	25 (5-5-5-5-5)
MSG	7 (1-2-3-2-3)
D-1	18 (4-3-4-3-4)
D-4	16 (3-4-3-4-3)
Proprietary (d) above	7 (1-2-1-2-1)

While the common salt prevailed as the prime reference standard, the formulations D-1 and D-4 acquired reasonably close second or approximation. Using the same salad preparation, the taste panel rated extensions of the D-1 and D-4 formulations in excess of range preferred with the following additional admixtures.

Seasoning	Ingredients: Ascorbic acid	Fumaric acid	Citric acid	Panel rating
Common salt				25
D-1	70	30	1	18
D-4	44.5	44.5	1.0	17
D-14	80	20	1	[1] 7
D-19	20	80	1	[2] 8

[1] Panel disclaimed salt-like flavor as excessively bland.
[2] Panel singled out sour-like flavor.

The above results indicate that a critical ratio or proportions prevail in this three-way formulation.

Example 3: A series of unsalted butter patties were seasoned with 40 mg of various formulations including common table salt for taste control or reference for the panel to rate the salt-like quality or saltiness applied to unsalted milk crackers with results indicated in the following tabulation.

Seasoning in Butter	**Panel Rating**
Common salt	25 (5-5-5-5-5)
Proprietary (d) in Example 1	15 (3-3-3-3-3)
D-1	16 (4-4-2-2-4)
D-4	14 (2-2-4-4-2)
D-19	6 (1-1-1-1-2)

The taste rating results demonstrate the substantial equivalency of the salt substitute of this process to the proprietary (d) product which is based on potassium salts and an active nerve ingredient which would be objected to by many physicians and would therefore not be suitable for the general purchase as an over-the-counter or food-store item.

Example 4: As in Example 1, the formulations were rated by another panel of five members by sprinkling over hamburg patties grilled over charcoal with preference rating summarized as follows, using packets of 40 mg for each patty.

Seasoning	**Panel Rating**
Common salt	25 (5-5-5-5-5)
D-1	12 (several ties)
D-4	16 (several ties)
Proprietary (d)	11 (several ties)
Monosodium glutamate	11 (several ties)

Formulation D-4 provided a close approximation of the common salt, with prospects that a little higher amount of sprinkling could provide a higher rating. The above examples thus demonstrate a consistent rating of the salt-like quality that is not only dependent upon the choice and ratio or proportions of the selected three, chemically defined ingredients, but also can be accepted from the standpoint of acceptable taste replicating in part that of the common salt.

Each of the ingredients provided in the formulations are based on chemical structures that exist in nature and to which man has been exposed as dietary substances. The formulations provide a ready need for salt substitutes to alleviate the well-known and widely accepted conditions of hypertension and other health dangers related to excess intake of salt, a feature that people can hardly ascertain until after the condition of danger appears.

The manner in which the unique three-component mixture provides the quality of saltiness is not clear. Numerous studies and theories point to the role of the cation as an excitatory factor on taste receptors and that the anion effect is of an inhibitory character. The factors involved direct much emphasis to the binding of the anions onto the cell membranes which activate a response which rates in turn the qualities of the principal senses of salty to the sweet range. It is therefore not evident from the speculated concepts that the specific mixture of this process could be predicted.

FLUORIDATED SODIUM CHLORIDE

R.R. Irani; U.S. Patent 3,406,077; October 15, 1968; assigned to Monsanto Company discusses fluoridation methods and in particular the production and use of a fluoridated salt. Prevention of tooth decay by the fluoridation of water is well-known. In many countries and communities a central water supply does not exist, therefore, prevention of tooth decay by water fluoridation is not feasible in those areas. Similarly fluoridated toothpaste has a significant beneficial effect upon the prevention of tooth decay. However, in certain areas of the world due to factors, such as the habits, economics, availability of toothpaste, etc., proper dental hygiene through the use of toothpaste is not practical.

It is also known that individuals have a relatively constant intake of certain food substrate, such as salt. Salt as used herein is sodium chloride or ordinary table salt unless otherwise specified. It has also been shown that table salt with a fluorine content of from about 0.01 to 0.03% by weight expressed as sodium fluoride has a significant beneficial effect upon tooth decay prevention.

Several problems exist in the introduction of these minute amounts of fluorine into the salt. For example, when the fluorine-containing compound is mixed directly with the salt, the fluorine-containing compound and the salt often segregate during handling and storage. Within a given large salt container, there can be portions free of fluoride ions (thus giving no beneficial fluorine effect when consumed) and other portions possibly can contain toxic amounts of fluorine. Under high humidity conditions, which are prevalent in many of the underdeveloped areas where dental care by fluoridated salt is most needed, the fluorine-containing substances tend to leach to the bottom of the salt containers.

Treating salt by spraying a water solution containing a fluorine-bearing substance on a heated salt bed which is sufficiently hot to evaporate the water from the solution gives a uniformly fluoridated salt which is not subject to segregation or leaching. Other methods of providing fluoridated salt, such as drying a water-soluble fluoride solution upon another additive, such as a flow conditioner, then mixing this additive containing the fluoride with the salt, solves the segregation and leaching problems. However, these processes are restricted to a water-soluble fluorine-bearing substance.

It has been found that fluorine-bearing materials in an extremely finely divided state can be coated with conditioning compounds which are also relatively finely divided to produce a fluoridated flow conditioner. The fluoridated flow conditioners contain intimate mixtures of the fluorine-containing substance and conventional flow conditioner, and do not separate into their individual components during handling and storage. Such fluoridated flow conditioners are useful when used to impart improved flow characteristics to edible compositions, such as salt, to produce an edible composition effective in the prevention of tooth decay.

Both water-soluble and water-insoluble fluorine-bearing compounds or mixtures can be used. Examples of the water-soluble fluorine-containing compounds which can be used include the simple fluoride salts, such as the alkali metal fluorides, such as sodium, potassium and lithium fluorides; stannous fluoride, cupric fluoride, ammonium fluoride, iron fluoride; complex salts containing fluorine, such as ammonium fluorosilicate, lithium fluorosilicate, sodium fluorosilicate, and potassium fluorosilicate; as well as mixed halide salts such as stannous chlorofluoride, cupric chlorofluoride and iron chlorofluoride.

In addition, the water-insoluble fluorine-containing compounds such as calcium fluoride, fluorapatite [$Ca_5F(PO_4)_3$], magnesium fluoride, barium fluoride and mixtures can be used. Mixtures of a water-soluble fluoride salt, such as mentioned above and fluorapatite can be used. Extremely finely divided flow conditioners and fluorine-containing compounds are combined to make a fluoridated flow conditioner. The term "flow conditioner" refers to a substance, the particles of which are very finely divided, and when used to coat particles of other materials which tend to agglomerate, reduce the agglomeration tendency of the other material thus improving the flow characteristics and reducing the natural caking tendency of the material coated.

For example, salt under high humidity conditions tends to agglomerate or cake, but when the salt is coated with a flow-conditioning agent such as tricalcium phosphate, the agglomeration is greatly reduced. Typical examples of suitable flow conditioners include the water-insoluble phosphates such as tricalcium orthophosphate, calcium pyrophosphate, calcium tripolyphosphate, calcium and magnesium disodium pyrophosphates as well as the aluminum silicates, such as sodium calcium aluminum silicate, potassium calcium aluminum silicate, and sodium magnesium aluminum silicate; calcium silicate hydrate; magnesium carbonate, calcium carbonate; silica aerogels; calcium sulfate and common clays.

It is necessary to prepare extremely minute particles of fluorine-containing compounds in order for the fluorine-bearing compound to be evenly dispersed throughout the conditioning agent and for the particles of the fluorine-containing compound to be retained in a dispersed fashion throughout the resulting fluoridated flow-conditioning agent. Although it is preferred that the average particle size of the fluorine-containing compounds be between about 5 and 50 microns, while particles with an average particle size smaller than 10 microns and essentially 100% by weight smaller than 40 microns (as measured by micromesh screen) are especially preferred, larger particle sizes may be used.

For example, fluorine-containing compounds with an average particle size of 30 microns with at most about 20 weight percent of their particles being between about 40 and 50 microns can be used. Particles with an average particle size larger than 100 microns will not become uniformly coated and mixed with the conditioning agent and will tend to segregate by the agitation received during normal handling.

The conditioner must also be relatively finely divided to give the proper coating to the particles of the material to be conditioned. Too large particles of conditioner will result in a fluoridated flow conditioner which will not uniformly coat the salt or other material and not effectively reduce the agglomeration tendency of the salt or other material to be conditioned. The conditioner tends to coat the fluorine-containing substance, therefore, the particles of the fluoridated flow conditioner are slightly larger than the particle size of the initial starting material.

It is preferred to have flow conditioners which have an average particle size between about 1 and 35 microns, while those materials having particles of about 10-micron size and essentially 100% by weight smaller than 40 microns as measured by micromesh screen are still further preferred. Suitable results, however, can be achieved with flow conditioners of a particle size smaller than a U.S. Standard 325-mesh screen or even when at most about 10 weight percent of their particles are as large as about 50 microns. Any conventional means of crushing, grinding or pulverizing can be used to produce the proper particle size of the fluorine-containing compound.

After the fluorine-bearing compound is in the physical form of relatively small particle size, that is about 100% passing through a U.S. Standard 170-mesh screen and with an average particle size smaller than about 50 microns, it is intimately mixed with the relatively finely divided conditioner particles to form the fluoridated flow conditioner. Any conventional means of mixing may be used. For example, a double helical mixer consisting of two helical ribbons, one a right-hand screw and the other a left-hand screw, mounted inside a housing of suitable volume will give excellent results. The time required to mix the components to give a composition of uniform fluorine content will be dependent upon several factors, such as the type of mixer used, the speed of rotation of rotating parts, the amount of conditioning agent to be fluoridated, the physical characteristics of the fluorine-bearing component and the physical characteristics of the flow-conditioner agent. In any event the components are to be intimately mixed so as to give a coating of flow conditioner on the fluorine-containing compound and the fluorine-containing compound is to be uniformly dispersed throughout the flow conditioner.

The amount of fluorine-containing compound used to fluoridate the flow conditioner will be dependent upon the amount of fluoridated flow conditioner that will be added to the material that is to be conditioned. The generally preferred range of fluorine-containing compound to the flow-conditioning agent is from about 0.1 weight percent to about 20 weight percent of the fluorine-containing compound in the fluoridated flow conditioner. The fluoridated flow conditioner, that is the flow conditioner containing the specified concentration of fluorine, can be used in various edible compositions. Any edible solid of relatively small particle size which can be uniformly coated such as salt, spices, sugar and the like, can be used.

The amount appropriate and type of flow conditioner for the particular end use desired and in turn the amount of fluorine in the flow conditioner can be varied to give the proper fluorine concentration and proper flow-conditioning benefits. Since medical science has studied the human intake of salt comprehensively and the annual intake has been firmly established, salt is the preferred edible composition to be used. Any commercial table salt may be used. The particle size preferred is that about 100 weight percent will pass through a U.S. Standard 16-mesh screen and less than about 10 weight percent through a U.S. Standard 70-mesh screen. Salt having larger or smaller particle sizes can be used if desired. Any conventional means of intimately mixing the salt prior to use with the flow conditioner to give a uniform coating of the flow conditioner on the salt particles may be used.

Example 1: Preparation of Fluorine-Containing Compound – One hundred pounds of fluorapatite with an initial particle size of about 100% smaller than a U.S. Standard 60-mesh screen and about 100% larger than a U.S. Standard 325-mesh screen are charged into a conventional ball mill. After 1 hr the particle size of the fluorapatite is reduced to about 100% smaller than 40 microns and with an average particle size of about 10 microns as measured by micromesh screens.

Example 2: Preparation of Fluoridated Flow Conditioner – Five hundred fifty pounds of tricalcium phosphate with about 100% smaller than 40 microns and with an average particle size of about 10 microns and 50 lb of fluorapatite $Ca_5F(PO_4)_3$, as prepared in Example 1 are charged into a double helical mixer. The helical screws are rotated at 150 rpm. The components are blended for about 1 hr. Nine samples are taken at three different horizontal points in the mixer bed and at three vertical levels in the mixer bed.

All samples show a uniform concentration of about 0.3% fluorine in the mixer mass. The standard sodium nitrate titration is used to determine the fluorine content. After packaging the mixer contents into standard 20 gal containers, the containers are subjected to vibration by moving the containers up and down for a distance of one-half inch in each direction at the rate of 30 times per minute for 1 hr. Analyses of samples of the flow conditioner taken at three different vertical levels inside the containers show the concentration remains at about 0.3% fluorine. Micromesh screen analysis of samples of the mixtures shows about 100% smaller than 50 microns.

Example 3: Conditioning Salt – One thousand parts of table salt, about 100% passing through a U.S. Standard 16-mesh screen and about 5% passing through a U.S. Standard 70-mesh screen, and 22 parts of a fluorapatite-tricalcium phosphate composition prepared as in Example 2 are charged into a conventional ribbon mixer. The mixer is allowed to blend for 30 min. The resulting mixture has excellent flow properties. Upon being subjected to high humidity conditions, the fluorine content remains uniform at about 0.013% fluorine expressed as NaF, when analyzed by the standard procedure in Example 2. Packages containing 1 lb of the mixture after being subjected to the vibration as given in Example 2, show no tendency to segregate into the individual components as determined by the standard fluorine analysis. The resulting composition is substantially more resistant to caking under high humidity conditions than similar salt without the flow conditioner.

Example 4: 1,000 parts of calcium pyrophosphate, about 100% smaller than 30 microns with an average particle size of 8 microns, and 10 parts of sodium fluoride, about 100% smaller than 40 microns and an average particle size of 30 microns are charged into a rotary drum rotating at 20 rpm. After mixing in the rotary drum for 30 min the sodium fluoride-calcium pyrophosphate mixture contains a uniform dispersion of sodium fluoride when samples are taken at random throughout the mixture in the drum. Fluorine content is determined by the standard method given in Example 2. The above mixture imparts excellent flow and noncaking qualities to salt when used in amounts from about 1 to 10%.

Other comparable fluoridated flow-conditioning agents can be prepared in the same manner to give the proper fluorine content, as described above, by replacing sodium fluoride with substantially similar amounts of other alkali metal fluorides useful in the practice of this process and in like manner other water-insoluble phosphates can be substituted for calcium pyrophosphate. For example, potassium fluoride and lithium fluoride and mixtures can be substituted for sodium fluoride in the foregoing composition and tricalcium orthophosphate, calcium tripolyphosphate, calcium disodium pyrophosphate and magnesium disodium pyrophosphate and mixtures can be substituted for calcium pyrophosphate and the resulting fluoridated flow-conditioning agent imparts excellent flow-conditioning properties to salt and shows no tendency for the fluoride to segregate in the resulting composition.

LOW-SODIUM MEAT PRODUCTS

Potassium Salt Cure

T.B. Zyss; U.S. Patent 3,782,975; January 1, 1974 explains how to use potassium phosphates in place of sodium salts in the injection curing of primal cuts (ham, bacon, corned beef, pastrami). In the curing of primal cuts by the injection process, the meat is pumped to about 110% of its weight with an aqueous curing solution which, typically, contains sodium chloride, sodium nitrite and/or nitrate, sugar and/or honey, sodium phosphate, sodium ascorbate and condiments. Bacon, corned beef and pastrami are injected with the curing solution by a procedure known as stitch pumping in which the solution is pumped directly into the tissues through a large number of closely spaced, hollow needles.

Arterial pumping is used to inject the curing solution into ham, although a procedure similar to stitch pumping could also be used. In arterial pumping, the solution is injected into the vascular system of the ham through a hollow needle inserted in, for example, the femoral artery. The curing solution is carried through all of the capillaries and establishes equilibrium with the tissue cells. Since ham shrinks about 10% by weight during the heating and smoking steps of the curing process, the curing solution is pumped into ham in an amount corresponding to about 10% of its green weight or raw weight.

If the curing solution is added to ham in an amount corresponding to 10 to 15% of its green weight, the ham is labelled water added; and if the curing solution is added in an amount from 15 to 20% or more of its green weight, the product is labelled imitation ham. After the pumping step, the primal cuts are maintained at a temperature usually between 32° to 42°F for about 3 to 5 days. During this period, the sodium nitrite reacts with the myoglobin of the meat to form a permanent nitrosomyoglobin. Sodium nitrate is a source for additional nitrite ion.

Following the cooling steps, the primal cuts may be further processed by heating and smoking. The resulting cured meats are refrigerated until sold or processed into canned products. The addition of the aqueous curing solution to the primal cuts imparts improved texture and tenderness to the meat structure which would otherwise be lacking if the brine was not added to the meat.

In order for the added water to be retained, it is necessary to have ingredients in the brine which assure good hydration and binding of the water with the meat. In the past, there has been debate as to whether the sodium ion or the chloride ion was the essential ion in the binding of water with meat where phosphates were present. In this connection, the compositions used in curing hams as well as compositions used in the preparation of sausage-type products contained sodium and/or chloride ions.

Although most people can readily injest primal cuts cured with a conventional sodium chloride/sodium phosphate brine, there are certain individuals, particularly those with heart disease and related ailments, who require a low-sodium diet. These persons have relatively little variety in their diet because of the small number of low-sodium foods. Additionally, other people who could benefit from a low-sodium diet, but who do not suffer severely from sodium chloride-containing foods, usually do not stay on the prescribed or recommended diet because of the poor selection of low-sodium foods. Thus, the production of low-sodium primal cuts would be of substantial benefit to those persons who are on a restricted diet because it would increase the variety of tasty foods available for their selection and consumption.

Potassium polyphosphates and potassium pyrophosphates can be used in conjunction with limited amounts of potassium chloride as a substitute for sodium chloride in the manufacture of dietetic sausage and meat loaf products. It is believed that sodium chloride is generally considered to be an essential ingredient for the reason that such products require a meat protein solubilizing agent (NaCl) so that on cooking, the ground particles have sufficient coherence to provide a product of suitable texture.

Potassium chloride also functions as a meat protein solubilizing agent and is not harmful to individuals on a low-sodium diet, but it is not satisfactory in that concentrations which are capable of providing sufficient meat particle coherence impart a bitter and unpalatable taste to these products. Potassium salts of polyphosphates and pyrophosphates can be cooperatively combined with low concentrations of potassium chloride to solubilize meat proteins and produce sodium salt-free sausage and meat loaf of satisfactory texture and taste. Thus, previous compositions used in the curing of hams, processing of poultry and preparation of dietetic sausage products contained either the sodium ion and/or the chloride ion.

In an improved process, a method is provided for curing fresh primal cuts of meat which does not require the presence of either the sodium ion or the chloride ion and which includes the step of subjecting a fresh primal cut to an aqueous curing solution containing water-soluble, nontoxic phosphate composition selected from the group consisting of orthophosphate, polyphosphate and pyrophosphate salts of potassium. The curing solution is free of sodium and chloride ions except for the nominal amount present in commercial water supplies used in making up the solution and, typically, is further formulated with special purpose additives such as potassium nitrite, potassium nitrate, sugar, potassium ascorbate and condiments.

The potassium phosphates used in the process include potassium orthophosphates, potassium polyphosphates, potassium pyrophosphates in their anhydrous and hydrated forms. Specific examples of potassium phosphates which can be used are as follows: potassium orthophosphates, KH_2PO_4, K_2HPO_4, K_3PO_4, K_2AlPO_4 (under orthophosphates for classification), and $K_2HPO_4 \cdot 2H_2O$; potassium polyphosphates, $K_5P_3O_{10}$; and potassium pyrophosphates, $K_4P_2O_7$.

The concentration of the potassium phosphate composition in the aqueous curing solution is generally in the range of about 4 to 12.5% by weight. When the primal cuts are pumped to about 110% of their weight with a curing solution containing about 4 to 12.5% by weight of phosphate, then the retained phosphate will be in the range of about 0.4 to 1.25% by weight. It has been found that particularly good results with respect to shelf life are obtained when the phosphate composition is a mixture of KH_2PO_4 and K_2HPO_4. The mixture generally contains about 25 to 75 parts by weight of KH_2PO_4 and 75 to 25 parts by weight of K_2HPO_4. A preferred phosphate mixture contains about equal parts by weight of KH_2PO_4 and K_2HPO_4.

In the laboratory, fresh primal cuts of ham, bacon, corned beef and pastrami, weighing about 454 g each, were injected with aqueous curing solutions consisting of potassium phosphate salts dissolved in distilled water and cured in the usual manner. The curing solutions were injected into the primal cuts in amounts corresponding to 10 to 30% of their weight and it was noted, upon completion of the curing process, that the added water was effectively bound to the meat.

The laboratory experiments demonstrated that water retention of the cured primal cuts was significant when the potassium phosphate level of the primal cuts was 0.1% by weight and that increased water binding or retention occurred up to a retained phosphate level of 1.25% by weight. The amount of water binding was substantially constant at a retained phosphate level from 1.25 to 2.0% by weight. It should be noted that at a retained phosphate level in the primal cut of 1.25% by weight the product will produce thirst in almost all users as well as a phosphate taste. However, the present legal limit for phosphates is such that the potassium phosphate composition is added to primal cuts to the level of 0.5% by weight.

Potassium nitrite and/or potassium nitrate are advantageously included in the curing solution in an amount from about 0.05 to 0.2% by weight. Potassium nitrite reacts with the myoglobin of the meat to produce the cured meat color. Potassium nitrate serves as a reservoir in the curing solution for additional nitrite ions. The legal limit for nitrite is 2 lb/100 gal of curing solution and the completely processed primal cuts may not contain more than 200 parts of nitrite per million parts of meat.

Sugar and/or honey may be included in the curing solution in an amount from about 1 to 8% by weight to impart a delicate taste to the cured product. The amount of sugar used is not critical but it relieves the bland taste and covers the phosphate which becomes noticeable for some individuals at the 0.7 to 0.8% retained phosphate level and above. Isoascorbic acid or potassium erythorbate can also be added to the curing solution in an amount up to 0.5% by weight to reduce any metmyoglobin to myoglobin. The legal limit for this ingredient is 75 oz/100 gal of curing solution.

The curing solution, which is usually at ambient temperature, is injected into hams by arterial pumping and into bacon, corned beef and pastrami by stitch pumping in an amount corresponding to about 5 to 20% of the weight of the primal cuts. Following the pumping step, the primal cuts are refrigerated for several days and then further processed by heating or heating and smoking.

Example 1: An aqueous curing solution was prepared having the following composition.

Ingredients	Quantity
Water	55 gal
Sugar	25 lb
KNO_2	10 oz
KNO_3	10 oz
Isoascorbic acid	11 oz
K_2HPO_4	4.5 lb
KH_2PO_4	4.5 lb

Two hams weighing about 10 lb each, a side of bacon, a round and a brisket for corned beef and a boned-out shoulder for pastrami were pumped with the curing solution to 110% of their green weight, which is the legal limit for added moisture. The hams were injected with the solution by arterial pumping and the other primal cuts were injected with the solution by stitch pumping. After pumping step, the primal cuts were refrigerated at 40°F for 3 days. Following refrigeration, the primal cuts were further processed by heating or heating and smoking. The hams were heated at 130°F for 4 hr without smoking after which one ham was heated at 140°F for 8 hr with heavy smoking and the other ham was heated at 165°F for 10 hr without smoking. The bacon slab was smoked at approximately 135°F for about 10 hr at which time the internal temperature had reached 128°F. In the case of pastrami, the primal cut was hung in the smokehouse and rubbed with a pastrami spice mixture in which no sodium chloride was present. The temperature was elevated in stages, with smoking to 125°, 150°, and 180°F in accordance with the usual procedure. Corned beef was further processed by heating it in a steam oven.

The hams, which had a green weight of 10 lb each, were pumped with 1 lb of curing solution to provide a total weight of 11 lb each. After processing was completed, the ham which was heated at 140°F with heavy smoking weighed 9.7 lb while the ham which was heated at 165°F without smoking weighed 9.4 lb. Two control hams which were processed without potassium phosphate weighed 9.1 and 9.0 lb. Thus, the control hams lost more weight than the hams cured with potassium phosphate and the texture of the control hams was somewhat tougher and drier. Like results with respect to water retention were obtained for the other primal cuts pumped with the potassium phosphate curing solution.

Example 2: Curing solutions were prepared having the following compositions.

	Quantity		
Composition	A	B	C
Ingredients:			
Water	55 gallons	55 gallons	55 gallons.
Sugar	25 pounds	25 pounds	25 pounds.
KNO_2	10 ounces	10 ounces	10 ounces.
KNO_3	do	do	Do.
Isoascorbic acid	41 ounces	41 ounces	41 ounces.
K_2HPO_4	9.0 pounds	18 pounds	38.75 pounds.
KH_2PO_4	do	do	Do.

Primal cuts corresponding to those identified in Example 1 were separately pumped to 110% of their green weight with Composition A, Composition B and Composition C. The pumping procedure and further processing of the pumped primal cuts were the same as in Example 1. The weights of the cured meats pumped with Composition A containing 9 lb of K_2HPO_4 and 9 lb of KH_2PO_4 were substantially equal to the green weights of the primal cuts while the weights of the cured meats pumped with Composition B containing 18 lb each of K_2HPO_4 and KH_2PO_4 were somewhat above the green weights of the primal cuts. The primal cuts pumped with Compositions A or B had excellent texture, a bland flavor and no trace of bitterness.

The taste of the primal cuts pumped with Composition C containing 38.75 lb each of K_2HPO_4 and KH_2PO_4 was sharper and drier than the primal cuts processed with lower levels of phosphate. Also, there was a certain production of thirst in those sampling the meat products cured with Composition C. By increasing the sugar and honey levels, the sharper phosphate taste could be covered in part and this could also be done by using more spice. In one other sample pumped experimentally with Composition C, water was retained to the 30% level but this was done by pumping on successive days in the chill room and the resulting product would have to be labeled imitation.

Example 3: This example illustrates the use of potassium tripolyphosphate, a mixture of tetrapotassium pyrophosphate and potassium aluminum orthophosphate, and a mixture of potassium tetrametaphosphate and dipotassium orthophosphate in the preparation of low-sodium, cured primal cuts. Curing solutions were prepared having the following compositions.

	Quantity	
Composition	A	B
Ingredients:		
Water	55 gallons	55 gallons.
Sugar	25 pounds	25 pounds.
KNO_2	10 ounces	10 ounces.
K erythorbate	41 ounces	41 ounces.
$K_5P_3O_{10}$	9.0 pounds	
$K_4P_2O_7$		4.5 pounds.
$KAlPO_4$		Do.
K_2HPO_4		

Primal cuts corresponding to those identified in Example 1 were separately pumped to 110% of their green weight with Compositions A and B. The pumping procedure and processing were the same as in Example 1. It was noted in each instance that the cured ham, bacon, corned beef and pastrami had excellent characteristics with respect to water retention, texture and taste.

Potassium and Phosphate Salts

In the manufacture of sausage products such as weiners, for example, and meat loaf products such as luncheon meat, sodium chloride is generally considered to be an essential ingredient. The reason for this is that these products require a meat protein solubilizing agent so that on cooking the ground or comminuted particles have sufficient coherence to provide a product of suitable texture and appearance.

This problem of meat particle coherence has prevented the making of sausage and meat loaf products which are acceptable for individuals with heart ailments (salt-free). In this regard, it should be noted that while potassium chloride also functions as a meat protein solubilizing agent and is not harmful to these individuals, it has not provided a satisfactory solution to this problem in that concentrations of potassium chloride which are capable of providing sufficient meat particle coherence impart a bitter and unpalatable taste to these products.

The use of low levels of potassium chloride in combination with phosphates is disclosed by *F.C. Olson and J.C. Trautman; U.S. Patent 3,447,932; June 3, 1969; assigned to Oscar Mayer & Co., Inc.* The potassium chloride concentration is, of course, subject to variance depending upon the type and amount of other spices used as well as the desired taste properties of the particular sausage or meat product being made. Generally, it is preferred that the concentration of potassium chloride be kept below 0.5%, although, under certain conditions, concentrations of up to as high as 5.0% can be used.

Examples of potassium polyphosphate and pyrophosphate salts which have been used in sausage and meat loaf production include potassium polyphosphate, potassium pyrophosphate, tetrapotassium pyrophosphate, potassium hexametaphosphate, potassium tripolyphosphate, tertiary potassium phosphate and monobasic potassium phosphate. However, other salts exhibiting protein-solubilizing characteristics similar to the abovementioned salts can be used. In this regard, it should be noted that while the potassium salts are preferred, other food-acceptable anion salts such as, for example, the ammonium and lithium polyphosphate and pyrophosphate salts can also be used.

The following examples serve to illustrate the application to the manufacture of both sausage and meat loaf products. Example 1 is particularly concerned with luncheon meat and Example 2 is directed to a weiner-type sausage. In Example 1 a standard luncheon meat formulation was used. This formulation consisted of the following basic ingredients.

	Pounds
80% lean port shoulder meat	100
Water	3
Potassium nitrite	0.016
Potassium nitrate	0.032
Ground black pepper	0.0625

In preparing the standard luncheon meat formulation, the pork was chilled to a temperature of 28°F and ground through a 3⁄16" plate. The ground emulsion was then vacuum mixed with the potassium nitrite, potassium nitrate, ground black pepper and the particular sodium chloride substitutes given in each of the examples.

Example 1: To the other ingredients in the standard luncheon meat formulation, 0.50 lb of potassium chloride and 0.75 lb of potassium tripolyphosphate were added and mixed for 12 min. The formulation was then placed in a cooler at 32°F for 48 hr and remixed under vacuum for 3 min. Some of the thus-prepared luncheon meat formulation was packed in 30½-oz cans, sealed under vacuum and processed for 130 min at 240°F. The remainder of the formulation was packed in 12-oz cans, sealed under vacuum and processed for 70 min at 225°F.

A cut out average of 97.5% was obtained for the 30½-oz can samples and of 98.4% for the 12-oz can samples. The cut out referred to above is the quotient obtained by dividing the drained weight of the finished product by the packed product and multiplying by 100. Cut outs in excess of 92% for 12-oz cans and of 90% for 30½-oz cans are generally considered acceptable as compared to products made with the usual quantities of sodium salt. Informal tasting of the above formulated luncheon meat product indicated that the product was considered acceptable by people familiar with low-sodium diets.

Example 2: The following example illustrates the applicability of this process to weiner-type sausages. The following ingredients were reduced to a uniform size by passing them through a standard meat grinder.

	Pounds
50% lean pork	54.80
88% lean beef	16.80
80% lean pork	15.00
Skinned pork jowls	10.00
Veal trimmings	3.40
Water	18.82
Spice, dextrose, corn syrup solids	2.00
Potassium tripolyphosphate	1.00
Potassium chloride	0.50
Ascorbic acid	0.05
Potassium nitrate	0.032
Potassium nitrite	0.016

All of these ingredients were added to a standard meat chopper and chopped until a temperature of 68°F was reached. The chopped batter was then mixed for one-half minute under maximum vacuum. The thus-prepared batter was then stuffed into cellulose weiner casings and subjected to the usual processing conditions in a continuous weiner-making operation.

During the chopping operation, it was observed that the meat product acted quite comparable to similar meat products with sodium salts. Thus the emulsion was extremely tacky and well knit together. The finished product showed no fat or water separation between the casing and the meat and the cellulose casing was easily stripped off. The finished weiners were of excellent texture and appearance, being quite firm on eating. The flavor thereof was judged to be acceptable by people familiar with a low-sodium diet.

COMPLETE DIETS

FORTIFIED LOW CALORIE FOOD

A liquid food produced as a water dispersion of carbohydrates, protein and fat contains 900 calories per serving. The product is described by *H.W. Howard, W.B. Jacobsen, J. Hsueh, R.C. Selby and T.W. Workman; U.S. Patent 3,219,454; November 23, 1965; assigned to The Borden Company*. The carbohydrate content consists of several water-soluble carbohydrates of varying rates of assimilation in the body, so that their energy is released over a widened period of time as shown by varying rates of intestinal absorption of the different carbohydrates. Lactose is used along with sucrose, water-soluble dextrins, dextrose and maltose in proportions in the order named, with lactose the largest.

The high proportion of lactose is considered to promote the development in the alimentary canal of flora which offset in part at least the shortage of bulking ingredients found in usual diets of this kind. The food contains usual types of flavoring materials, mineral and vitamin supplements, and for best results guar gum and carrageenin, either alone or mixed, or other viscosity increasing or thickening agents commonly used in food products. The liquid is stable when refrigerated or frozen. It is substantially nonsettling under refrigeration for several days at least. It can be spray dried to powder form for use after reconstitution with water.

In a blind taste panel test, the vanilla flavored liquid food was compared with a commercial food of the same class and like flavor containing substantially less carbohydrate for 100 parts of fat and containing also more and different kinds of protein. Of the panel, 87% preferred the new food product on the basis of flavor, viscosity, general appearance, sweetness level and absence of objectionable aftertaste; only 3% preferred the other commercial food; and 10% had indeterminate findings. With the chocolate flavored materials the figures in order were 99%, 0% and 10%. A test group, using a quart a day (900 calories) lost on the average a half pound a day in weight for 2 weeks.

The fat used is any of the common edible fats. Butterfat is preferred, although other fats may be substituted as, for example, soybean or corn oil either in refined or partially hydrogenated condition or both if the particular advantages of butterfat are not desired in the finished product. The protein used must be water-dispersible so as to prevent sedimentation in the liquid food and facilitate reconstitution in case the product is distributed in powder form. Furthermore the protein should be relatively high in content of lysine and other essential amino acids some of which are low in cereal proteins, in order that the relatively small amount of the protein, as compared to that of other compositions of this type, will provide the necessary amounts of the amino acids essential for human nutrition. No protein was found economically available that is as satisfactory for this purpose and as acceptable in palatability as the proteins of milk, that is, casein, lactoglobin and lactalbumin as they occur naturally in milk. While other animal proteins such as those of meat or egg would meet the requirement as to the essential amino acids, there are other disadvantages in substituting such other proteins. Additional proteins, e.g., soy or peanut protein, if fortified where necessary by certain essential amino acids to make the proportions of such acids about the average found in body proteins, would meet also the one requirement but not the palatability and other features such as lack of settling necessary for preferential acceptance.

If the soy or peanut proteins are used they are in any case first degraded or partially hydrolyzed to water-soluble or dispersible form and used in conjunction with at least an equal weight of milk protein. The flavoring, vitamins, minerals, and any other enriching components are of usual kind and in amounts meeting authoritative dietary standards.

Table 1 shows a composition recommended for commercial use, to which there are to be added the flavoring material, vitamins, mineral supplements and stabilizing gums. The edible stabilizing gum used to decrease separation and settling from the food in aqueous dispersion is supplied in extracted form as in a commercial food grade of carrageenin, guar, or locust bean gum or of further processed material such as sodium alginate and sodium carboxymethylcellulose.

TABLE 1

Components	Grams used	Gms. for 100 Fat	Calories Provided	
			No.	Percent of Total
Protein	58	341	230	25.6
Carbohydrates, total	130	765	520	57.7
Lactose	85	500	340	37.8
Sucrose	28	165	112	12.4
Dextrins	13	76	52	5.8
Dextrose and maltose	4	24	16	1.8
Fat	17	100	150	16.7
Total	205		900	100

The gum is used in the amount corresponding to about 0.3 to 0.5 gram of the gum per quart of the food weighing 1,020 grams of solids content including minerals about 220 grams. Satisfactory results are obtained in the use of the gum when carrageenin and guar gums are used, either alone or jointly in amount to raise the viscosity of the food to about 75 to 250 cp at 45°F and suitably to within the range 100 to 150 cp for the food made with chocolate flavor and 110 to 170 for the vanillin flavored foods. In all cases the high ratio of total carbohydrate to fat and the gradation of the carbohydrates, between the quickly absorbable and the more slowly absorbable materials such as the dextrins, will be noted.

The minor components or supplementary materials, when used, are added in any usual nontoxic form of food grade. Thus the metals, except for the iron in the amount shown, the linoleic and arachidonic acids, and some of the vitamins may be supplied by skim milk. Others may be added in compounds that are nontoxic, edible, and permissible in foods. Soluble compounds are selected when available. When no soluble source of the supplement can be had, as in the case of vitamins A and D, then it is used in conjunction with any known and permitted solubilizing or emulsifying agent. Examples of metal salts that meet the requirements are sodium (selected metal) pyrophosphate, e.g., sodium ferric pyrophosphate, other phosphates, chlorides, citrates and lactates, all in amounts not above the officially set tolerances. Examples of other source materials are thiamine hydrochloride, vitamin A palmitate with about an equal weight of mixed glycerol monooleate and propylene glycol, lecithin or other emulsifiers and butterfat, the latter providing the unsaturated linoleic and arachidonic acids.

TABLE 2: SUITABLE CONTENT OF MINOR COMPONENTS

Component used:		Amount per quart of liquid food (900 calories)
Calcium	mg	2,000
Magnesium	mg	200
Phosphorus	mg	1,600
Potassium	mg	2,300
Sodium	mg	970
Cobalt	mg	0.001
Copper	mg	0.50
Fluorine	mg	0.27
Iodine	mg	0.35
Iron	mg	12
Manganese	mg	0.03
Molybdenum	mg	0.01
Zinc	mg	6.4
Linoleic acid	mg	270
Arachidonic acid	mg	170
Ascorbic acid	mg	75
Biotin	mg	0.06
Choline	mg	218
Folic acid	mcg	3.8
Inositol	mg	218
Niacin	mg	20
Pantothenic acid	mg	5.9
Pyridoxine	mg	0.8
Riboflavin	mg	3.0
Thiamine	mg	1.5
Vitamin B_{12}	mcg	9.4
Vitamin A	I.U.	5,000
Vitamin D	I.U.	400
Vitamin E	mg	0.27
Vitamin K	D.G. u.	460

Example 1: A quart of liquid food of caloric value 900 per quart, designed to supply the necessary components for a day's diet in conjunction with the fat to be metabolized from the body of the human consumer, is made on the following formula, skim milk being the medium for the dispersion.

TABLE 3

Component	Amount	Percent of Daily Human Requirement as Established by U.S. Food and Drug Administration
Milk protein	58 gm	
Fat	17 gm	
Carbohydrate	130 gm	
Minerals (ash)	12.7 gm	
Vitamin A as palmitate	5,000 units	125
Vitamin D	400 units	100
Vitamin C as ascorbic acid	75 mg	250
Thiamine as hydrochloride	1.5 mg	150
Riboflavin	3.0 mg	250
Niacin	20.0 mg	200
Iron as Fe*	12.0 mg	120
Calcium	2.0 gm	267
Phosphorus	1.6 gm	213
Carrageenin	0.26 gm	
Guar gum	0.14 gm	

*Iron supplied as sodium ferric pyrophosphate.

Chocolate base flavoring (8% fat) was used in the amount of 14 grams, supplying 1.12 grams fat. The remainder of the fat (15.88 grams) was supplied by milk fat as cream containing in glyceride form 0.27 gram of linoleic acid and 0.17 gram of arachidonic acid. The minerals, vitamins and other minor components are of kinds and in the proportions stated in Tables 1 and 2.

The carbohydrate moiety in the food of the formula in Table 2 was approximately as follows: lactose 59%, sucrose 33%, soluble dextrin 6%, and maltose and dextrose 2%. The sucrose was introduced in granulated form. The dextrose, maltose and soluble dextrin were provided by spray dried starch conversion syrup solids. The milk protein, some of the lactose, the calcium, phosphorus and the necessary water are supplied by skim milk and the fat by cream. Additional spray dried milk solids (nonfat) were mixed to provide additional lactose and protein.

The water-insoluble vitamins A and D are preemulsified together with a minimum amount of skim milk and an emulsifying proportion of glycerol monooleate or other emulsifying agent of food grade. The order, manner and technique and machinery for mixing, homogenizing and pasteurization are conventional in making solutions or dispersions of vitamins and mineral supplements, fat and carbohydrates in fluid milk as in the ice cream or other dairy industry.

Example 2: The procedure and composition of Example 1 are used except that the 58 grams of milk protein are replaced by 38 grams of milk protein and 32 of enzyme hydrolyzed soy protein. The product so made is satisfactory from the standpoint of calorie content but not as acceptable in other regards as the product made in Example 1.

Example 3: In making other flavors of the product such as vanillin, eggnog, banana, raspberry, or strawberry, the flavoring mixtures that are usual in establishing these particular flavors are substituted for the chocolate in the composition and procedure of Example 1, in amount to give the depth of flavor desired. Correction is made for the fat content (as in cocoa) or absence of fat content in the flavoring composition to be substituted.

Example 4: In making the food with tomato flavor, a tomato pulp as in the form of a puree or paste is incorporated in amount to give the flavor desired, in place of the chocolate flavoring of Example 1. Changes are necessary, however, in making the tomato product. The pectin content of the tomato is reduced, either by selection of the tomato to be processed to the paste or puree or by any known method for separation of pectin from tomatoes.

The pectin is reduced to an amount corresponding to not more than about 0.002 to 0.02 part for 100 parts of the liquid food of which the flavoring is to be a part. While no pectin at all is required with the tomato flavoring, it is permissible to use any amount up to that shown as for instance 0.01 to 0.1% of the dry weight of the food. Also the major (calorific) components of the composition are modified somewhat in proportions when the tomato flavoring is the one selected, proportions of such major components in a suitable formula for a quart (220 grams dry weight) of the food being as follows:

Components	Grams Used	
Protein		58
Carbohydrates, total		130
Lactose	74	
Dextrins	23	
Dextrose and maltose	8	
From tomato	25	
Fat		17
Total		205

The tomato, protein, and fat are included with the components of those classes from all sources in the table above thus reducing correspondingly the amounts required from other sources. The minerals of the tomato are included with minerals from other sources which constitute most of the difference between 205 and 220 grams dry weight per quart. The carbohydrate of the tomato is listed separately in the table above. Also the carrageenin, guar, or like gum, is either reduced in amount or omitted entirely. The total content of stabilizing gum mixed as such and of pectin supplied with the tomato is selected so that the gelling tendency is avoided and the viscosity of the food is established within the range stated, namely, 75 to 200 cp at 45°F for the cold stored food.

If these changes from Example 1 are not made particularly in the reduction of the pectin content of the tomato flavoring used, the finished fluid food is not acceptable to the trade because of incipient gelling or unsatisfactorily low flowability when the liquid food is refrigerated previous to use. While the compositions described are adapted particularly well for distribution to the consumer or the retail store under refrigeration similar to fluid milk, the compositions have been found to preserve their desirable taste and not to coagulate or precipitate during or after subjection to high temperature sterilization. For this reason they are adapted to sterilization at such temperatures for short times, cooling and then canning in previously sterilized containers under aseptic conditions.

Example 5: In practicing this sterilization and canning procedure, the composition of Example 1, for instance, is used as follows. It is subjected to heat sterilization at temperatures and times that produce sterilization equivalent to F_0 10 to 20, all by conventional technique, cooled to approximately room temperature and then canned under aseptic conditions, the temperature of the sterilization being preferably in range of 270° to 290°F. There is no deterioration such as results from like treatment of fluid milk. The canned product remains stable when held for several months' storage at room temperature.

Example 6: The procedure of any of the Examples 1 through 4 are followed except that the finished compounded liquid food is subjected to quick freezing under conditions that are conventional in quick freezing fruit juice or the like. The quick frozen material may be stored in frozen form and thawed at the time of use, with preservation of the desirable properties and flavor of the product.

COMPLETE BALANCED FOOD

E. Halin; U.S. Patent 3,338,717; August 29, 1967 prepares a stabilized food having high assimilability which contains soybean meal, vegetable oil, arrowroot starch, saccharide, hydrogenated oil, vitamins and other additives. The process consists of several interrelated steps. Step 1 involves introducing a predetermined amount of vitamin C into a stated amount of edible hydrogenated oil heated sufficiently to be substantially in the liquid state. Generally temperatures of about 45°C are adequate for this purpose, depending largely on the particular oil used. Suitable hydrogenated oils include hydrogenated cottonseed oil, hydrogenated soybean oil, hydrogenated safflower oil, hydrogenated olive oil, and hydrogenated peanut oil, among others.

Also suitable is coconut oil, which is a naturally occurring saturated oil and is, in that sense, already the equivalent of the oils that occur naturally in the unsaturated condition and have to be hydrogenated for this purpose. The mixture of vitamin C and the edible hydrogenated oil is stirred slowly to keep the vitamin C in substantially uniform suspension and then cooled until it solidifies. After complete solidification, the mixture is cut, chipped or shredded, as desired, in order to get it into the form of reasonably small particles. In this particulated form, it is introduced into a blender, along with a weighed portion of finely particled arrowroot starch.

In place of the arrowroot starch, there may be used potato starch, if desired. The specificity of this component is based on the presence of a wide range of desired nutritional factors, along with a high degree of ready assimilability. The mixture of the Step 1 components is then ready for blending with any or all of the components of the other four steps of the process.

In Step 2, a weighed portion of 2-methyl-1-naphthoquinone (vitamin K) is added to and dissolved in a defined amount of an edible vegetable oil, such as olive oil, cottonseed oil, soybean oil, or peanut oil. There is added to the oil solution a predetermined amount of vitamin A and vitamin D dissolved in a weighed portion of an edible hydrogenated oil, which has been heated to about 45°C in order to convert it into the liquid state. This edible hydrogenated oil is the same kind as used in Step 1.

The mixture is stirred in order to promote solution of the various parts. The system is then chilled to about 5° to 10°C and the solution stored until it completely solidifies. After solidification, this material is cut, chipped or shredded, as desired, into reasonably small particles and introduced into a blender. A weighed portion of arrowroot starch is also added to the blender and thoroughly mixed with the previously described material. The component of Step 2 is then ready to be blended with components of Step 1.

In Step 3, there are intimately ground together, in a mortar or similar device, predetermined amounts of vitamin B_1 (thiamine hydrochloride), vitamin B_2 (riboflavin) and vitamin B_5 (nicotinamide or niacin). In most instances, particularly when the final food product is intended for adults, there are also incorporated defined amounts of vitamin B_6 (pyridoxine), vitamin B_4 or H (biotin) and vitamin B_{12} (pantothenic acid or cobalamin). To these vitamins is added a defined amount of finely comminuted calcium phosphate. If it is desired to incorporate iron into the formulation, there is, also, added a weighed portion of finely divided ferrous sulfate or ferrous gluconate.

In many instances, particularly when the requirements of adults are contemplated, there are also incorporated other mineral sources, such as manganese sulfate, potassium chloride, zinc sulfate, and magnesium sulfate. The vitamins and salt or salts are added to a blender and intimately mixed with weighed amounts of iodized salt and a saccharide, preferably a monosaccharide, such as glucose. The glucose may be conveniently obtained in the form of dried corn syrup solids which are readily available commercially. Also satisfactory in this respect are disaccharides, such as sucrose. All of these components are blended to a uniform consistency and distribution in a blender, such as the ribbon type. The components of Step 3 are then ready for mixing and blending with the components of the other steps.

In Step 4, there is introduced into a preheated pressurizable vessel at 10 lb steam pressure such as an autoclave or other pressure regulable cooking vessel, a predetermined amount of soybean meal or flour. The soybean meal is heated with agitation at about 210° to 220°F with live steam introduced into the inner jacket of the autoclave, and with steam at lower pressure, about 4 to 5½ lb introduced inside the autoclave for about 12 to 15 minutes. In order to remove the condensed steam, a vacuum is applied, 25 to 27 inches of vacuum being satisfactory for this purpose, for about three minutes.

The soybean flour is then cooled in the autoclave and passed through a screen, about 100 mesh. The portions held on the screen are processed through a mill or grinder to comminute finely the larger particles of the processed soybean meal and thereby provide the soybean meal in a substantially uniform particle size. The soybean flour is then ready to be blended with the other components of the other steps and can conveniently be readied for the final blending by storing in a bin or like enclosure and then the quantity required is used as desired.

Step 5 consists of introducing a weighed amount of an edible vegetable oil into a mixing or blending vessel. This edible vegetable oil is the same kind as defined under Step 2. As indicated previously, the components of the various five steps may be introduced in any order into the final mixer or blender. It is best to introduce first the edible vegetable oil of Step 5, followed by a blend of the components of Steps 1, 2 and 3, and finally by the processed soybean meal of Step 4 or the latter two may be reversed. In any case, the components of Steps 1 through 5 are intimately mixed and blended. The total blend is then screened to provide a uniform particle size in the product, processed, if desired, through an Entoleter or similar device and then finally packaged.

By observing the times required in each step, it is possible to develop a satisfactory sequence of steps so that the total process is efficiently consummated either on a substantially continuous or batch basis. The process can be readily applied to a wide range of food material to be produced varying from laboratory adaptations for about 25 lb or less to full commercial development of 2,500 lb or more. The various components, previously delineated are as follows. Sufficient vitamin and mineral supplements are added to supply the known amounts required for contemplated users.

	Percent
Processed soybean meal	30 - 62
Edible vegetable oil	12 - 40
Arrowroot starch or equivalent	9 - 25
Saccharide	6 - 10
Edible hydrogenated oil	1 - 5
Calcium phosphate	3.16 - 4.5
Iodized salt	0.75 - 1.5

A specific formulation for children from infancy to about 10 or 12 years of age is, in parts by weight, as follows:

Processed soybean meal	30
Edible vegetable oil	15
Arrowroot starch or equivalent	16
Saccharide	18
Calcium phosphate	3.2
Iodized salt	1
Vitamins	1

For older persons beyond about 40 years of age the same specific formulation will be used except for variation in the vitamin content, and particularly an increase of the ascorbic acid which is important in the development of connective tissues. For persons in the age range from about 12 to 40 years of age the same specific formulation can be used but with the vitamin content modified. The relationship among vitamins for a daily administration of useful therapeutic formulation is as follows.

Vitamin A	25,000 USP units
Vitamin B	1,000 USP units
Thiamine hydrochloride	10 mg
Riboflavin	5 mg
Nicotinamide	100 mg
Pyridoxine hydrochloride	5 mg
Cobalamin (B_{12})	20 mcg
Panthothenate	20 mg
Ascorbic acid	200 mg

COMPREHENSIVE DIET

Instead of a comprehensive diet in a liquid form, *W.F.J. Cuthbertson and G.A. Childs; U.S. Patent 2,961,320; November 22, 1960; assigned to Glaxo Laboratories Limited, England* prepared the food in a stable dry form to be made up with water into liquid form. All the constituents of such a solid dietary preparation should be firstly readily dispersible in aqueous media and secondly readily digestible. Those requirements place a considerable limitation upon the constituents which may be used and give rise to particular difficulty in providing that the food should have the necessary calorific value.

Carbohydrates are a principal source of calorific value, but virtually the only carbohydrates which are both readily dispersible and readily digestible are the sugars, and to supply the whole of the necessary calorific value with sugars would make the product so sweet as to be unpalatable and would probably also lead to digestive disturbances. To supply some or all of the calorific requirements with fats leads to difficulties in obtaining a dry, stable product; furthermore the fat tends to separate out on dispersion of the product in aqueous media giving an unattractive and unpalatable product. The use of wetting or emulsifying agents to overcome this last difficulty is objectionable in that medical authorities consider that foodstuffs should not contain such substances.

It was found that pulverulent, dispersible readily assimilable proteins of low bulk density (volume/weight), are able to absorb and retain a substantial proportion of certain suitable fats and that when these proteins containing absorbed fats are dispersed in aqueous media, a chemically and physically stable dispersion is obtained from which the fats have no substantial tendency to separate under normal conditions of use. It appears that in such dispersions the protein acts as a wetting or emulsifying agent ensuring dispersion of the oil. Thus it is possible to formulate a comprehensive diet preparation in which a substantial proportion of the calorific requirement is provided by a fat thus reducing the proportion of sugar required.

In this way the palatability of the resulting dietary preparation is considerably improved. Furthermore the preparations do not under normal conditions of usage exude fat or appear greasy, although under physical stress (i.e., conditions of unusual heat or pressure) some exudation of fat will of course occur. The solid product remains relatively free-flowing and is readily dispersed in aqueous media. The resultant liquid further exhibits substantially no oiliness as regards taste or appearance.

The palability of the preparation may be further improved by replacing a further quantity of sugar with a water-soluble dextrin, which is able to supply a proportion of the calorific requirement without adding to the sweetness of the product.

Preferred fats for the preparations are those which have a melting point below 15°C such as arachis oil, sesame oil, olive oil and cottonseed oil; of these arachis oil is preferred for reasons of palatability. Fats having melting points higher than 25°C are not suitable as they are difficult to incorporate in the protein and tend to separate out on dispersion of the food in aqueous media.

Various proteins may be used, but such proteins must be of low bulk density (volume per weight) such that they will absorb the fat. It is best to use a protein the bulk density of which is less than 6.5 cc/g, although material having a somewhat higher bulk density can be used but with increasing difficulty in handling; however in general, it is best that the bulk density of the protein be not less than 4 cc/g, again for ease of handling. Protein having a bulk density less than 4 cc/g when used may also give a product having a more greasy appearance and with poorer flow properties.

The protein must of course be water-dispersible and readily assimilable; furthermore, it should have a substantial proportion of its particles in the form of flakes in order to ensure stable absorption of the fat. Examples of proteins which can be used are the assimilable and water-dispersible caseinates, particularly the metal caseinates such as calcium and potassium caseinates; other proteins which can be used include soluble egg albumin.

Calcium caseinate is particularly valuable in that it can be readily prepared in a pulverulent flaky form having low bulk density by simple roller drying. Such a roller dried product may be obtained with a bulk density of about 5.5 cc/g, and is able to absorb up to twice its weight of a fat.

In general it is best to use a proportion of fat to protein somewhat less than the maximum proportion of fat which the protein will absorb, so as to ensure that there is no substantial tendency for any of the fat to be exuded. Thus while calcium caseinate will absorb up to twice its weight of a suitable fat, in fact with this protein it is best to use a proportion of fat to calcium caseinate of 1.25:1.

In view of the fact the proteins of the stated type are relatively costly as compared with the other ingredients, it is best to use only so much of such proteins as is required to absorb the fat used, and to provide the balance of the protein requirement by a cheaper source of protein, which should, of course, be stable, readily assimilated and dispersible in aqueous media. A suitable additional protein source is, for example, a vegetable protein or milk protein, e.g., in the form of dried whole cream milk or dried skim milk. The dietary preparations also include a sugar, such as sucrose or glucose; the whole of the calorific requirement of the patient can be supplied by the fat and the sugar.

The composition also includes vitamins (preferably all those necessary for human nutrition) and also trace elements; the vitamins required in general include vitamins A, the B group, C D and E; the trace elements should include sodium, iron, manganese, copper and molybdenum. It is best that certain of the trace minerals which should be included in the comprehensive diet should be present as salts which are not soluble in water but are soluble in the digestive juices. Thus copper, iron and manganese in the form of their water-soluble salts would tend to inactivate certain of the vitamins during storage of the preparation, which defect appears to be reduced if insoluble salts are used.

Iron, copper and manganese are used in the form of salts which are insoluble in water but soluble in the digestive juices; such salts are for example the carbonates. The absorption of the fat in the pulverulent protein should be carried out under careful conditions; thus if too vigorous conditions of mixing are used only a small proportion of the fat is absorbed and the remainder separates to give a greasy dough as it becomes squeezed out of the protein particles. When the proportions are correctly chosen, for instance less than 2 parts of fat are used per part of roller-dried calcium caseinate, and the mixing is carried out gently, the product remains dry and the mixing should be carried out in such a way that a dry powder is obtained.

The comprehensive diet is prepared by adding the oil-soluble vitamins to the fat and carefully mixing the fortified fat with the pulverulent protein derivative; this mix is then added to an intimate mixture of the other constituents and the whole thoroughly blended. The following procedure is used for the preparation of a 2,000 lb batch of comprehensive diet, and for this procedure three stock mixtures are required, namely (1) vitamin mixture, (2) mineral mixture and (3) oil-soluble vitamins-arachis oil mixture; the preparation of these stock mixtures (in each case sufficient for 10,000 lb comprehensive diet) is first described.

Vitamin Mixture	Grams
Ascorbic acid	500.00
Nicotinic acid	320.00
Calcium D-pantothenate (Ca content between 8.2 and 8.6%)	50.00
Thiamine hydrochloride	45.00
Pyridoxine hydrochloride	10.00
Vitamin B_{12}	0.04
Calcium caseinate (Casilan brand)	777.5

After blending the whole mix is micropulverized and stored until required.

Mineral Mixture	Grams
Potassium dihydrogen phosphate	29,750
Sodium chloride	13,100
Ferrous carbonate, saccharated	740
Manganous carbonate	420
Copper carbonate	100
Sodium molybdate	4.0
Cobalt oxide	2.0
Icing sugar	1,284

In the above mineral mixture manganese content is not less than 42% manganese, iron content is not less than 33% total iron and 26% ferrous iron and copper content is not less than 52% copper. The above ingredients are roughly mixed, e.g., in a Hobart or Peerless mixer, then micropulverized to effect final mixing and stored until required.

Oil-Soluble Vitamins-Arachis Oil Mixture

Vitamin A palmitate	120 x 10^6 I.U.
Vitamin D_2	15 x 10^6 I.U.
α-Tocopherol acetate	240 g
Menaphthone	50 g
Arachis oil	30 lb

The menaphthone is dissolved, with the aid of a little heat, in the arachis oil and then the vitamin A palmitate, vitamin D_2 and α-topopherol acetate added. The whole is then thoroughly mixed using a slow speed stirrer and stored away from direct sunlight at a temperature not below 15°C. This stock solution is used for fortifying the arachis oil and care should be taken to see that no separation has taken place during storage. The 2,000 pound batch of comprehensive diet was prepared in the following stages:

Stage 1: 10 lb of castor sugar are mixed for 10 minutes, e.g., in a small Hobart mixer with 340 g (¾ lb) of vitamin mixture.

Stage 2:

	Pounds
Soluble dextrin	220
Castor sugar	100
Mineral mixture	20
Sugar-vitamin mixture from Stage 1	10¾

These ingredients are added, in the order given, to a blender, e.g., a Savage blender and blended for 10 minutes. Spray dried skim milk 1,090 lb is then added and blending continued for a further 20 minutes.

Stage 3:

	Pounds
Arachis oil	306
Oil-soluble vitamins-arachis oil mixture	6

The two ingredients are mixed by stirring with a slow speed stirrer for 10 minutes.

Stage 4:

Calcium caseinate (Casilan brand)	10 x 25 lb
Fortified arachis oil from Stage 3	10 x 31 lb 3 oz (10 x 3.45 gal)

25 lb of calcium caseinate are introduced into a large Hobart or Peerless mixer and with the paddle revolving at its slowest speed, 31 lb 3 oz (3.45 gal) of fortified arachis oil are slowly added. When all the fortified arachis oil has been added, stirring is continued for not more than 5 minutes. The blended material is transferred to the blender containing the Stage 2 mixture, but mixing in the blender was not commenced until nine further calcium caseinate-fortified arachis oil premixes had been prepared as described above and transferred, in each case, to the blender.

Stage 5: The blender then contains:

	Pounds
Blended material from Stage 2	1,442
Calcium caseinate oil premixes from Stage 4	562

This mixture is blended for 15 minutes, before commencing discharge of the material to packing.

CHEMICALLY DEFINED DIETS

Special Palatable Diets

Complete diets require the presence of protein, vitamins, minerals, fats, and a source of calories generally in the form of carbohydrates. Natural foodstuffs, such as meat, fruit, grains, and leafy vegetables provide these constituents. It has long been the aim of nutritionists to develop complete synthetic diets that would incorporate these constituents, and such diets have been indeed developed.

Proteins are high molecular weight, highly complex polymers composed of a variety of the so-called essential and nonessential amino acids. Utilization of protein by the animal organism requires that the protein be degraded by the proteolytic enzymes of the gastrointestinal tract to the constituent individual amino acids because the amino acids can be absorbed through the gastrointestinal tract only in the free, uncombined form.

For a dietary regimen to be considered adequate for the support of all normal physiological functions, it should contain the essential amino acids in the appropriate levels and in the proper proportion of one to the other. The function of the nonessential amino acids is to provide a source of metabolizable nitrogen required by the animal organism for the biosynthesis of proteins, purines, nucleic acids, and other metabolites. Examples of nonessential amino acids include alanine, cysteine, cystine, glycine, proline, glutamic acid, tyrosine, aspartic acid, and serine. Proper nutritional balance requires that these nonessential amino acids be provided in sufficient quantity and within a range of proportions to each other that is less restrictive or critical than the balance required for the essential amino acids.

M. Winitz; U.S. Patent 3,701,666; October 31, 1972; assigned to Morton-Norwich Products, Inc. states that a "defined diet" refers to a diet formulation consisting essentially of highly purified nutrients, essential vitamins, and minerals. If the nutrients of the defined diet are all present in the form of compounds whose precise molecular configuration is known, this defined diet is termed a "chemically defined diet." Nutrients whose precise molecular configuration may not be known and whose inclusion may remove the diet from the chemically defined diet category, include, for example, proteins, peptones, starches, dextrins and fats. On the other hand, nutrients whose molecular configurations are precisely known and which may be a part of chemically defined diets include, for example, amino acids and their simple precursors, such as purified peptides, mono- and disaccharides and esters of pure fatty acids.

Various attempts have been made to formulate synthetic diets for human consumption to supply all of the requirements of the essential amino acids and nitrogen needed for growth and sustenance of normal physiological activity. Although nutritionally adequate amino acid mixtures containing in addition carbohydrates, fats, minerals, and vitamins have been formulated and have been successfully used in metabolic and nutritional studies with experimental animals and with humans, long term provision of defined diets to humans has been thwarted by virtue of the unpalatable nature of the diets. Usefulness has generally been limited to experimental projects where the desire of the subjects, as in the furtherance of research, overcame the problem of the unpalatability.

It is believed that in many instances the unpalatability of previous chemically defined diet compositions has been caused wholly or in part by the chemical breakdown of amino acids and/or by the chemical interaction of various components in the formulation, and not by the inherent nature of the amino acid. It has been found that an amino acid formulation palatable to human taste can be provided if the ingredients and the conditions of the formulation process are closely regulated, particularly by controlling formulation temperatures and orders of addition in order to minimize the formulation of amino acid derivatives and the interaction of components to form undesirable reaction products. Amino acid derivatives are considered to be products which evolve by decomposition of the amino acids or by interaction of an amino acid with another component of the formulation.

Palatability in an aqueous solution formulation is harmed if the amino acids and any derivatives contribute lower alkyl mercaptan, such as methyl mercaptan, in a total amount greater than about 15 miligrams per liter of solution. If nonessential amino acids are used to supply the metabolizable nitrogen necessary for the human body to perform its natural metabolic functions, then there are two other criteria considered important to preserve the palatability of the formulation. The amino acids and any derivatives should not contain sulfhydryl groups in an amount greater than about 0.05 gram per liter of solution. Also glutamic acid, or alkali metal salts, such as sodium glutamate, should not be present in an amount greater than about 1.43 grams per liter of solution.

Palatable amino acid compositions for human consumption can be formulated and used in a diet together with carbohydrates, fats, vitamins and/or minerals. The process is generally concerned with making such formulations from substantially pure amino acids. However, it is also concerned with the overall process for formulating substantially complete diet.

By following the foregoing criteria, it is possible to choose the ingredients of a formulation to avoid initially employing amino acids which will cause the limits set forth above to be exceeded, thereby destroying palatability. The amount of sulfhydryl (SH) groups and disulfide linkages (which form SH froups upon reduction in solution by other components, such as ascorbic acid) in the diet is regulated by limiting the amount of such groups added to the formulation in the form of cystine, which contains a disulfide linkage, and cysteine, which contains an SH group.

Furthermore, edible commercial grades of methionine possess an off-taste by virtue of the contamination thereof with other impurities, such as lower alkyl mercaptans. Thus, commercial methionine is further purified before its inclusion in a dietary mixture, as by recrystallization from water or a water-ethanol mixture, or from some other suitable solvent or mixture of solvents. Moreover, methionine is subject to thermal degradation, and it has been found that the temperature of the formulation after dissolution of methionine should be controlled to prevent such degradation.

In chemically defined diets, the amino acid components, in the form of free essential and nonessential amino acids or suitable precursors such as peptides, are provided as total levels sufficient to satisfy normal physiological requirements for nitrogen. Accordingly, the balance among the various amino acids is selected to meet the normal metabolic needs of the human subject and to maintain the desired nitrogen balance. Because of strong interdependencies between the required level of a given amino acid and the level of one or more of the other amino acids present in the diet, it is not practicable to establish a precise range of levels for each of the amino acids. Generally, the ratios of the levels of amino acids in the diets should approximate those of a high quality protein, such as meat, eggs, or milk, for example.

The following table illustrates a diet formulation containing these essential amino acids at relative levels with respect to one another which are at about the centers of these preferred ranges. Deviation in the amount of any particular essential amino acids with respect to the others in the table should not be more than plus or minus about 5% by weight in order to remain within the preferred ranges.

The diet formulation set forth in the following table is entitled a base diet because, if it were to be administered for any substantial length of time, e.g., more than about two weeks, a source of cobalt and molybdenum ions would be added, and the values of folic acid and menadione (vitamin K) would be increased.

Diet Formulation (base)

Amino acids:		
L-alanine	g	1.76
L-arginine·HCl	g	3.89
L-aspartic acid	g	3.75
L-glutamine	g	6.19
Glycine	g	2.87
L-histidine·HCl·H_2O	g	1.08
L-isoleucine	g	1.65
L-leucine	g	2.61
L-lysine·HCl	g	2.45
L-methionine	g	1.69
L-phenylalanine	g	1.88
L-proline	g	2.35
L-serine	g	1.21
L-threonine	g	1.65
L-tryptophan	g	0.51
L-tyrosine ethyl ester·HCl	g	2.80
L-valine	g	1.82

Minerals:		
Acetic acid	mg	5.95
Calcium chloride·$2H_2O$	g	2.93
Cupric acetate·H_2O	mg	6.12
Ferrous ammonium sulfate·$6H_2O$	g	0.070
Magnesium oxide	g	0.258
Manganous acetate·$4H_2O$	mg	12.49
Potassium chloride	g	2.33
Potassium hydroxide	g	1.27
Potassium iodide	mg	0.19
Sodium chloride	g	1.21
Sodium glycero-phosphate·5½ H_2O	g	8.13
Sodium hydroxide	g	1.13
Zinc chloride	mg	0.85

Water-soluble vitamins: [1]		
p-Aminobenzoic acid (K salt)	mg	354.9
Ascorbic acid	mg	70.0
d-Biotin	mg	0.20
d-Calcium pantothenate	mg	10.0
Choline chloride	mg	85.0
Folic acid	mg	0.10
Inositol	mg	116.5
Niacinamide	mg	13.3
Pyridoxine·HCl	mg	2.0
Riboflavin phosphate (Na salt)	mg	1.64
Thiamine·HCl	mg	1.2
Cyanocobalamin	mcg	5.0

[1] Vitamin levels given represent amounts to be present in finished product.

Carbohydrates:		
Glucono-delta-lactone	g	6.31
Glucose	g	400.5

Fats and fat-soluble vitamins:		
Safflower oil [2]	g	1.33
Vitamin A acetate	mg	1.72
Vitamin D_2	mcg	10.0
Alpha-tocopherol acetate	mg	20.0
Polysorbate 80 (emulsifier)	mg	66.7
Menadione	mg	0.12

[2] Or equal amount of ethyl linoleate.

Flavoring: Synthetic flavoring agents plus water are added.

The diet formulation in the above table is suitable for supplying all the nutrient needs of the human species. In this respect, the formulation is considered to provide, in the weight amounts shown, the complete daily requirements for the average human. The illustrated composition provides a caloric intake of about 1,800 calories and metabolizable nitrogen in an amount of about 5.88 grams. The illustrated composition is intended to be administered orally as primarily an aqueous solution, with the fat and fat-soluble vitamins being present as an emulsion. Although the amount of water used need only be sufficient to solubilize the water-soluble ingredients, it is preferably administered in a more dilute form at a level of about one calorie per milliliter. Such a caloric level is provided by diluting the weight formulation shown with sufficient distilled or deionized water to bring it up to about 1.8 liters.

As previously indicated, the nonessential amino acids provide a source of metabolizable nitrogen. However, in addition, some of the nonessential amino acids can be used as substitutes in part for requirements of the essential amino acids. For example, a portion of the requirement for methionine can be substituted for by either cystine or cysteine and a portion of the requirement for phenylalanine can be substituted for by tyrosine. Moreover, in general there is more latitude in the relative levels of the nonessential amino acids; however, they are preferably provided at the relative levels present in the formulation shown in the above table, plus or minus about 5% by weight. It should also be recognized that another source of nonessential nitrogen could be used, for example, an organic ammonium salt, such as ammonium citrate.

As previously indicated, the carbohydrates in the diet are used as the prime source of energy, and in the diet illustrated in the above table, the carbohydrates are the primary source of calories. Although carbohydrates may generally be used in various forms, as also previously indicated, it is considered advantageous to provide as the sole source of carbohydrate in the diet either glucose, maltose, starches or dextrins. It has been found that human subjects who consume a defined diet of this type with such a sole source of carbohydrate exhibit a marked reduction in their blood serum cholesterol level. It has been found that it is important to closely control the process steps in combining the various ingredients to form such a complete diet, or various portions in order to prevent degradation of individual ingredients and/or interactions between ingredients which might well give rise to off-tastes.

In general, certain amino acids are not very water-soluble, and such relatively insoluble members are used in the form of the hydrochloride salts and/or esters thereof, which are nutritionally the same. Tyrosine is preferably provided in such forms. In order to promote the dissolution of various of the amino acids in water, the water is usually maintained at a temperature of about 90° to 100°C. However, it has been found that two of the amino acids of the formulation are too suceptible to thermal degradation to permit their inclusion into the aqueous solution at such temperatures without seriously harming the efficacy and palatability of the resultant solution.

These two ingredients are the essential amino acid tryptophan and the nonessential amino acid glutamine. Tryptophan is extremely susceptible to partial deterioration; and accordingly, before tryptophan is added to the aqueous solution, the temperature of the solution is reduced to at least about 70°C and preferably to at least about 40°C. In general, it is not considered necessary nor practical to reduce the temperature of the solution lower than to about 30°C. Likewise, glutamine is not added to the aqueous solution until the temperature has been lowered as specified above.

Particular carbohydrates from the group of monosaccharides, disaccharides, starches and dextrins, which may be used have various degrees of water-solubility, and good solubility is desirable in formulating the food composition. Moreover, it has been found that the favorable endothermic heat of solution of carbohydrates, such as glucose, can be used to achieve the temperature lowering of the solution preparatory to the addition of tryptophan and glutamine while concurrently accomplishing dissolution of a substantial portion of the carbohydrates component in the aqueous solution at temperatures where solubility is greater.

Consideration in the process may also be given to methionine which, as previously indicated, is subject to degradation that results in the creation of lower alkyl mercaptans, particularly methyl mercaptan. The time for which methionine is subjected to the relatively high temperatures may be reduced by adding the methionine just before the temperature lowering is carried out, in which instance it should not be exposed to elevated temperatures for a sufficient period of time to harm palatability.

The nonessential amino acid, aspartic acid, is difficultly soluble in water even at temperatures in the range of about 90° to 100°C. However, it has been found that aspartic acid can be fairly readily dissolved in alkaline water having a pH between about 8 and 14. Thus, the dissolution is facilitated by separately dissolving the aspartic acid in alkaline water having a pH of about 8 or above and then adding this presolution to the main solution.

Another consideration in making a nutrient composition of this type is that the growth of microorganisms, such as bacteria should be prevented. To achieve this end, from the time the temperature is lowered by the first dissolution of the carbohydrate, the amount of water present in the solution, relative to the amount of carbohydrate and solutes, is regulated so that the solution is hypertonic. Hypertonic is defined as having an osmotic pressure higher than that within microorganisms so that undesirable microorganisms, for example, *E. coli,* cannot grow in the solution.

The amounts of water added subsequently throughout the process are similarly balanced with the amounts of additional solute so that the hypertonicity of the solution is maintained, at least up to and through the addition of all the water-soluble ingredients. Depending upon the intended use of a particular batch of the food composition, the product may be dried, in which case the solution should remain hypertonic to and through the drying process. Also, the product might be shipped in concentrated solution to another location where it would be likely diluted before administration, in which instance there is a considerable advantage in maintaining the solution hypertonic during shipping. On the other hand, if it is the intention to package the solution in a form for immediate and simple administration it should remain hypertonic until final dilution, which would be normally carried out just shortly before a pasteurization or sterilization step that should immediately precede packaging.

At the conclusion of the addition of all of the water-soluble ingredients, an emulsion is made with the fat and fat-soluble ingredients. A suitable edible emulsifier, such as a polyoxyethylene sorbitan monooleate, is added in sufficient amount. In sequence thereafter, the fat component, followed by the fat-soluble vitamins, are individually added. With all of the ingredients in the single batch, a suitable emulsifier or homogenizer is employed to create a stable emulsion. A purified naturally occurring fat, such as purified safflower oil or corn oil, may be used.

If the composition is to be packaged for immediate administration, water is added after emulsifying to increase the volume of the product to the desired dilution, followed preferably immediately by pasteurization or sterilization. Again, temperature considerations should be recognized, and pasteurization is carried out by heating to a temperature of about 90°C for about one minute, and not substantially longer. Then the emulsion is rapidly cooled to a temperature below about 40°C. During the time of raising the emulsion to pasteurization temperature, the time of pasteurization, and the time of cooling, the temperature should not be above about 60°C for longer than five minutes total. It has been found that pasteurization in this particular manner achieves the intended purpose while preserving the palatability of the food composition.

If a dried product is desired, the pasteurization step need not be carried out if hypertonicity of the relatively concentrated emulsion is to be relied upon to prevent bacterial growth. It has been found that by freeze-drying this product, preferably to a moisture level or less than about 5% by weight, a dry product is formed which is free from bacterial growth and taste-stable during storage at room temperature. Moreover, freeze-drying produces a dried product that is easily reconstitutable to the form in which it will normally be administered, i.e., a water solution.

The freeze-dried product easily redissolves as an aggregate, and if the product which was freeze-dried was in the form of an emulsion, an emulsion is again formed upon redissolution. It is considered to be definitely advantageous to be able to provide a product of this type which can be stored and/or transported in the dry form and which can be easily dissolved as an aggregate to form a palatable diet composition, especially one containing all of the daily necessary nutrients for the human species.

Example: A palatable chemically defined diet is formed in the manner described below. The amino acids are used in the form of the L-isomers. The amino acids are first recrystallized from water, or from a water-ethanol solvent mixture, in the presence of activated charcoal in order to remove potentially undesirable impurities which may contaminate the individual amino acids. The batch to be prepared will provide about 1,000 liters of a diet composition having a concentration such that each milliliter contains approximately 2.7 calories.

Before beginning the actual dissolution of the components, preparation of various presolutions to be used is carried out. Into about 30 liters of distilled or deionized water, 1.71 kg of reagent grade sodium hydroxide is initially dissolved, followed by 5.63 kg of L-aspartic acid, stirring to achieve complete solution at room temperature.

Into another 30 liter quantity of distilled or deionized water, 4.06 kg of reagent grade potassium hydroxide is dissolved. To this solution is added 17.67 kg of glucono-delta-lactone, stirring to dissolve, followed by 385 grams of reagent grade magnesium oxide. The magnesium oxide is stirred to form an initial suspension, and stirring is continued at room temperature to accomplish complete dissolution.

Into six separate, 100 milliliter quantities of distilled or deionized water, the following quantities of salts are individually added: hydrated ammonium molybdate, 430 mg; hydrated cobaltous acetate, 1.7 grams; hydrated cupric acetate, 2.55 grams; zinc chloride, 1.28 grams; potassium iodide, 0.255 gram; and hydrated manganous acetate, 18.3 grams. These six individual solutions are then added in sequence to 3 liters of distilled or deionized water, while stirring, to make up a composite solution which is hereafter referred to as the trace mineral presolution, and to which sufficient additional water is added to dilute it to 4 liters.

The batch preparation is begun by heating 400 liters of distilled or deionized water to 90°C, in a steam-jacketed kettle equipped with an efficient stirrer. The amounts of the amino acids set forth immediately herebelow are added to the water while the temperature is maintained at about 90°C.

	Kg		Kg
L-alanine	2.64	L-methionine	2.54
L-arginine·HCl	5.84	L-proline	3.52
L-aspartic acid presolution glycine	4.30	L-serine	1.82
L-histidine·HCl·H_2O	1.62	L-threonine	2.47
L-isoleucine	2.47	L-phenylalanine	2.82
L-leucine	3.92	L-tyrosine ethyl ester·HCl	4.20
L-lysine·HCl	3.67	L-valine	2.73

Each of the ingredients is added separately and dissolved before the next is added. The methionine may be added last, so as to reduce the time during which it is exposed to the 90°C temperature. Upon completion of the dissolution of the last of the amino acids, the application of heat to the kettle is stopped, and about 400 kg of glucose is added in a substantially continuous stream with continuous stirring. The addition of the glucose is performed sufficiently rapidly so that a temperature of about 40°C is reached within ten minutes after the glucose addition begins. Thereafter, heat is supplied as necessary to maintain the temperature of the solution between about 30° and 35°C, while stirring is continued, until all of the glucose is dissolved.

When the glucose dissolution is completed, 770 grams of L-tryptophan are dissolved, followed by the addition of 9.29 kg of L-glutamine after the tryptophan dissolution is complete. After the glutamine is dissolved, the glucono-delta-lactone-magnesium oxide presolution, which measures about 30 liters, is slowly added at a fairly constant rate over a period of five to ten minutes while stirring.

Following the addition of this presolution, 5.36 kg of hydrated sodium glycerophosphate is added, followed by 5.48 kg of sodium chloride, and by 6.98 grams of hydrated ferrous ammonium sulfate, dissolution of each component being completed before addition of the next. At this time, about 192.5 kg of glucose are dissolved in the solution, to make a total amount of dissolved glucose of 592.5 kg. Stirring is employed, and the temperature of the solution is maintained between 30° and 35°C until the dissolution of the glucose is completed.

The following USP grade vitamins are then added in this sequence, insuring that each is dissolved before the next is added: 639 grams p-aminobenzoic acid, 96 grams of ascorbic acid; a presolution of 1.28 grams of d-biotin in distilled water; 12.8 grams of d-calcium pantothenate; 354.8 grams of choline bitartrate; 2.56 grams of folic acid; 1.28 grams of inositol; 15.36 grams of niacinamide; 2.56 grams of pyridoxine·HCl; 2.31 grams of riboflavin; 1.53 grams of thiamine·HCl; and 2.56 grams of a 1:1000 trituration of cyanocobalamin with mannitol.

Following the dissolution of the vitamins, 2.5 kg of hydrated calcium chloride is slowly added in increments of about 250 grams. Upon complete dissolution of the hydrated calcium chloride, and with the solution at a temperature not above 30°C, the four liters of the trace minerals presolution is slowly added, with stirring, over a period of 5 minutes.

Next, the solution is given an orange flavoring and coloring by adding the following ingredients in the amounts stated, which are designed to color and flavor the 1,000 liters of concentrate: 3.36 kg of orange juice flavor; 83.8 grams FD&C Yellow No. 5; 50 grams FD&C Yellow No. 6 and 16.8 kg of citric acid.

To the colored and flavored aqueous solution are added in sequence: 100 grams of Polysorbate 80 emulsifier; 2 kg of pure ethyl linoleate; 3.64 grams of vitamin A acetate; 5.73 grams of alpha-tocopherol acetate; 4.58 grams of menadione; and 57 mg of vitamin D. All of the vitamins employed meet USP standards. The solution with the fat and the fat-soluble vitamins added is then homogenized at room temperature for a sufficient time to create a uniform emulsion of the fat phase and the liquid phase. Additional distilled or deionized water is used to wash the homogenization apparatus and make certain of the quantitative use of the diet composition. This water is added to the emulsion. About 200 to 300 liters of water can be used for this purpose and subsequently added to the main batch in order to end up with a final volume of about 1,000 liters.

The emulsion is pasteurized by passage through a heat exchange-type pasteurizer at a rate so that the emulsion is exposed to a temperature of about 90°C for about one minute. The product exiting from the pasteurizer is cooled using a suitable cooling device. The pasteurization treatment is such that the emulsion is exposed to a temperature of about 60°C or higher for no longer than a total of five minutes. The product from the pasteurizer is conducted directly to canning apparatus wherein it is filled into ordinary cans in a manner usual for the food processing industry. After canning, the product can be stored at room temperature for months prior to administration without adverse affect.

Purified Amino Acid Components

It has been found that a formulation containing amino acids is unpalatable if it contains lower alkyl mercaptans, such as methyl mercaptan, in such a total amount that an aqueous solution of the formulation would have a concentration greater than about 15 milligrams of such mercaptans per liter of solution, when the pH of the solution is about 3.7. The tolerance level decreases slightly with increasing pH, with the tolerable concentration being about 7.5 mg/liter at a pH of about 5.7. It has also been found that the sulfhydryl group of the amino acid cysteine likewise detracts from palatability, and there should not be present a sulfhydryl group concentration greater than about 0.05 gram per liter of solution at a pH of about 5.7. The tolerable sulfhydryl concentration decreases at lower pH's, and at a pH of about 3.7 it is about 0.038 g/liter.

M. Winitz; U.S. Patent 3,698,912; October 17, 1972; assigned to Morton-Norwich Products, Inc. reports also that glutamic acid imparts a strong characteristic flavor to an amino acid formulation, which flavor can render a formulation unpalatable. The flavor is also incompatible with the taste of the other ingredients likely to be used in chemical diets and renders the entire diet formulation unpalatable if it is present in an amount of more than about 1.43 grams per liter. A process is used in which protein hydrolysates containing some or all of the palatability offenders are treated to remove the offenders and replace them at least in part with nutritionally equivalent nonoffensive components and thereby provide nutritionally balanced amino acid formulations.

The protein hydrolysates may be prepared from fairly abundant high-quality proteins, such as casein, lactalbumin, soybean protein, fish protein and bacterially produced protein from petroleum. However, other suitable proteins may also be used, such as gelatin, egg albumin, whole blood protein, wheat gluten and zein, for example. As a starting material, one may hydrolyze protein with a mineral acid or by means of proteolytic enzymes to form a mixture of amino acids wherein the amino acids are almost exclusively of the L-configuration, with little or no D-amino acids.

After an aqueous solution of the protein is formed, substantially all of the following amino acids are removed from the solution: tyrosine, cystine, aspartic acid and glutamic acid. These four amino acids may be removed in any order, and if the particular protein hydrolysate being treated should not contain one or more of these amino acids, or should contain one or more of these amino acids in only very minor amount, provision for removing such an amino acid need not be made. The cystine and tyrosine may be removed separately but it is preferred that they be removed together. This pair of acids is precipitated from the main hydrolysate solution while it is maintained at a temperature below about 50°C and a pH between about 5 and 7.

It is also best to remove the glutamic acid and aspartic acid together by absorption onto anion exchange resins. Removal of these four amino acids as pairs of two provides a simple and expedient method of removing these components from the main hydrolysate solution, which may then be treated as a unit in preparing it to serve as the base of a palatable amino acid formulation adapted for human consumption.

One pair of amino acids, i.e., tyrosine and cystine, includes a major contributor to unpalatability, i.e., cystine, and also includes a nonessential amino acid of limited water solubility, i.e., tyrosine. The other pair of amino acids includes another palatability offender, i.e., glutamic acid, as indicated before, and another nonessential amino acid of limited solubility in water, i.e., aspartic acid. Removal of glutamic acid facilitates its elimination or reduction in the final product.

Following removal of substantially all of the aspartic acid, tyrosine, glutamic acid and cystine from the protein hydrolysate solution, water-soluble forms of these components, or nutritional equivalents are added to the hydrolysate solution, or to the dry mixture of amino acids remaining after evaporation of the hydrolysate solution to dryness, to alleviate the deficiency that now exists because of their absence. Tyrosine and aspartic acid are supplied to the hydrolysate solution or to the dry mixture in forms having a water solubility at least equal to that of leucine.

It is preferred that the tyrosine which was recovered from the protein hydrolysate solution is converted to a water-soluble tyrosine derivative (preferably tyrosine ethyl ester hydrochloride), that the aspartic acid is converted to a water-soluble salt, such as a sodium salt, and that these water-soluble components are then returned to the main protein hydrolysate solution or to the dried counterpart. Of course, these two amino acids need not be returned to the same protein hydrolysate from which they were removed or in precisely the same amounts because water-soluble L-isomers of these two amino acids are commercially available. However, there are economic advantages to converting and returning the removed components.

In amino acid formulations which are designed to provide all of the metabolic requirements for nitrogen, and in diet compositions which might be used, for example, as the sole daily source of nutrition, the balance among the various amino acids should be such as to meet the normal metabolic needs of the human subject and to maintain the desired nitrogen balance. There are strong interdependencies between the required levels of any given amino acid and of one or more of the other amino acids. It is known that various of the nonessential amino acids can be used to partially supply a portion of essential amino acid requirements, and vice-versa, essential amino acids can be provided in lieu of certain nonessential amino acids to perform the same nutritional function.

Inasmuch as naturally occurring proteins of high quality, such as meat, eggs, milk and certain grains, include the essential and nonessential amino acids in suitably balanced relationships to one another to satisfy human nutritional requirements, amino acid formulations made from protein hydrolysates are considered generally to be adequate to supply the metabolic nitrogen needs of the human species if they approximate the amino acid balances of these proteins and contain these amino acids in the L-isomer form.

Before the addition or return of amino acids to the protein hydrolysate containing the majority of the amino acids from the protein, it may be further treated to assure its freedom from impurities and contaminants which may result from the hydrolysis process. Purification in this manner may be rather simply carried out by boiling the main solution for about one minute after adding adsorptive activated carbon and filtering the carbon from the solution while it is still hot. Purification should be such as to assure the removal of substantially all of the lower alkyl mercaptans.

To facilitate dissolution of methionine into the main protein hydrolysate solution, which is added to replace the cystine that is removed, the solution is maintained at an elevated temperature, i.e., above about 70°C. However, when the protein hydrolysate solution contains tryptophan, a lower temperature plus vigorous stirring is used. The water-soluble salts of tyrosine and aspartic acid may also easily be added at the elevated temperature if desired. However, methionine is subject to thermal degradation to produce lower alkyl mercaptans. Glutamine is not added to the solution at temperatures above about 70°C, and thus is usually dissolved after cooling of the solution. When a dry formulation is being prepared, the glutamine is usually added by blending after the protein hydrolysate has been dried.

Generally, it is desirable to use these amino acids as a part of a more complete dietary composition including at least a carbohydrate component, particularly a monosaccharide, disaccharide, dextrin, starch, or mixtures of any of the foregoing. It has been found that the favorable endothermic heat of solution of such carbohydrates, for example glucose, can be used to achieve the desired temperature lowering of the solution prior to the addition of tryptophan and glutamine.

However, if the carbohydrate is used to accomplish the lowering of the temperature, interaction may occur at elevated temperatures between the amino acids, particularly lysine, and aldehyde or ketone groups which may be present in the carbohydrate component, for example, glucose. Such chemical interaction results in the formation of undesirable condensation products which are brown in color and which have a flavor resembling caramel. Precautions should be taken to minimize the extent of the time-temperature integral over which amino acids and such carbohydrates are both present in the solution in order to thereby minimize the extent of caramelization that may occur.

This objective is achieved by adding the carbohydrate sufficiently rapidly while agitating the solution to enhance its dissolution. In this respect, the addition of the carbohydrate component, particularly glucose, should be sufficiently rapid to drop the temperature to about 40°C within ten minutes from the initial addition. Not only does rapid dissolution avoid possible unpalatability resulting from interaction between amino acids and aldehyde groups of the carbohydrate, but it also reduces the time at which the methionine is exposed to the relatively high temperatures.

In the formulation of complete dietary compositions which provide all of the nutritional needs for the human subject when taken as a sole source of nutrient for days, weeks or months it is desirable to add to the protein hydrolysate, after the removal of aspartic acid, tyrosine, glutamic acid, and cystine, the following nutrients either singly or in combination:

- minerals in the form of the mineral salts of sodium, potassium, magnesium, calcium, manganese, zinc, copper, iron, cobalt, iodine, phosphorus, chlorine and/or molybdenum;
- water-soluble and/or fat-soluble vitamins such as vitamin A, vitamin B_{12}, vitamin C, vitamin D, vitamin E, para-aminobenzoic acid, biotin, pantothenic acid, inositol, niacin, pyridoxine, riboflavin, thiamine and/or materials with equivalent vitamin activity;
- fats, such as safflower oil, corn oil, ethyl linoleate and/or equivalent fat substances;
- a source of carbohydrate;
- any one of the amino acids or a combination of amino acids necessary to achieve a well-balanced dietary composition.

Example: Sixty grams of casein are suspended in 300 milliliters of concentrated hydrochloric acid in a glass bottle, and the bottle is placed in an autoclave for about five hours at about 120°C. The resultant hydrolysate is concentrated to a thick syrup under reduced pressure, and the excess HCl is removed by evaporation twice, after the addition each time of about 75 milliliters of water. The final residue is dissolved in 150 ml of hot distilled water, treated with Norit (a highly absorptive activated carbon) and then filtered while hot. The filtrate is evaporated in vacuo to a thick syrup.

A more dilute aqueous solution is formed by the addition of about 500 ml of water to the residue, and this solution is then treated with concentrated ammonium hydroxide to a pH of about 5. The resultant solution is chilled to about 4°C and maintained at that temperature for about twenty-four hours. Under these conditions the cystine and tyrosine precipitate from the solution. The precipitate is filtered using suction, and the filtrate is saved. The precipitate is then washed with a small amount of warm water, which water is added to the filtrate.

The filtrate plus wash water is evaporated to dryness under about one-half atmosphere pressure, and the residue is dissolved in water. Pretreated Amberlite IRA resin (a strong base-type anion exchange resin) is added, while stirring, until the pH of the solution reaches about 6.5. The Amberlite IRA resin is pretreated successively with HCl, water, sodium carbonate, and water again. The resin is filtered from the solution, and the filtrate acidified using hydrochloric acid to a pH of about 2.8. The acidified filtrate is again treated with a sufficient fresh resin to bring the pH to about 6.5. The resin is again filtered from the solution, and the two resin portions are combined. The aspartic acid and the glutamic acid are adsorbed onto the resin.

The filtrate is heated to boiling, Norit is added, and the solution is filtered while hot. The volume of the solution is adjusted to 400 milliliters by either the addition of an appropriate amount of water or by evaporation. This solution contains the following amino acids from the protein: threonine, isoleucine, leucine, lysine, methionine, phenylalanine, valine, arginine, histidine, alanine, glycine, proline and serine.

The combined cystine-tyrosine precipitate is dissolved in a minimal amount of aqueous 10% ammonium hydroxide at a temperature of about 95°C. After dissolution of the precipitate, the solution is chilled to a temperature of about 4°C and maintained there at that temperature for several hours. Under these conditions, tyrosine precipitates from the solution. The precipitate is filtered from the solution and then washed successively with cold aqueous 10% ammonium hydroxide and cold water. The washed precipitate is dried under vacuum and then suspended in 30 ml of absolute ethanol.

The suspension is saturated with dry hydrogen chloride gas, and the resulting solution is permitted to stand for about sixteen hours at room temperature. The solution is evaporated to dryness under vacuum, and 30 ml of absolute ethanol is added to the resulting residue. The evaporation is repeated, and the residue which is now in the form of tyrosine ethyl ester hydrochloride is dissolved in hot ethanol. The solution is treated with Norit and filtered. The filtrate is evaporated to dryness under reduced pressure of about one-half atmosphere. The residue is recrystallized from an ethanol-ether mixture and yields 3.7 grams of pure L-tyrosine ethyl ester hydrochloride.

The combined ion exchange resin portions are extracted with 1.0N HCl until the pH of the extract is less than 2. The extracts are treated with Norit, filtered and evaporated to a low bulk. The solution is then saturated with dry hydrogen chloride gas, and chilled to a temperature of about 0°C whereat the glutamic acid precipitates in the form of L-glutamic acid ·HCl. The filtrate from these crystals is evaporated in vacuo to a syrup to remove excess hydrochloric acid, diluted with water and then boiled. It is then treated with an excess of copper carbonate, and the excess is filtered while the solution is still hot.

Upon cooling to room temperature, the blue copper salt of L-aspartic acid crystallizes in substantially quantitative yield. This salt is then suspended in water and treated with hydrogen sulfide gas. The resultant copper sulfide is filtered from the solution with the aid of Norit, and the filtrate is evaporated to a low bulk. Ethanol in an amount about four times the volume of the solution is added, which causes the precipitation of L-aspartic acid. A yield of approximately 3.6 grams of L-aspartic acid is obtained, which is then converted to the sodium salt thereof by dissolution in 30 ml of distilled water containing 1.1 grams of sodium hydroxide.

The 400 ml volume main solution containing the majority of amino acids is heated to 90°C. Successively added thereto are the sodium L-aspartate solution, the 3.7 grams of L-tyrosine ethyl ester hydrochloride, plus one gram of pure L-methionine, stirring being employed continuously to facilitate dissolution.

The heat is removed and the solution is rapidly cooled to 40°C or below by the addition of about 400 grams of glucose, in increments of about 50 grams each, using continuous stirring to facilitate dissolution. Under these conditions, the temperature reaches 40°C about ten minutes subsequent to the initial addition of glucose. Upon dissolution of all of the glucose, 0.84 gram of L-tryptophan is added. Subsequent to complete dissolution of the tryptophan, 6.1 grams of L-glutamine are dissolved in the solution.

Analysis of the solution shows that amino acids are contained in approximately the following amounts: L-trypthophan 0.84 g, L-threonine 2.3 g, L-isoleucine 3.5 g, L-leucine 5.3 g, L-lysine 4.2 g, L-methionine 2.6 g, L-phenylalanine 2.9 g, L-tyrosine ethyl ester hydrochloride 3.7 g, L-valine 3.9 g, L-arginine 2.2 g, L-aspartic acid 3.6 g, L-glutamine 6.1 g, L-histidine 1.6 g, L-alanine 1.8 g, glycine, 1.1 g, L-proline 6.2 g, and L-serine 3.5 g.

The above analysis shows that substantially quantitative yields of most of the amino acids are obtained from the 60 grams of casein that were hydrolyzed, with the exception of cystine, glutamic acid and tryptophan. The solution is fairly clear, and tasting shows that although it exhibits a relatively sweet taste characteristic of the glucose, it is palatable to the human taste. The formulation is considered to be suited to provide a complete chemical nutrient composition for the human species by the addition to the solution of additional carbohydrate and appropriate amounts of vitamins, minerals, and fat.

Glutamic Acid and Methionine Limits

In another description by *M. Winitz; U.S. Patent 3,697,287; October 10, 1972; assigned to Morton-Norwich Products, Inc.* a series of chemically defined and palatable diets are given, each meeting the following basic requirements:

(a) the inclusion in the formulation of amino acids or substantially simple amino acid derivatives containing sulfhydryl (SH) groups (e.g., cysteine, homocysteine) or other groups which are convertible to sulfhydryl groups (e.g., cystine, homocystine);

(b) the inclusion in the formulation of amino acids or substantially simple amino acid derivatives whose taste is incompatible with the flavor of the mixture of the other ingredients (e.g., glutamic acid or its alkali metal salts, such as sodium glutamate); and

(c) the use of amino acids or substantially simple amino acid derivatives that are not of sufficient purity. Derivatives are considered to be products which evolve by decomposition of the amino acids or by chemical conversion of amino acids to other discrete chemical entities.

The off-taste caused by the sulfhydryl group of amino acids or simple amino acid precursors is avoided when the concentration of such amino acids which contain the sulfhydryl group, or which contain groups that are convertible to the sulfhydryl group, is limited in the food composition.

Where a nutritionally balanced diet requires either glutamic acid or its nutritional equivalent, the off-taste problem can be obviated by providing an equimolar amount of the nutritional equivalent, glutamine, in lieu of all of the glutamic acid. If, for example, a particular protein hydrolysate would contain glutamic acid and/or sodium glutamate in an amount so that the desired end solution would exceed about 1.43 grams of glutamic acid per liter, an objectionable taste is avoided by removing at least a portion of the glutamic acid and substituting a sufficient amount of glutamine so that the glutamic acid (and/or sodium glutamate) concentration will be less than about 1.43 grams of glutamic acid per liter. It appears that the tolerance for glutamic acid is independent of pH in the pH range of interest.

Another potential cause of off-taste in chemically defined diet compositions is the presence of impurities in the individual amino acids (or simple precursors). One of the principal impurities that can cause off-taste is a lower alkyl mercaptan, such as methyl mercaptan, which is a contaminant of most commercial edible grades of methionine. Removal of the simple mercaptan impurities can be achieved by purifying the methionine just prior to its inclusion in a dietary mixture. Purification can be effected by recrystallization from water, or a water-ethanol mixture, or other suitable solvents or mixtures of solvents. The objectionable off-taste imparted by simple mercaptan impurities is avoided by limiting the concentrations of simple mercaptans in a diet composition for human consumption.

It has been found that the tolerable concentration of mercaptan is pH dependent, as was the case with the tolerance for sulfhydryl groups. However, the mercaptan level is generally inversely proportional to the pH. For a food product having a pH of about 5.7 the methyl mercaptan concentration should not exceed about 7.5 milligrams per liter to assure palatability. Moreover, if a composition should exceed this mercaptan level upon initial formulation, the taste worsens upon storage for any appreciable length of time, becoming completely obnoxious to taste and smell.

Certain limits have been set forth above for the limits of concentrations of various chemical moieties to provide for palatable food compositions for human consumption. These limits pertain to the total concentrations of these moieties in the food composition, without regard to the source of such moieties, whether from a given component, any other component, or from their generation as derivatives during compounding or storage.

As previously indicated, the amino acid components of a diet are selected to meet the normal metabolic needs of the subject and to maintain the desired nitrogen balance. In chemically defined diets, the amino acid components, in the form of free essential and nonessential amino acids or suitable precursors thereof, such as peptides, are provided at total levels sufficient to satisfy normal physiological requirements for nitrogen.

Accordingly, the balance among the various amino acids is selected to meet the normal metabolic needs of the human subject and to maintain the desired nitrogen balance. Because of strong interdependencies between the required level of a given amino acid and the level of one or more of the other amino acids present in diet, it is not practicable to establish a precise range of levels for each of the amino acids. Generally, the ratios of the levels of amino acids in the diets should approximate those of a high quality protein, such as meat, eggs, or milk, for example.

Preferred levels of the essential amino acids have been established which can be adapted for supplying the metabolic needs of the human species. A composition using the essential amino acids in these levels with respect to one another is given in the following table. The diet is designed to be taken in two parts, with the fats and fat-soluble vitamins being separated from the remainder.

Amino Acids

L-lysine·HCl	3.58 g	L-tyrosine ethyl ester·HCl	4.10 g
L-leucine	3.83 g	L-aspartic acid	5.50 g
L-isoleucine	2.42 g	L-threonine	2.42 g
L-valine	2.67 g	L-proline	3.42 g
L-phenylalanine	2.75 g	Glycine	4.20 g

(continued)

Amino Acids

L-arginine·HCl	5.70 g	L-serine	1.78 g
L-histidine·HCl·H_2O	1.58 g	L-tryptophan	0.75 g
L-methionine	2.48 g	L-glutamine	9.15 g
L-alanine	2.58 g		

Vitamins

Thiamine·HCl	1.00 mg	d-Biotin	0.83 mg
Riboflavin	1.50 mg	Folic acid	1.67 mg
Pyridoxine·HCl	1.67 mg	Ascorbic acid	62.50 mg
Niacinamide	10.00 mg	Cyanocobalamin	1.67 μg
Inositol	0.83 mg	p-Aminobenzoic acid	416.56 mg
d-Ca pantothenate	8.33 mg	Choline bitartrate	231.25 mg

Minerals

Potassium iodide	0.25 mg	Ammonium molybdate·$4H_2O$	0.42 mg
Manganous acetate·$4H_2O$	18.30 mg	Potassium hydroxide	4.00 g
Zinc chloride	1.25 mg	Magnesium oxide	0.38 g
Cupric acetate·H_2O	2.60 mg	Sodium hydroxide	1.68 g
Cobaltous acetate·$4H_2O$	1.67 mg	Ferrous ammonium sulfate·$6H_2O$	0.68 g
Sodium glycero-phosphate·5½ H_2O	5.23 g	Calcium chloride·$2H_2O$	2.44 g
Potassium sorbate	1.00 g	Sodium chloride	5.35 g

Carbohydrates

Glucose	570 g	Glucono-δ-lactone	17.20 g

Fats and Fat-Soluble Vitamins

Ethyl linoleate	2.000 g
Vitamin A acetate	3.640 mg
Vitamin D	0.057 mg
α-Tocopherol acetate	57.29 mg
Menadione	4.58 mg

In one series of experiments, twenty-five normal adult subjects were confined for a period of twelve weeks to the above diet as the sole source of sustenance. Ingestion of anything else other than distilled water was prohibited. The aqueous portion of the diet was provided ad libitum to each experimental subject who ingested amounts of the water-soluble components sufficient to meet his subjective needs. Caloric intake varied from subject to subject, but averaged 2,100 calories per day per subject for the group as a whole. The fat compound was given to each subject as a separate daily supplement. The diets were considered to be palatable to the subjects.

Each subject followed a daily activity schedule that included regulated moderate physical exercise. The subjects exhibited no unusual symptoms and showed a marked reduction in their blood serum cholesterol level. This reduction is considered to be due to the use of glucose, maltose, starches and/or dextrins as the sole source of carbohydrate. As an example, a liter of solution of a dietary composition might include purified safflower oil or ethyl linoleate in an amount of 0.74 gram, plus or minus about 5%, which may be formed into an emulsion with the aqueous phase using about 37 mg of polyoxyethylene sorbitan monooleate.

Example 1: A composition containing the following essential and nonessential amino acids plus the vitamins, minerals, carbohydrates and flavoring and coloring agents shown in the above diet formulation is prepared. The amounts of the amino acids are altered slightly to the amounts listed below.

	Grams		Grams
L-lysine·HCl	3.58	Sodium L-aspartate	6.40
L-leucine	3.83	L-threonine	2.42
L-isoleucine	2.42	L-proline	10.33
L-valine	2.67	Glycine	1.67
L-phenylalanine	1.75	L-serine	5.33
L-arginine·HCl	2.58	L-tyrosine ethyl ester·HCl	6.83
L-histidine·HCl·H_2O	1.58	L-tryptophan	0.75
L-methionine	1.75	L-glutamine	9.07
L-alanine	2.58		

Each of the amino acids listed above is first recrystallized from water or water-ethanol or other suitable solvent mixture in the presence of activated charcoal. The purified amino acids in the amounts indicated above are then dissolved successively in 500 milliliters of distilled or deionized water at a temperature of 90°C. The composition of the remainder of the formulation is identical with that shown in the above table. The glucose component is added to the amino acid solution at this temperature and dissolved. The mineral components are added either as the individual dry salts or as the aqueous presolutions described below.

Presolution A consists of the potassium hydroxide, magnesium oxide, and glucono-delta-lactone components dissolved in water. Presolution B consists of the trace-mineral salts dissolved in water. Presolution B consists of the trace-mineral salts dissolved in water. After cooling the batch to 30°C, the water-soluble vitamins are then added either individually or as a premix. A peach flavoring agent is then added, plus the citric acid indicated, which brings the pH of the solution to about 3.7. The chemical diet prepared in this manner has been shown in taste-panel tests to be palatable and highly acceptable for human consumption.

Example 2: A composition of the following essential and nonessential amino acids is made by dissolving them in the specified quantities in 500 milliliters of distilled water:

	Grams		Grams
L-lysine·HCl	3.58	Sodium L-aspartate	6.40
L-leucine	3.83	L-threonine	2.42
L-isoleucine	2.42	L-proline	10.33
L-valine	2.67	Glycine	1.67
L-phenylalanine	1.75	L-serine	5.33
L-arginine·HCl	2.58	L-tyrosine ethyl ester·HCl	6.83
L-histidine·HCl·H_2O	1.58	L-tryptophan	0.75
L-methionine	1.75	L-glutamine	9.07
L-alanine	2.58	L-cysteine ethyl ester·HCl	0.92

The preparation procedures for purification of the amino acids and the sequence and conditions for adding the amino acids, carbohydrate, minerals, vitamins, and flavoring for this example are identical with those specified in Example 1. The composition of Example 2 only differs from that of Example 1 in that 0.92 gram of L-cysteine ethyl ester hydrochloride is included in Example 2. The total dilution of each of these compositions is such that the final aqueous solution measures about 2.5 liters. This amount of cysteine ethyl ester hydrochloride is equivalent to about 0.37 gram per liter.

HIGH PROTEIN WAFER

A baked wafer-like product that contains sufficient protein, fat, carbohydrate, vitamins and minerals to replace a full meal was designed by *J.L. Gabby and G.K. Ashby; U.S. Patent 3,185,574; May 25, 1965; assigned to Mead Johnson & Company.* The first consideration in formulating the product is to provide a composition which will supply from 58 to 100 grams of dietary protein per day. This amount of protein fulfills the accepted requirement for maintenance of nitrogen balance in an individual depending upon the level of physical activity in which he partakes. Seventy grams per day is considered an average and adequate amount of protein for regular daily intake except for manual laborers who may require more. This product provides this amount of protein in a daily dietary unit of from ½ to 1 lb of the finished baked pieces.

In order to provide this amount of protein, the baked pieces contain from 15 to 40% by weight of protein. For wafers adapted for weight losing diets, the amount by weight of fat may be as low as 5%. For the preparation of a high calorie wafer for weight graining regimens, as much as 20% by weight of edible fatty glyceride may be included in the baked piece. In addition, from 5 to 25% the weight of the baked piece is made up of moisture, flavors, fillers, nutritionally inert ingredients, etc. These ingredients, when they include vitamins and minerals, are present in sufficient amounts to provide a nutritionally balanced product. The remainder of the composition is made up of digestible carbohydrate, which is at least 35% by weight of the finished baked piece.

Such wafers can be baked in essentially any desired size, but it has been found that a rather thick baked piece (about ½ centimeter) of relatively small diameter (about 5 centimeters), has an appealing appearance and provides about 25 calories per wafer in formulations containing about 9% fat. Wafers of higher caloric content are prepared by increasing the proportion of fat, since, on a weight basis, both carbohydrate and protein have approximately the same caloric equivalent, while that of fat is substantially higher.

The protein additive mixture contains about 31 parts by weight of calcium caseinate, 60 parts by weight of soy protein concentrate, and 2 parts by weight of defatted cottonseed flour per 100 parts by weight. Use of an amount of defatted cottonseed flour of up to 5% of the finished wafer weight is particularly advantageous since the shortening characteristics of the fatty glyceride components are enhanced.

The fat content of the finished wafer is within the range 5% to 20% by weight, and the protein content thereof 15% to 40%. Both of these amounts are related to the amount of water to be used in preparing the dough in the course of manufacturing. The fat content can be provided by a mixture of three parts by weight of coconut oil and seven parts by weight of corn oil, not considering the minor contribution to the fat content provided by other ingredients, such as the flavors, vitamins, etc.

From 5 to 25% of the weight of the finished baked piece is made up of ingredients which cannot be classified as proteins, fats or carbohydrates. This class of materials includes the vitamins and minerals which are added to provide a nutritionally adequate and balanced product. It also includes the fibrous constituents, which usually amount to 4% of the finished wafer weight, and which are inherently associated with the various ingredients, particularly those from cereal sources.

It also includes a certain amount of moisture, usually from 3 to 5%, since it is not possible to remove all of the moisture on baking to provide an absolutely anhydrous wafer. Furthermore, such is not to be desired, since about 3% moisture contributes to the tenderness and texture of the finished baked piece. Within the limits of this range other nutritionally inert materials, such as fillers, binders, and extenders, including indigestible material, such as cellulose derivatives, resin, etc. may be included.

The balance of the baked piece is then made up of digestible carbohydrate components which should be at least about 35% of the finished baked piece weight. If a lower proportion of carbohydrate is used, a crumbly, undesirable baked piece is obtained. Of the digestible carbohydrate content, it is preferred that approximately one-half be sugar, preferably sucrose, and the remainder starch.

As is customarily the procedure in wafer or cookie manufacture, the sugar and shortening ingredients are creamed together, as the first step, along with minor constituents such as the lecithin, egg white, salt, flavors, and others. In the process the vitamins may be incorporated during the creaming operation or in the second step with the wheat gluten. The wheat gluten component, preferably cookie grade wheat flour, and the mineral premix are added to the creamed mixture. A dry, crumbly mixture results to which a predetermined weight of water is added to provide a fluid, batter-like liquid completing the second step.

If desired the water may be added to the creamed mixture prior to the wheat gluten and minerals. This offers no advantage, however. The second step is preferably conducted by first adding the wheat gluten and minerals since the resulting dry mass may be stored for relatively prolonged periods before mixing with water to form the batter and completion of the process. The batter-like fluid formed after addition of the water is prone to microbial contamination and concomitant decomposition if not handled according to procedures accepted by the baking industry.

The third step involves adding the protein additive of low water solubility described above to the batter-like fluid of step 2 and mixing just sufficiently to provide a homogeneous, moldable dough. Five or six minutes of mixing is usually sufficient. Molded dough units are then formed and baked in the usual fashion, e.g., 4 to 8 minutes at 370° to 420°F as the fourth and final step.

The weight of water referred to in step 2 is empirically predetermined by routine experimentation so that just sufficient is used to provide a moldable dough. If insufficient water has been used in forming the batter prior to mixing in the protein additive in step 3, a crumbly dry mixture is obtained rather than a moldable dough. A second trial batch is then prepared using a somewhat larger proportion of water until the proper balance is obtained.

When somewhat too much water has been used in step 2 to permit formation of a dough in step 3, a sticky mixture is obtained which cannot be molded. This difficulty can sometimes be remedied by allowing the mixture to stand for a period at room temperature until the excess water has been absorbed by the protein additive. This period should not exceed about one hour, however, since excessive hydration of the proteins occurs, resulting in a hard, brittle product on baking, or in extreme cases in a dough from which the water cannot be removed under usual baking conditions.

A corollary to this, of course, is that a properly formed dough should be baked within a reasonable time after molding to avoid excessive hydration of the protein. To this end, if the process is to be interrupted prior to baking, it is advisable to withhold adding the protein additive until the baking ovens are ready to receive the molded wafers. Once the protein additive is mixed with the batter-like mass, execution of the remainder of the operation within about one hour is advisable.

The results of a series of experiments of this type which were conducted to determine optimum amounts of water for dough batches having various compositions are graphically arranged in Figure 8.1 which is a plot on rectangular coordinates with the percent fat, that is, edible fatty glyceride, values arranged along the ordinate and the percent protein along the abscissa. These percentages refer to the percent by weight of the edible fatty glyceride and total protein ingredients in the finished baked piece.

The overlapping geometric figures positioned between the coordinates enclose those fat and protein compositions for wafers. The smaller area outlined in heavy print and enclosed within the larger area encloses the preferred compositions for use in weight losing diets. It will be apparent that the present wafers may contain from 15 to 40% protein by weight and from 5 to 20% fat. The protein and fat limits for the preferred compositions for weight loss are 25 to 40% and 5 to 15%.

It is apparent from inspection of Figure 8.1 that certain fat and protein combinations within these limits do not lend themselves to formation of a satisfactory product.

FIGURE 8.1: HIGH PROTEIN WAFER

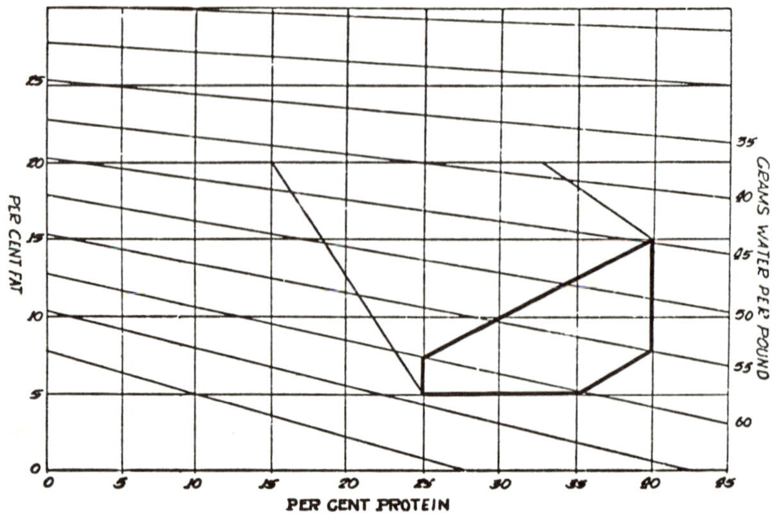

Source: J.L. Gabby and G.K. Ashby; U.S. Patent 3,185,574; May 25, 1965

Thus, when the protein content exceeds 35%, the minimum fat content should exceed 5% by 0.5 times the difference between the actual protein content of the baked piece and 35%. When the protein content is less than 25%, the minimum fat content exceeds the 5% value by 1.5 times the numerical difference between the actual protein content of the finished wafer and 25%. As to the maximum fat content of the finished wafer, it is apparent that the maximum value is 20% except when the protein content exceeds 32.5%. The maximum fat content then is less than that figure by ⅔ of the difference between the actual protein content of the baked piece and 32.5%.

The minimum values of fat content for wafer compositions preferred for weight loss coincide with those given above. The maximum values for fat content of such compositions are uniformly less than the above maximum values. For a wafer containing 40% protein, the fat content may range from a minimum of 7.5% to a maximum of 15%, and for the weight loss compositions which contain less than 40% protein the maximum fat content is less than 15% by ½ the difference between 40% and the actual protein content of the finished baked piece.

Example 1: A vitamin premix is prepared by blending the following ingredients in a suitable apparatus, e.g., a twin shell blender. The amounts given provide sufficient premix for a 1,000 lb batch of wafers.

Vitamin A	17.5 million units*
Vitamin D	1.75 million units
Ascorbic acid	5,000.00 grams

(continued)

Thiamine mononitrate	12.00 grams
Riboflavin	6.60 grams
Niacinamide	33.00 grams
Pyridoxine hydrochloride	4.40 grams
Vitamin B_{12} concentrate, 0.1% (w/w) in gelatin	4.40 grams
Calcium pantothenate	24.00 grams
D-α-tocopheryl succinate	18.18 grams
Powdered cane sugar	1,630.38 grams

*Preferably supplied as a granular gelatin product.

In a similar fashion a mineral premix is prepared from the following ingredients.

Calcium phosphate	2,373.50 grams
Potassium phosphate, dibasic	1,911.00 grams
Potassium carbonate	2,240.00 grams
Magnesium carbonate	1,077.00 grams
Ferrous sulfate	33.60 grams
Manganese sulfate	12.40 grams
Copper carbonate	5.60 grams

For a 1,000 lb batch of cinnamon-molasses flavored wafers, the following ingredients were weighed and transferred to a horizontal mixer and mixed at high speed to form a cream.

Coconut oil	22 lb 8 oz
Corn oil	52 lb 8 oz
Soy lecithin	6 lb
Powdered cane sugar	200 lb
Vitamin premix	from above
Egg white, spray dried	8 lb 13 oz
Salt, iodized	8 lb 13 oz
Lemon oil, USP	1¼ oz
Butter flavor	10 oz
Vanilla flavor	2 lb
Cinnamon, ground	5 lb 10 oz

After a smooth cream is formed, the speed of the mixer is reduced, and 238 lb of wheat flour is blended into the mass until the batch is homogeneous and dry in appearance. Ammonium carbonate, 13 lb 4 oz; water 108 lb, the temperature of which is adjusted to that of the room; and molasses, 82 lb, is then added and the entire batch mixed to a smooth batter. About 4 to 6 minutes is required. The following materials are then weighed and added all at once to the batter without previous blending with one another.

Wheat bran	11 lb
Yeast, torula	20 lb
Cottonseed flour	11 lb
Mineral premix	from above
Calcium caseinate	106 lb
Soy protein concentrate, $^{16}/_{35}$ mesh	254 lb

The batch is then mixed just sufficiently to form a moldable dough, care being taken not to overmix. An additional portion of soy lecithin, 5 lb, is then added and mixing is continued just long enough to coat the dough with the lecithin. The dough batch is then transferred to the hopper of a rotary die cookie molding machine and formed into wafer-like pieces. Each piece weighs 0.22 oz. The wafers are baked 4½ minutes in a band oven, for the first half of the cycle at 420°F and the second at 370°F. The wafers are packed 36 to a ½ lb box which provides 900 calories. These wafers contain on a weight basis 30.8% protein (nitrogen analysis multiplied by 6.25), of which 6.7% is wheat gluten, 54.9%

the proteins of soy flour, 2.1% the proteins of cottonseed flour, and the remainder is supplied by the other protein-containing ingredients, 8.8% fat, and 48.5% carbohydrate, 52% of which is sucrose and the balance starch provided by the various cereal ingredients.

Example 2: This example will illustrate the result of not adhering to the order prescribed for mixing the ingredients. In this example the water is added after combining the protein additive with the other dough components rather than before. The following materials were combined by creaming in a mechanical mixer.

	Grams
Lecithin	3.5
Corn oil	25.3
Coconut oil	10.2
Sugar	121.0
Dried egg white	4.0
Salt	4.0

The mixing speed was then reduced and the following ingredients blended therewith.

	Grams
Wheat flour	111.0
Mineral premix (Ex. 1)	8.0
Soy protein concentrate, $^{16}/_{35}$ mesh size	115.0
Calcium caseinate	48.0
Cottonseed flour, defatted	5.0
Torula yeast	9.0
Wheat bran	5.0

A crumbly dry mass resulted, not suitable as a dough. An amount of water as calculated from Figure 8.1, 57 grams, was added to the mixture and mixing continued. A damp meal rather than a dough formed. If the water was added before the protein additive components (calcium caseinate, defatted cottonseed flour, and soy protein concentrate), satisfactory results would be obtained.

An attempt to improve the texture of the damp mealy mass was made by adding two additional 20 gram portions of water. Thus with a total of 97 grams of water, a mixture barely of the consistency of a dough was obtained which was firm and crumbly, and difficult to roll and cut. It was necessary to bake it 10 minutes instead of the usual 7 minutes. The resulting baked pieces were hard and brittle around the edges and tough in the center. It was considered an unsatisfactory product.

INFANT FORMULAS WITH FATS

High β-Palmitic Fats

Ready-to-feed infant formulations have been prepared by the appropriate blending of ingredients, including varying proportions of selected animal fats to yield a fatty acid composition approximating that of human milk fat. However, the fatty acids of such artificial formulations have not been as well absorbed by the infant, or by experimental animals, e.g., the laboratory rat, as are the natural fatty acids in human milk.

It has now been found that the superior absorbability of human milk fat resides to a great extent in its high content of β-palmitic acid, that is palmitic acid in the 2-position of the triglyceride molecule. It is believed that the positional distribution of palmitic acid is significant because in the intestinal tract, pancreatic lipase digests the triglyceride to form two free fatty acids and a β-monoglyceride. That is, the lipase specifically releases the fatty acids in the α- and α'-positions of the triglyceride molecule leaving substantially intact the fatty acid at the β-position. Also the long chain saturated fatty acids, e.g., palmitic and stearic acids, which are present after digestion of the triglyceride are known to be relatively poorly absorbed, in contrast to the better absorption of these acids when present as the β-monoglycerides. Fats having palmitic acid in the β-position have been found to be well absorbed. Because of the specificity of hydrolysis by pancreatic lipase on the fatty acids in the α'- and α'_1-positions of the triglyceride molecule, a fat with a relatively high proportion of β-palmitic acid will be better absorbed than a fat blend with a similar fatty acid composition but with a high percentage of palmitic acid esterified in the α,α'-position and a lower percentage as β-palmitic acid.

A special fat blend is used by *R.M. Tomarelli and F.W. Bernhart; U.S. Patent 3,542,560; November 24, 1970; assigned to American Home Products Corporation* to improve the fat absorption of infant formulas to be equivalent to human milk. This can be achieved by manufacturing an infant formulation containing a fat blend in which from about 45 to 100% by weight of the palmitic acid present is β-palmitic acid. Especially advantageous results have been obtained with β-palmitic acid concentrations above about 80%. The total palmitic acid content is preferably at least 25% of the total fat content as is further described below in Table 1. The remainder of the formulation may be nonfat milk, such as cow's milk, soy lecithin, carrageenin, vitamins, minerals and water.

It has been found the total fat absorption increases directly with an increase in β-palmitic acid content. A ready-to-feed infant formulation more closely simulating the fat content of human milk may be prepared as follows. Natural animal fats may be selected to supply the desired proportion of β-palmitic acid. Such natural animal fats include lard, porpoise fat, and whale fat.

Many other animal fats and oils may be blended in various proportions to achieve a formulation having the desired characteristics. The formulation may be prepared by blending fats having a relatively high proportion of β-palmitic acid, such as lard and butter oil, with other known fats and oils which have a relatively low proportion of β-palmitic acid, such as corn, soybean, palm, peanut, coconut, olive, babassu, cottonseed, oleo and tallow. Lard has about 85% of β-palmitic acid and butter oil has about 44% of β-palmitic acid. Such blends are shown in Table 1.

TABLE 1: TYPICAL FAT BLENDS HAVING HIGH β-PALMITIC ACID CONTENT

Fatty acid moiety	Weight percent fatty acids					
	70 lard, 10 coconut oil, 20 palm oil		70 lard, 10 coconut, 20 butter oil		Human milk fat	
Lauric	5.5		6.0		4.7	
Myristic	5.5		5.3		7.8	
Palmitic	27.8	(59*)	26.0	(71*)	27.3	(68*)
Stearic	10.3		11.8		9.7	
Oleic	38.8		34.4		34.5	
Linoleic	10.0		8.4		9.5	
Linolenic	0.7		0.9		1.5	
Total	98.6		92.8		95.0	

*Percent of β-palmitic acid based on total palmitic acid content.

An infant food formulation may also be prepared by adding to a β-palmitic acid deficient formulation a concentrate containing triglycerides, diglycerides or monoglycerides, or a combination of them, which contain about 30 to 100% by weight of β-palmitic acid. The concentrates may be of natural or synthetic origin.

A complete ready-to-feed formulation may have the composition shown in Table 2, where w/w is percent by weight of ingredient per total weight of formulation and w/v is percent by weight of ingredient per total volume of the formulation.

TABLE 2: TYPICAL READY-TO-FEED FORMULATION

	Undiluted		
	Liquid, w/v	Powder, w/w	Diluted, w/w
Fat	2 - 10	8 - 40	1 - 5
Carbohydrate	8 - 20	32 - 80	4 - 10
Protein	1.8 - 6.6	7.2 - 26.4	0.9 - 3.3
Ash	0.30 - 1.6	1.20 - 6.4	0.15 - 0.8
Total solids	16 - 34	95 - 100	8 - 17
Water	84 - 66	0 - 5	92 - 83

Example: This experiment shows the role of 2-palmitic triglycerides in the absorption of fats in an animal. Experiments were conducted with standard laboratory animals to explore the relationship of triglyceride structure to fat absorption. The fats studied included human milk fat, butter oil, lard, oleo (beef fat), experimental fat mixtures made up of oleo, palm, peanut and corn oil, identified as OPPC, and an experimental fat mixture made up of lard, butter oil and coconut oil, identified as LBC. The OPPC has a percent by weight fatty acid composition similar to human milk but has a low proportion of β-palmitic acid. The LBC has a percent by weight fatty acid composition similar to human milk fat and has a high proprotion of β-palmitic acid.

Part 1: Determination of the Positional Distribution of the Fatty Acids – The proportion of each fatty acid in the β-position of the triglycerides of the fats under study was determined by the procedure of Brockerhoff and Yurkowski, *J. Lipid Res.* 7, 62 (1966). In brief, the fat samples were digested by pancreatic lipase to a predetermined point; the products of digestion were extracted from the mixture and the monoglycerides separated by thin layer chromatography. The fatty acid composition of the monoglycerides and of the original triglycerides was determined by gas-liquid chromatography.

The percent of the total amount of each fatty acid that occupies the β-position was calculated from the formula:

$$\frac{\text{mol percent of palmitic acid in monoglyceride}}{3 \times \text{mol percent of palmitic acid in triglyceride}}$$

using palmitic acid for an example. An important feature in this composition of the various fats is that palmitic acid is predominantly in the β-position only in human milk, lard, and in the LBC mixture.

Part 2: Determination of Fat Absorption – Young male Sprague-Dawley rats were maintained on a fat-free basal diet consisting of casein, 220 grams per kilogram of diet (g/kg); glucose, 738 g/kg; salt mixture, 40 g/kg; vitamin mixture, 1 g/kg; and choline chloride, 1 g/kg. Each rat received a daily supplement of 0.1 ml of corn oil except for 3 days before the start of an assay and during the 6 day assay period.

At the start of the assay the rats were divided into groups with equal average body weights, 6 rats per group. Each group was fed the test fat, incorporated into the basal diet at the 15% fat level replacing an equal amount of glucose. The fat diets were fed for 3 days and then the fat-free diet for the next 3 days. All feeding was ad libitum. Feces were collected for the entire 6 days. The fat intake was determined from a measurement of food consumption.

In each experiment, a group receiving only the fat-free diet for the 6 days, served as a control for the estimation of endogeneous fecal fat excretion for rats of the same strain, age and body weight. The total fatty acid content of the food and the feces was determined by the procedure of van de Kamer (Hawk, *Physiological Chemistry*, 14th Ed, p 539). Percent of fat absorption based on fat intake was calculated from the following formula.

$$\text{Fat absorption} = 100 - \left[\frac{\text{meq fecal fatty acids} - \text{meq endogenous fatty acids}}{\text{meq food fatty acids}} \times 100\right]$$

The absorption of specific fatty acids was calculated in the same manner, i.e., meq of each fatty acid in the feces was corrected for endogeneous excretion of that fatty acid.

The LBC mixture, resembling human milk fat in total fatty acid composition but containing a slightly higher proprtion of β-palmitic acid, was slightly better absorbed than human milk fat. In contrast, the OPPC mixture also with a fatty acid composition simulating human milk but with a low proportion of β-palmitic acid was definitely less well absorbed.

A direct relationship exists between absorption and percent in the β-position for palmitic acid but does not exist for stearic acid. The superior absorption of the β-palmitic acid of LBC in relation to its proportion in the β-position compared with lard is believed to be due to the higher stearic acid content in the lard which interferes with β-palmitic acid absorption.

From the foregoing results, it may be concluded that the absorbability of a fat is dependent on a number of factors. Of major importance is its content of stearic and palmitic acids, the less well absorbed constituents. The content of oleic and linoleic acid is known to be of importance. For instance, it is known that the absorption of stearic and palmitic acids is improved by the addition of unsaturated oils to a fat mixture. It has been found from the foregoing experimentation, with regard to palmitic acid absorption, that positional distribution of the fatty acid in the triglyceride molecule plays a role in fat absorption. Particularly, it has been found that a high β-palmitic acid content increases the total fat absorption.

In analyzing the data of the present study, it appeared that, at least with the six fats studied, the most important factors interfering with total fat absorption in the rat are the content of palmitic acid in the α-position and total stearic acid content.

Assimilable Fat Blend

In another process *F.W. Bernhart; U.S. Patent 3,649,295; March 14, 1972; assigned to American Home Products Corporation* specifies other fat compositions to "humanize" infant feeding formulas. Specific difficulties with fat mixtures used in infant formulas have been previously pointed out. It was found that the fat in an artificial formula with approximately the same fatty acid composition as breast milk fat was not well absorbed in young infants. Of greater importance was the finding that the lowered fat absorption resulted in low and in some infants negative absorption of calcium. The low-absorption results from binding of calcium by unabsorbed fatty acids, particularly palmitic and stearic acids, forming highly insoluble soaps.

Another variety of fat compositions used in humanized infant formulas is characterized by levels of fatty acids, greatly different from those present in human milk fat. A syndrome consisting of edema, skin lesions, an elevated platelet count and morphologic changes in erythrocytes has been described in premature infants fed formulas containing fat with a relatively high content of linoleic acid. The objective in formulating a "humanized" fat composition for use in infant formulas is to provide a fat mixture which is well absorbed from the intestines, and at the same time provide a physiological level of linoleic acid.

The reason for not obtaining both objects is due to the fatty acid composition of available fats and oils. Fatty acids are absorbed to different extents by the newborn infant. Stearic acid and palmitic acid are poorly absorbed from the infant intestine. All the other fatty acids are absorbed to a considerably greater extent than stearic and palmitic acid. Oleic acid is intermediate in the extent of absorption. The fats which are used in existing infant formulas are grouped into four classes as follows: palmitic acid oils, palm oil; lauric acid oils, coconut and babassu; animal fats, oleo oil (beef fat); and seed oils, corn, peanut, soybean and cottonseed oil.

Among the palmitic acid oils and lauric acid oils, the ones listed above include all members of these classes. The other commercial animal fats (butter, lard, mutton tallow) are not used because of economic considerations (butter), religious (lard) and relative unavailability (mutton tallow). The other available seed oils are not considered suitable for various reasons: linseed because of containing over 50% linolenic acid; olive oil because of uneconomical price, safflower because it contains over 70% linoleic acid; sesame seed oil because of the presence of phenolic compounds and sunflower oil because of unavailability.

Example 1: Compositions of edible fat blend are provided as set forth in the table below, as follows:

Oil	A	B	C	D	E	F	G	H	I	J
Oleic oil	25	25	45	15	20	43	42	39	30	27
Oleo oil	33	33	10	26	35	10	10	30	45	17
Soybean oil	14	15	--	--	10	--	7	--	--	24
Corn oil	--	--	25	12	--	--	--	--	15	--
Cottonseed oil	--	--	--	12	--	15	7	--	--	--
Coconut oil	27	27	--	35	34	15	--	30	10	31
Babassu oil	--	--	20	--	--	15	34	--	--	--
Soy lecithin	1	--	--	--	1	2	--	1	--	1

Example 2: Using the fat composition of the table above, Column A, the following ready-to-feed in infant formula is provided, amounts set forth for 100 ml.

Fat composition	3.6 grams
Lactose	7.2 grams
Protein (electrodialyzed whey and nonfat milk)	1.5 grams
Vitamin A	264 USP units
Vitamin D	42.0 USP units
Vitamin E	0.64 I.U.
Vitamin B_1 (thiamine·HCl)	0.071 mg
Pantothenic acid	0.21 mg
Vitamin B_2 (riboflavin)	0.11 mg
Vitamin C (ascorbic acid)	5.3 mg
Vitamin B_4 (pyridoxine·HCl)	0.042 mg
Vitamin B_{12} (cyanocobalamin)	0.11 mcg
Niacinamide	0.53 mg
Folic acid	3.2 mcg
Calcium	42 mg
Phosphorus	33 mg
Sodium	15 mg
Potassium	55 mg
Chloride	41 mg
Magnesium	6.0 mg
Iron	0.85 mg
Copper	0.04 mg
Zinc	0.32 mg
Iodine	6.9 mcg
Water (qsad)	100.0 ml

The fat-soluble vitamins are dissolved in the fat composition and the remaining ingredients are dissolved in the water. The fat composition and water solution are then mixed and homogenized. Adequate amounts of other trace minerals are present in the whey and nonfat milk. The process is repeated using the other fat composition of Example 1.

INFANT FORMULAS–ELECTRODIALYZED WHEY

Balanced Calcium and Phosphorus

One of the techniques used in processing cow's milk in attempting to simulate human milk product is the ion-exchange desalting method in which cow's milk is processed through beds of ion exchange resins. This technique is costly to operate and difficult to maintain under proper sanitary conditions. More importantly however, such technique does not properly adjust the anion content of the milk, to the level necessary for a human milk substitute. For example, the phosphates, which are undesirably high in cow's milk, are not adequately removed by the technique described. While the technique does affect a degree of ash reduction, certain components are removed disproportionately, thus adversely affecting the essential balance of such components in the final product. This technique, moreover, results in a product having its protein content altered in a manner that results in heat instability when sterilization is carried out on a final product incorporating the desalted milk.

A number of attempts have been made to construct or more particularly compound simulated human milk from individual purified ingredients that is, precipitated washed casein, precipitated washed whey proteins, lactose, fats, vitamins and minerals. This procedure is very costly. Furthermore the desirable water-soluble components such as vitamin B and nonprotein nitrogen components are lost in the protein preparation according to such techniques. In addition the purified proteins cannot be properly resolubilized in the formula to give an uncoagulated feeding preparation like breast milk.

It is, of course, well known that the mineral components of milk, in proper balance, are essential in insuring growth and the proper functioning of the infant. For example, an excess of minerals which must be excreted in the urine imposes an undue burden on the organs of the infant responsible for this function. Accordingly, it is essential that any combination serving as a substitute for human milk provide the proper balance of all the components of human milk.

The difficulties faced in producing a simulated human milk from bovine milk will be more readily understood if attention is given to the table below in which representative compositions of bovine and human milk are given. All values are given on a dry basis.

	Bovine	Human
Total protein, g/100 g	26.8*	12.3*
Casein, g/100 g	22.0	5.0
Whey protein, g/100 g	4.8	7.3
Lactose, g/100 g	38.9	57.4
Fat, g/100 g	28.6	28.6
Calcium, mg/100 g	1,000	264
Phosphorus, mg/100 g	790	123
Potassium, mg/100 g	1,130	450
Sodium, mg/100 g	476	123
Chloride, mg/100 g	846	352
Magnesium, mg/100 g	99	33
Total ash, g/100 g	5.7	1.7
Calories/100 g	540.0	540.0

It will be observed from the table that human milk is richer than bovine in whey protein and lactose, has the same fat content and the same caloric value as bovine milk but has a relatively lower content of all other constituents shown. Processing bovine milk in an economical manner directly to a product having the composition of human milk on reconstitution in the same weight amounts has been a most difficult problem.

The use of electrodialyzed whey to reduce and balance the ash content of cow's milk is discussed by *R.E. Clark, E.R. Spence and R.C. Stribley; U.S. Patent 3,320,072; May 16, 1967; assigned to American Home Products Corporation.*

It has been found that if bovine milk is subjected to electrodialysis in a manner that reduces the ash content to an amount approximately 30% to 50% that of normal nonfat bovine milk, and electrodialyzed whey is added which has been desalted so that the ash of whey is about 10% that of normal whey, edible lactose and edible fat provides quite unexpectedly a satisfactory simulated human milk product. Such a product is characterized in having a reduced total mineral content substantially consistent with the mineral content of human milk, an ash content on a reconstituted basis of from 0.21 to not more than 0.27%, clearly consistent to the total ash content of human milk and a whey protein to casein ratio substantially similar to that of human milk. The product will also have a whey protein to casein ratio of about 1.46:1 and a substantially correct protein to phosphorus ratio when compared to human milk. Additionally and importantly, such a product has the advantage of being heat stable during the required high temperature sterilization processing.

In preparing the simulated milk compositions utilizing electrodialyzed milk, it is necessary that minor amounts of edible lactose and fats be added in preparing the final composition. The edible fats suitable for this purpose include the animal and vegetable fats as well as blends of the same. The fats used are preferably low in C_4, C_6 and C_{12} to C_{14} saturated fatty acids. A fat blend suitable for the defined purpose contains a major amount of an oil such as oleo oil together with lesser amounts of coconut oil, corn oil or soybean oil and the like.

Though the process requires the use of electrodialyzed milk, it is more specifically and preferably directed to a simulated milk product of the type and composition described utilizing a substantial portion of electrodialyzed skim milk the ash content of which has been reduced to 34 to 45% of normal skim milk. As previously indicated, it is necessary that the ash of whey be reduced to not more than 10% that of normal whey. This will bring the phosphorus level to about 0.14% (dry basis). It is preferred however that the phosphorus level be less than 0.1%, therefore the degree of whey desalting preferred is that which reduces the ash of whey to the range of 2 to 4% that of normal whey.

In addition to the essential and important components above-identified for the simulated human milk product it is of course understood that trace amounts of minerals and/or vitamins can be added to provide the final composition with these elements as desired, for example, metals such as ferrous iron, copper, potassium, sodium and calcium in their salt forms such as the citrate, chloride or bicarbonate may be used.

The most important will be the calcium salt the addition of which may be necessary to meet a specification standard, that is to provide a desired calcium phosphorus ratio when a particular batch of the electrodialyzed milk being used is deficient in calcium. This of course can occur when slight variations in the electrodialysis processing occur. While the continuous sampling of the dialysis product will minimize operating conditions which temporarily exceed predetermined standards for a specified product, variations are readily corrected by addition of these salts as suggested.

The electrodialyzed milk and electrodialyzed whey supply significant quantities of the B vitamins which substantially reduce the cost of the added vitamins. In the preparation of the formula, vitamin levels are then further adjusted by addition of vitamins to the formula as necessary to provide the required levels. Such levels are recommended for a daily allowance basis by the National Research Council.

The infant feeding composition of this process contains: fat blend, 1.0 part; lactose, 0.45 part; electrodialyzed skim milk, 0.56 part and electrodialyzed whey, 1.5 parts. Ca is present in the final composition by reason of the specific treatment of the electrodialyzed whey and electrodialyzed skim milk in an amount of from 140 to 160 mg/100 g of formula solids. Phosphorus however is present in an amount of from 117 to 130 mg/100 g substantially identical with human milk. The addition therefore of only 100 to 130 mg Ca per 100 g provides the same Ca/P ratio as in human milk.

The fat blend would typically consist of the following: oleo oil 55%, coconut oil 17.5%, corn oil 10.0%, soybean oil 16.5% and soybean lecithin 1.0%. To this mixture, there are then normally added the critical and required minerals such as ferrous iron, copper, calcium, potassium and sodium, preferably in the salt form and totalling in the composition described, in the same part weight ratio basis not more than 0.0015 part. Finally there are added to the composition the vitamins identified below, which are on the basis of a reconstituted liquid quart (2.1 lb containing 12.4% total solids):

Vitamin A	2,500 USP units
Vitamin D	400 USP units
Vitamin E	6 I.U.
B_2 (thiamine hydrochloride)	0.67 mg
B_3 (riboflavin)	1 mg
C (ascorbic acid)	50 mg
B_6 (pyridoxine hydrochloride)	0.4 mg
B_{12} (cyancobalamin)	1 mcg
Niacinamide	5 mg

Example: Partial Desalting of Nonfat Milk by Electrodialysis – The equipment used in providing the desired electrodialyzed nonfat milk was an electrodialysis stack consisting of 20 cell pairs (one desalting cell and one concentrating cell comprise a cell pair) 9" x 10" and approximately 1 mm thick together with standard electrodes and end plates. The dialysis membranes used were the homogeneous, permselective type designated as Ionics CR 61, cation permeable, and AR-111A, anion permeable. The stack is provided with a suitable source of direct current, and is equipped with flow, pH, conductivity, and electrical meters, as well as appropriate gauges, pumps, piping and tanks.

In carrying out electrodialysis 45 pounds of raw skim milk was heated to 165°F, held for 30 seconds, and cooled to 120°F. The milk circuit of the dialysis stack was standardized by pumping 100 ppm chlorine solution through the stack before use. The chlorine solution was rinsed out with distilled water and the stack filled with water. Prior to initiation of dialysis, a sample of the skim milk in the stack was analyzed as shown hereafter. The brine solution used was prepared from 32 gallons of water heated to 120°F to which 175 grams NaCl and 80 ml concentrated HCl were added.

The electrode rinse solution was prepared from 5 gallons of water, heated to 120°F to which was added 165 grams K_2SO_4 and 35 ml concentrated H_2SO_4. A starting current of 0.4 amp was delivered to the stack. During operation, samples were removed at 30-minute intervals for solids and ash determination. The pH gradient across the stack was held between 0.1 and 0.2 pH units. The average gradient across the stack for the run was -0.19 pH units. The composite milk averaged 0.0023 pH units drop per minute of operation. The run was stopped after 2:45 hours at which time the milk had a pH 6.12.

The dialyzed milk, after being diluted with rinse water, had 7.29% solids and 0.4498% nitrogen (protein/solids–39.369%). Commercially prepared electrodialyzed whey (EDW) powder having the following composition was used. The fluidized electrodialyzed whey was neutralized to pH 6.8 with KOH after electrodialysis.

Composition of EDW Solids	Percent
Total proteins	14.5
Casein	0
Whey proteins	14.5
Lactose	84.4
Ash	1.1
Calcium	0.12
Phosphorus	0.13
Sodium	0.01
Potassium	0.3
Chloride	0.03
Magnesium	0.01

The electrodialyzed whey, electrodialyzed skim milk, lactose and fats were combined to give 10 lb of total solids (25% TS). The mix was agitated and heated to 160°F. The minerals were then added. The formula was clarified, high-temperature-short-time pasteurized (180°F, 10 seconds), cooled and spray dryed.

Special Concentration Process

In another description, *R.E. Clark, E.R. Eckhardt, R.R. Spence and R.C. Stribley; U.S. Patent 3,201,245; August 17, 1965; assigned to American Home Products Corporation* describes the critical step of concentrating the whey prior to electrodialysis.

Example 1: A quantity of sweet cheese whey was preheated to 167°F in a high velocity tubular preheater held at 167°F for about 5 seconds and evaporated to approximately 28% total solids in a double effect evaporator in which the maximum product temperature in the first effect was about 165°F. The concentrated whey was electrodialyzed in an electrodialysis stack equipped with ion-selective membranes to an ash content of 1% dry solids basis and spray dried under conditions which did not result in significant heat denaturation of the proteins. The electrodialyzed whey powder had a composition of 14% proteinous material, 81% lactose, 1.0% mineral salts and 3.5% moisture.

A quantity of this electrodialyzed whey powder was combined with 9% solids nonfat milk, edible grade lactose, a blend of oleo oil, coconut oil, corn oil, soybean oil, and soybean lecithin, vitamins, minerals and water to give a preparation having about 24.1% total solids. The mixture was then cooled to 45°F and introduced into cans and heat sterilized at 265°F. An analysis of the product appears in the table below.

Example 2: A quantity of cheese whey is heated to a temperature of 155°F and immediately flashed down to 140°F by admitting the same to the first stage of a double effect evaporator. The sweet cheese whey is concentrated to approximately 28% total solids content and is passed to an electrodialysis apparatus and dried as in Example 1. The electrodialyzed whey then has an ash content less than 2% dry basis. The electrodialyzed whey (1 lb) is mixed with a nonfat milk (3.5 lb) together with 0.25 lb of lactose and 0.5 lb of the fat blend of Example 1 above. The mixture is stirred and heated to a temperature of 165°F, homogenized at a pressure of 3,000 psi and heat sterilized at a temperature of about 260°F. An analysis of the product is set forth in the table below.

Example 3: Another test was carried out according to this method of the previous example in which a triple effect evaporator was used at a temperature of 158°F. Final blend of product was carried out in the amounts and conditions previously stated. The product was sterilized at a temperature of 240°F. The analysis of the characteristics of this product is shown in the table below.

Example 4: Following the procedure of the previous example a whey was pretreated such that this whey, prior to electrodialysis, was evaporated in a single stage evaporator at a temperature of 130°F. A product was prepared in the amount and manner stated in Example 2 and sterilized at a temperature of 250°F. The analysis of this product following sterilization is set forth in the table below.

Whey Preheat Temperature	Viscosity of Sterilized Infant Formula	Visual Appearance of Infant Formula
Example 1 – 167°F	Not measurable	Coagulated
Example 2 – 155°F	6.0 cp	Smooth*
Example 3 – 158°F	7.5 cp	Smooth*
Example 4 – 130°F	7.2 cp	Smooth*

*Smooth appearing upon draining from a spoon and also gives a smooth cloud-like appearance when dropped in clear cold tap water.

From the foregoing data, it is apparent that the improvement which is directed to controlling the temperature of the cheese whey prior to and during concentration or evaporation in the manner specified results in a most satisfactory product even when sterilized at high temperature as is required. On the other hand, it will be noted that when this temperature exceeded 165°F an unsatisfactory product resulted.

INFANT FORMULAS—*Lactobacillus bifidus*

N-Acetyl-d-Glucosamine Enhancement

F.W. Zilliken; U.S. Patent 3,274,003; September 20, 1966; assigned to American Home Products Corporation proposes the addition of N-octanoyl, N-benzoyl or N-carboethoxy-d-glucosamine to infant feeding formulas to enhance the growth of *Lactobacillus bifidus*. A typical commercially available infant feeding composition, derived from cow's milk but having major physical and chemical properties essentially the same as human milk, contains approximately 1.5% of protein, 7.0% of carbohydrate, 3.5% of fat, 0.38% of ash and a mineral constituent, such as iron, present in a relatively minor amount as, for example, approximately 0.0005%, as the food is made up as a liquid ready to feed to the infant.

Such a composition is, in many respects, similar to human milk, will supply approximately 20 calories per fluid ounce, and contains, in addition, various vitamin constituents, such as vitamin A, thiamine, riboflavin, vitamin C, vitamin D and niacin, all present (with the exception of vitamin C) in amounts considerably in excess of the amounts in which those vitamins are present in an equal volume of human or breast milk. It is to be understood, of course, that the composition described is but one of many available infant feeding formulas, all based on, or derived from, cow's milk. Such compositions are usually supplied, either in a concentrated liquid form which is adapted to be diluted with water before use, or as a solid powder or granular composition which is dissolved in water in specified proportions to provide solutions of specified concentrations.

While some of these commercially available infant food compositions are similar to human milk in their major physical and chemical properties, there are, however, certain important differences in the predominating microorganisms present in the intestinal bacterial flora characteristic of an infant fed exclusively with human or breast milk, as compared with the intestinal flora of an infant who is fed with one of the commercial food formulas based on, or derived from, cow's milk. The intestinal flora of the breast fed infant is characterized by the presence of the microorganism *Lactobacillus bifidus* as the predominating microorganism. As seen in smears when freshly isolated from the infant's feces, the *L. bifidus* organism appears as gram-positive, straight or curved rods that are nonmotile and that do not form spores. One or both ends may appear to be split longitudinally to give the effect of two short branches, this appearance leading to the term "bifid." One end of the organism may be bulbous or racket-shaped.

While this microorganism predominates in the intestinal tract and feces of infants fed with human milk from lactating mothers, it is not characteristic of the predominating microorganism in the intestinal tract of infants fed with other foods, such as cow's milk, or with the commercially available food preparations derived from, or based on, cow's milk. Infants so fed show a less uniform intestinal flora with an appreciable proportion of gram-negative rods and cocci. The fact that a stable *L. bifidus* flora is characteristic only of the intestinal tract of breast-fed infants is now fully understood and appreciated by pediatricians and other specialists in the medical, biological, and nutritional sciences.

In the infant colon in which a stable *L. bifidus* flora is lacking, it is now generally believed that the infant is more likely to suffer from disorders attributable to the growth of other bacteria, some of which are harmful to the infant, since competition for the available food supply in the infant intestinal tract where *L. bifidus* flourishes reduces the possibility that other bacteria, particularly the harmful *E. coli,* may grow in considerable numbers or in extensive colonies.

In any event, where the gram-positive bacillus *L. bifidus* is present in large numbers in the infant intestine, which is now found only when the infant is breast-fed, there is considerably less growth of other microorganisms, particularly of the gram-negative *E. coli.*, the Clostridia, and certain air-borne microorganisms such as the Sarcinae bacteria. While the reasons for the superior resistance to disease of breast-fed infants as compared with infants who are fed on cow's milk, or on the usual infant formulas in which cow's milk is a principal constituent, are not fully understood, pediatricians are generally agreed that in some measure, at least, the superior resistance is attributable to the presence of an extensive *L. bifidus* flora in the intestines of nurslings, this bacterium being absent in any considerable numbers from the intestines of infants who are not fed with human milk. Extensive research work has been carried out on the identification and isolation of the active biological principle which is responsible for promoting the growth of *L. bifidus* in the intestines of infants who are nourished with human milk, which active principle is evidently not present in sufficient amount to insure the growth of *L. bifidus* as the predominant microorganism in the intestinal tract of infants who are fed with cow's milk or formulas based on cow's milk.

This research has demonstrated that the specific growth-promoting substances or factor is probably also present, at least to some extent, in cow's milk as well as in human milk, but the amount contained in cow's milk is so small as to be only a very small percentage of the amount contained in an equal volume of human milk. It is far insufficient in quantity to provide the infant with enough of the growth-promoting substances or factor to insure dominance of the *L. bifidus* microorganism in his colon and feces.

While the concentration of the specific growth promoting factor for *L. bifidus* in cow's milk varies to some extent with various samples of milk, and to some extent with the particular strain or variety of the *L. bifidus* microorganism which is selected for the test, ordinarily the specific growth-promoting factor is present in cow's milk only in about $\frac{1}{30}$ the concentration (approximately 3.33%) in which it is present in human or breast milk. This concentration is evidently too low to insure that *L. bifidus* will be the predominating microorganism in the intestinal flora of infants fed with cow's milk.

As only about one-half of the solid content of the commercial infant food preparation of which the analysis is given above is cow's milk, the *L. bifidus* growth promoting factor is actually present in such an infant food formula only to the extent of approximately $\frac{1}{60}$ (approximately 1.67%) of the amount in which it is present in an equal volume of human milk. Thought has therefore been given to the possibility of adding to the infant food compositions some material or substance which would supply the specific *L. bifidus* growth promoting substances or factor in greater amount without adversely affecting other characteristics of the composition, so that this growth-promoting factor could be present in the infant's nourishment to an extent sufficient to insure the predominance of the microorganism *L. bifidus* in his intestinal tract.

It has been discovered that the addition of one of a very limited group of N-substituted-d-glucosamines to an infant composition of the type described provides the growth promoting factor to support the growth of *Lactobacillus bifidus.* The compound found useful for these purposes is one of the group consisting of N-octanoyl-d-glucosamine, N-benzoyl-d-glucosamine or N-carboethoxy-d-glucosamine. These compounds are available commercially or can be prepared by known methods.

As has been suggested, the compounds used in the infant feeding composition of the process are included therein in small amounts. These compounds have been found to be unexpectedly more effective than other somewhat related compounds used for this purpose. The growth-promoting activity of the compounds may be expressed as the number of micrograms of the material which are necessary to give a standard response E, when tested tested with strains of *L. bifidus,* such as *L. bifidus* var *penn.,* which response E is that given by 0.06 milliliter of human milk of average growth-promoting factor. The growth-promoting activity of *L. bifidus* of breast milk varies to some extent, and the standard unit, E, is based on average conditions.

It is recognized that N-acetyl-d-glucosamine possesses growth-promoting activity that averages about 2,800 micrograms/E. However, it has been found that the selected compounds have a growth-promoting activity on an average of ten times that of the N-acetyl-d-glucosamine referred to when tested on a comparative basis. More importantly, however, the selected compounds are each less toxic than the N-acetyl-d-glucosamine compound.

For example, in rat studies N-octanoyl-d-glucosamine is substantially nontoxic, having an LD_{50} of less than 10 g/kg rat. Similarly, N-benzoyl-d-glucosamine and N-carboethoxy-d-glucosamine have toxicities of 7 g/kg and 8 g/kg respectively. This is to be contrasted with the N-acetyl-d-glucosamine which has a toxicity of 2 to 3 g/kg when tested under identical conditions. In the table below is set forth the measure of the growth-promoting activity of the compounds as compared with N-acetyl-d-glucosamine. The data for all compounds is based on the same test conditions.

Compound	Micrograms/E
N-acetyl-d-glucosamine	2,800.0
N-octanoyl-d-glucosamine	292.0
N-benzoyl-d-glucosamine	194.0
N-carboethoxy-d-glucosamine	172.0

The selected compounds are almost completely unabsorbed from the intestinal tract thus permitting them to reach the lower parts of the intestine and produce a better bifidus flora. This is evidenced by the data set forth in the table below, based on a series of tests carried out according to the inverted sac method, in which 5 inches of jejunum (rabbit) are washed, suspended, inverted, filled with the compound to be tested and sealed in a beaker glass maintained at 30°C.

	Original Concentration, mg	- - -Outside Concentration*- - - After 3 hr, mg	After 6 hr, mg
N-acetyl-d-glucosamine	500	250	200
N-carboethoxy-d-glucosamine	500	480	450
N-benzoyl-d-glucosamine	500	460	420
N-octanoyl-d-glucosamine	500	490	475

*As measured by the Morgan-Elson test and total reducing capacity, using D-glucose as a reference.

It has been found that the amount of the N-octanoyl-d-glucosamine, N-benzoyl-d-glucosamine or N-carboethoxy-d-glucosamine, which should be incorporated in an infant feeding composition, where that composition is based on cow's milk and is of the type generally described above, may range from about 0.5% to 5.0% by weight, this percentage being based on the total weight of solids present in the composition. Generally speaking, an amount of the selected compound of about 1.0% will be satisfactory.

Lactite and Lactulite Enhancement

It is known that lactulose has a satisfactory bifidus-activity, that is to say, that when infants not nursed at the breast are given a food preparation having a lactose (lactulose) protein quotient greater than 2.6 and containing 0.5 to 7 g of lactulose per 70 calories, a pure or substantially pure bifidus flora is produced in the intestinal tract of the newborn children.

F. Petuely; U.S. Patent 3,239,358; March 8, 1966; assigned to North American Philips Company, Inc. found that not only with lactulose but also with its hydration product, which is a mixture of lactite and lactulite, with either of these substances food preparations for infants may be made which enable a pure or substantially pure bifidus flora to be produced in the intestinal tract.

The mixture of lactite and lactulite can either be produced by hydrogenation of an aqueous solution of lactulose with Raney nickel and hydrogen or by mixing lactite with lactulite.

The Formula 1 (β-galactosidosorbite) can be assigned to lactite and the Formula 2 (β-galactosidomannite) to lactulite. Formula 3 is attributed to lactulose.

(1)

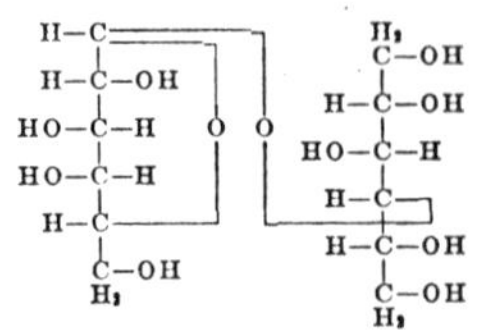

(2)

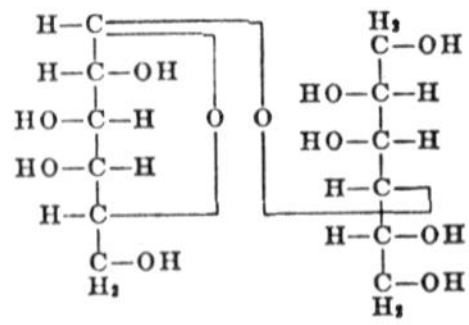

(3)

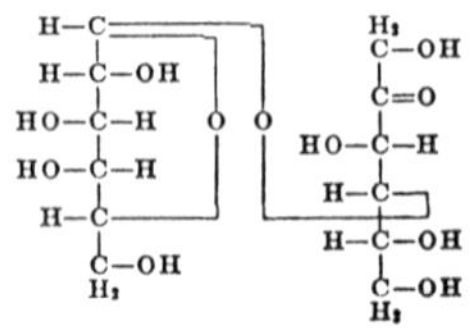

Lactite is a known compound the production of which is described in a thesis of the Technische Hochschule at Brunswic "Darstellung von Cellobiulose und Melibiulose" (Hoepke), 1958. The compounds can be produced by hydrogenation of lactose, for example with Raney-nickel and hydrogen, at a pressure of about 160 atm.

Lactulite is produced by hydrogenation of lactulose. By separating off the lactite formed in the same reaction, the lactulite can be obtained in the pure state. The hydrogenation may be carried out in the same manner as that described by Hoepke for the production of lactite. The separation of lactite from lactulite may be performed by the known methods for the separation of sugars or sugar-like compounds. Satisfactory results were obtained with chromatographic adsorption.

For the production of the preparations, the addition to a conventional food preparation suitable for children and having a lactose-protein quotient exceeding 2.0, of from 0.5 to 3 g of lactite or from 0.2 to 2 g of lactulite or a mixture of 0.1 to 2.5 g of lactite and 0.1 to 1.8 g of lactulite per 70 calories is sufficient.

The finished food preparation should have a lactose-protein quotient between 2.0 and 8, preferably between 2.5 and 5. In general, a pure bifidus flora is produced by the daily administration to an infant of an amount of a food preparation corresponding to 100 to 150 calories, preferably to about 120 calories, per kg of body weight.

In most cases a daily dose of about 12 g of lactite or 6 g of lactulite is sufficient to produce and maintain a pure bifidus flora. This small dose of lactulite is an improvement upon the use of lactulose which usually has to be used in daily amounts exceeding 6 g. The lactite has the practical advantage that it can be more readily prepared than lactulose, inter alia owing to the formation of a smaller number of by-products.

Example 1: Method of Producing Lactite – 250 g of lactose were dissolved in 170 cc of distilled water at a temperature of 80°C and then introduced in an autoclave together with 12.5 g of Raney-nickel. Hydrogen was then introduced into the autoclave at a pressure of 120 atm, the (rotating) autoclave being heated to a temperature of from 140° to 150°C. During heating, the pressure increased to about 160 atm. During the treatment the pressure decreased necessitating the introduction of more hydrogen.

After about 4 hours the reaction was completed, as was apparent from the fact that the pressure remained constant. The reaction liquid was cooled and filtered. The nickel salts were removed from the green filtrate by passing it over ion exchangers. The cation exchanger used was Lewatit K.S.B. and the anion exchanger was Lewatit M1. The yield was 94%.

Example 2: Method of Producing Lactulite – In the same manner as described in Example 1, 15 g of lactulose dissolved in 150 cc of water were reduced with a Raney-nickel catalyst. The maximum temperature used was 90°C, while the hydrogen pressure was maintained at about 120 atm.

Example 3: Method of Producing a Bifidus-Active Food Preparation – 100 cc of pasteurized whole milk having a fat content of 3.8% were mixed with 50 cc of boiled water, 8 g of lactose and 2 g of lactite. The mixture had a calorific value of about 150 calories. Infants from 0 to 6 months were given in the usual manner daily amounts of this food such that they received about 120 calories per kg of body weight. Within from 2 to 4 days, a substantially pure bifidus flora was found in the feces of all the children.

Example 4: In exactly the same manner as described in Example 3, a food preparation was made from 100 cc of pasteurized whole milk, 50 cc of boiled water, 8 g of lactose and 1.5 g of lactulite. When this mixture was used to feed infants, a substantially pure bifidus flora was produced.

Ion-Exchange Enrichment Process

It is known that the microorganism Bifidus (synonyms: *Bifidibacterium bifidum, Lactobacillus bifidus, Actinobacterium bifidum*) is to be found preponderantly in the alimentary canal of suckling children. In the feces of suckling children nourished with mother's milk the *Lactobacillus bifidus* represents 90% of the intestinal bacteria. When changing the feeding from mother's milk to the usual milk a complete change of the intestinal bacteria takes place.

After this the *Lactobacillus bifidus* vanishes completely or otherwise represents only a little part of the intestinal bacteria of the suckling child nourished with cow's milk and a still smaller part of the intestinal bacteria of children fed with mixed nourishment and of adults. Tests have been made for a long time to change the composition of the food of suckling children not nourished with mother's milk so as to ensure that the *Lactobacillus bifidus* remains in the feces. Compositions of foodstuffs effecting this are called "bifidogen." Such bifidogen foodstuff can be obtained by:

(1) Food additives, the reaction mechanism of which is unknown,
(2) Additions to cow's milk of growth factors which are necessary for the growth of *Lactobacillus bifidus.*

Included in the first group is the method which consists in the increase of the ratio lactose/protein and in the addition of lactulose. In vitro this method shows no effect on *Lactobacillus bifidus.* In order to realize the second possibility it is necessary to know the growth factors which are needed by *Lactobacillus bifidus.*

Certain special mutants need a factor (factor bifidus 1) the special activity of which can be met by a series of derivatives of N-acetyl glucosamine, in particular by the 4-0-galactosidyl-N-acetyl glucosamine and the β-alkyl-N-acetyl glucosaminide. It is supposed that this mutant fills up a nonidentified gap in the chain of reactions which give rise to the mucopeptide in the coat of the stomach and of the intestine.

It has been shown that nearly all of the bacteria recoverable from the feces of suckling children do not need this factor, but only grow in the presence of another substance, called "bifidus 2" (BF_2). The presence of the factor has been exactly verified. *Lactobacillus bifidus* grows in a nutrient medium containing various amino acids, p-aminobenzoic acid, biotin and calcium pantothenate only if a preparation containing the factor bifidus 2 is added to the mixture. The nutrient medium used consists of the following ingredients and has a pH of 6.4.

Asparagine	5 g
Amino acids obtained by acid hydrolysis of casein (Casamino-acids Difco)	10 g
Cystine	0.200 g
Lactose	10 g
Sodium acetate·$6H_2O$	5 g
$MgSO_4 \cdot 7H_2O$	0.5 g
$Na_2HPO_4 \cdot 12H_2O$	2.33 g
KH_2PO_4	0.90 g
p-Aminobenzoic acid	1 mg
Biotin	0.010 mg
Calcium pantothenate	1 mg
H_2O, qs	500 ml

2.5 ml of this medium are used together with a standardized inoculum resulting in a total volume of 5 ml. One unit BF_2 corresponds to that amount which after 5 days anaerobic fermentation at 100°F results in the growth of that amount of germs which result in the production of such amount of acids (acetic and lactic acid) which is equivalent to one ml N/10 acid per test tube of 5 ml contents. This method verified that the factor BF_2 is not identical with any of the known growth factors. In particular, it cannot be replaced by any purine (adenine, guanine, hypoxanthine), further it cannot be replaced by streptogenin-rich preparations.

It has been found that the factor BF_2 is present in enzyme hydrolysates of proteins, in particular of casein, in different amounts depending on the reaction conditions. It has been possible to produce BF_2-rich enzyme hydrolysates with 5,000 to 20,000 units/g. By administering these preparations to suckling children it has been found that with the addition of the factor BF_2 to a bifodogen foodstuff, the amount of *Lactobacillus bifidus* in the feces is essentially increased. If sufficient amounts of BF_2 were used *Lactobacillus bifidus* represented 40 to 95% of the intestinal bacteria.

H.G. Hostettler; U.S. Patent 3,556,801; January 19, 1971; assigned to Ursina AG, Switzerland describes a procedure to obtain a material rich in the factor BF_2. In the first step of the chromatographic process the protein hydrolysate is passed through a column of a strong acid cation exchange resin in the acid form, for instance Dowex 50 X 4 (mesh size 50 to 100) in the acid form.

100 g protein hydrolysate containing the factor BF_2 are dissolved in 1 N sulfuric acid (e.g. 100 g in 1,000 ml) and any insoluble fraction is removed by filtration. The pure solution was placed on top of a column of Dowex 50 X 2 (mesh size 50 to 100); for instance, a column of 2.5 cm diameter and 35 cm height can be used. The fraction passing through the column without being adsorbed is disregarded.

The fraction being adsorbed by the resin is eluted at first with 2 N ammonia and then with water (e.g., 1,000 ml ammonia, 1,000 ml water). This fraction is concentrated under reduced pressure removing the greatest part of the ammonia. Instead of the Dowex 50 other strong acid cation exchange resins can be used for the first chromatography, e.g., Amberlite IR 112 and 120, Chempro C 20, Duolite C 25, Ionac C 240, Lewatit S 100, Nalcite HCR, HGR and HDR, Permutit Q and RS, Wofatit KPS 200 and Zeokarb (Zerolit) 225.

Then the concentrated eluate from the first chromatography is processed through a column of a weak acid cation exchange resin, such as Amberlite IRC-50 (mesh size 50 to 100) using as above a column of 2.5 x 35 cm filled with 500 g of Amberlite IRC 50. The resin has been buffered to pH 4.7 by passing a 0.1 M solution of ammonium acetate through the column. The concentrated solution obtained by the first chromatography is placed on top of the second column and the fraction not adsorbed by the resin is collected. The elution is accomplished by passing an ammonium acetate buffer solution through the column. Both the not adsorbed fraction and the eluate are evaporated under reduced pressure and are dried by sublimation in vacuo or by lyophilization. The product is designated as fraction N in the following part. In the above example the volume of the solution placed on the top of the column during the first chromatography was 1,000 milliliter buffer solution. The BF_2 activity of fraction N was 20,000 to 25,000 units per gram. The fraction contains 12 to 16% of the original protein hydrolysate. The fraction N can be used as a half pure preparation of BF_2 in foodstuffs for suckling children. Instead of the Amberlite IRC-50 used on the above process other weak acid cation exchange resins of the carboxyl type can be used for the second chromatography; examples for such resins are Duolite CS 101, Permutit H 70 and C, Wofatit CP 300 and Zeokarb (Zerolit 226).

The factor BF_2 can be obtained in a more concentrated form according to the following technique: A preparation is subjected to a further chromatography process using Dowex 50 W X 2 (mesh size 200 to 400). For chromatographing a solution of 2 g of fraction N in 100 ml water, 750 g of the resin in a column of 2.5 x 15 cm can be used for example. At first the resin is treated with 0.2 N ammonium acetate buffer solution of pH 5.46. The elution is accomplished with the same buffer solution which is passed through the column at a constant rate (e.g., 120 ml/h). The effluent fraction containing the factor BF_2 is collected. In the above example the desired factor is found in the fractions from 3,200 to 7,500 ml.

The solution is evaporated in vacuo and dried by sublimation or lyophilization. Using the most favorable reaction conditions, from 100 parts by weight of the original hydrolysate 0.56 part by weight of the product is obtained with 28% of the original BF_2 activity. The product has an activity of 500,000 BF_2 units per gram. According to another technique the fraction N is filtered through dextran gel (e.g., Sephadex G_{25}) in an aqueous solution as concentrated as possible (e.g., 20 mg/ml). The fraction not adsorbed by the gel is collected; it contains nearly the total factor BF_2 (yield 90 to 100%) and shows an activity of 25,000 to 50,000 units per gram (fraction N_s). This fraction can be used as a bifidogen preparation.

According to another technique the fraction N (or the fraction N_s) can be chromatographed in a basic anion exchange resin, such as for instance Dowex 1 x 2; however, any other basic ion exchange material producing the desired concentration of the bifidogen factor can be used.

The elution is accomplished with pyridine acetate or with collidine acetate containing increased amounts of acetic acid. The eluted products are identified by the ninhydrin reaction. They are eluted as a series of altogether about 20 peaks. The number of peaks varies dependent on the kind of the hydrolysate used for the preparation of fraction N (or N_s). The fraction enriched with BF_2 corresponds to a single peak.

Example: The column used for the chromatography had a diameter of 2 cm and a height of 150 cm and was filled with Dowex 1 x 2; the temperature was 95°F and the rate of flow 40 ml/h; the eluate was collected in 10 ml fractions.

At first the column is eluted with a pyridine acetate solution (1%; pH 8.0) then with a mixture of 1% collidine plus 1% pyridine plus enough 0.1 N acetic acid that the pH is 7.5. Then 2 N acetic acid is used for the elution (fractions 150 to 250) and finally glacial acetic acid.

A solution containing 100 to 300 mg of fraction N is placed on the top of the column. The BF_2 activity is eluted with peak No. 17, the other peaks contain inactive material. The so-obtained product is designated as N_p or $N_{s\text{-}p}$. Its activity is 500,000 to 1,000,000 BF_2 units per gram (yield of factor BF_2 from 20 to 70%). This product can also be used as bifodigen preparation.

If any of the various semipurified or highly purified preparations described above are administered together with the usual foodstuff to a suckling child, in the feces of whom no

Lactobacillus bifidus is to be found, this bacterium is developed. After the administering a period of 1 to 15 days will elapse until *Lactobacillus bifidus* occurs in the feces dependent on the dose. By administering 5,000 to 10,000 BF_2 per day *Lactobacillus bifidus* occurs after 10 to 15 days in the feces and comprises from 10 to 40% of the total intestinal bacteria. By administering more than 100,000 units BF_2 per day the flora of bifidus develops very rapidly (in 1 to 3 days) and comprises from 50 to 95% of the total bacterial flora.

SPECIAL FORTIFICATION PROCESSES

ENCAPSULATED FAT-SOLUBLE VITAMINS

Silicon Dioxide Method

A number of the oil-soluble vitamins such as vitamin A, and water-soluble vitamins such as the B group tend to develop unpleasant odors, a problem which must be overcome in order to render the vitamin-containing products acceptable for human consumption. Various techniques have been attempted to render these vitamins odor-free, such as by coating the vitamins with a water-insoluble or low water-soluble material such as waxes, glyceryl monostearate, stearic acid, gelatin, etc. These techniques often result in vitamin compositions which have an acceptable odor for most uses. However, for some purposes, such as where the vitamins are included directly in foods, the vitamins must be substantially odorless in order to meet with adequate consumer acceptance.

Odor-free vitamins are prepared according to the process of *B. Borenstein and F.W. Jahns; U.S. Patent 3,497,589; February 24, 1970; assigned to Hoffmann-La Roche Inc.* which includes two essential steps: (a) a coated vitamin or coated vitamin mixture containing at least one of the following: thiamine, riboflavin, niacin, vitamin A, or a vitamin A-active material, wherein the coating is of a material having a low water solubility, e.g., a cold water solubility of less than about 1%, and (b) at least about 0.01, preferably from about 0.01 to 10.0 parts by weight of silicon dioxide per part by weight of the coated vitamin material, the silicon dioxide having a particle size in the range of from about 0.010 to 0.040 micron, an internal surface area of less than 20% of the total surface area, wherein the total surface area is in the range of from about 150 to 350 m^2/g, and which contains no more than about 3% water.

The silicon dioxide can be prepared by the high temperature oxidation of $SiCl_4$. This type of silicon dioxide has been found to exhibit a remarkable ability to adsorb vitamin odors when used in this process, without excessive adsorption of water vapor. The above compositions can be prepared by intimately mixing ingredients (a) and (b) above, according to standard mixing techniques, such as with a planetary mixer.

It has been found that the quantity of silicon dioxide required to substantially reduce or eliminate unpleasant odors from the coated vitamins is generally greater when a foodstuff is present than in compositions containing only coated vitamins and silicon dioxide. The minimum amount of silicon dioxide which should be provided in producing a vitamin supplemented foodstuff is expressed by the formula, $MS = 0.02V + 0.0001F$ where MS is the minimum weight of silicon dioxide to be employed; V is the weight of coated vitamins;

and F is the weight of foodstuff. Thus, one can add the coated vitamin-silicon dioxide composition to the foodstuff and, if necessary to achieve the calculated minimum, subsequently adding to the mixture such additional quantity of silicon dioxide as may be needed. There is no maximum amount of silicon dioxide that can be used. Obviously, there would be no practical purpose in using amounts of silicon dioxide considerably in excess of the minimum amount required. Generally, quantities of silicon dioxide ranging from the minimum amount given by the above formula up to about ten times the minimum amount are employed. Vitamin A-active materials that can be used in this process include vitamin A esters, e.g., vitamin A palmitate, vitamin A acetate, etc.

The coated vitamins are well known, e.g., B vitamins encased in a coating of a mixture of monoglycerides and diglycerides of naturally occurring fatty acids having from 16 to 18 carbon atoms, niacinamide coated with naturally occurring saturated fatty acids having from 16 to 18 carbon atoms, gelatin coated vitamin A-containing compositions. Normally from about 50 to 80% of coating material is present based on the total weight of the coated vitamin product. In addition to individual coated vitamins, any multivitamin composition which includes one or more of the above-named vitamins has a coating as above that can be used in the process.

Example 1: A coated multivitamin beadlet mixture composed of the following ingredients

Vitamin B_1, (33% coated with a mixture of mono- and diglycerides of C_{16} to C_{18} fatty acids)	15 mg
Vitamin B_2, (33% coated with a mixture of mono- and diglycerides of C_{16} to C_{18} fatty acids)	18 mg
Vitamin B_6, (33% coated with a mixture of mono- and diglycerides of C_{16} to C_{18} fatty acids)	7 mg
Niacinamide, (33% coated with stearic acid)	75 mg
Iron, reduced	15 mg
Potassium iodide	0.7 mg
Vitamin A palmitate, (500,000 IU/g, gelatin coated)	2000 units

was intimately mixed with 5.0 mg of silicon dioxide having a particle size of about 0.015 micron, a total surface area of about 190 m^2/gm, and containing about 1% water. The above mixture was added to, and mixed with, 490 grams of sugar and 495 mg of silicon dioxide having a particle size of about 0.015 micron, a total surface area of about 190 m^2/g and containing about 1% of water. The vitamin enriched sugar composition, thus obtained, was then stored in a closed container at room temperature for 4 months. At the end of this time, there was essentially no vitamin odor. An identical sugar composition was prepared and stored as described above except that no silicon dioxide was present. After 1 week, the sugar composition had an objectionable odor.

Example 2: 50 grams of beadlets of vitamin B_1 (33%) coated with mono- and diglycerides of C_{16} to C_{18} fatty acids, are mixed in a mortar with 1.5 grams of silicon dioxide having a particle size of about 0.015 micron, a total surface area of about 190 m^2/gm, and containing about 1% water. The mixture is then stored in a bottle at room temperature. In a separate bottle, 50 grams of the above beadlets of vitamin B_1 unmixed with silicon dioxide, is stored at room temperature. After 6 months' storage, the beadlets mixed with silicon dioxide do not have an offensive odor, while the beadlets not mixed with silicon dioxide have a decidedly disagreeable odor.

Gelatin-Ascorbic Acid-Sorbitol Emulsion

It has been an almost uniform practice in the dairy industry to use liquid vitamin containing compositions in the fortification of fluid whole or skim milk. At the present time, however, there appears to be preference in the industry toward the use of vitamin containing products in the form of powders or beadlets. There are presently available commercially a large variety of dry free-flowing vitamin containing compositions. For the most part, however, the products which are now available are not satisfactorily useful in the fortification of fluid milks since they do not have one or more of the properties required for a milk supplement. For example, one attribute which must be possessed by a product used in fortifying fluid whole or skim milk is an ability to be dispersed readily and completely in cold as well as warm milk. In many instances, the dry vitamin products which are now available do not possess the requisite dispersibility characteristics. Moreover, to be satisfactory for use, the vitamin containing composition must be such that it does not impart an unfavorable flavor to the fluid milk, either when initially added to the milk or upon refrigerated storage of the fortified fluid milk during distribution to and consumption by the user.

A free-flowing fat-soluble vitamin composition is prepared by *M.A. Cannalouga and N. Cohen; U.S. Patent 3,558,323; January 26, 1971; assigned to Hoffmann-La Roche Inc.* and the product is particularly suitable for use in fortification of fluid milk or spray-dried skim milk. The essential ingredients of the process are (1) from about 47 to about 92% by weight of the reaction product between (a) gelatin, (b) ascorbic acid or a mixture of ascorbic acid and citric acid, or a mixture of citric acid with sorbose, (c) sorbitol and (d) water, the nature and production of which will be described hereinafter; (2) from 5 to 15% by weight of an edible highly saturated organic oil, e.g., modified coconut oil or coconut oil; (3) from about 5 to about 50% by weight of a fat soluble vitamin; (4) and from about 0.1 to about 3% by weight of α-tocopherol. As optional ingredients, the compositions may contain antioxidants, plasticizers, buffering agents, coloring agents, etc.

Example 1: In this example, 10.38 kg of a 70% aqueous sorbitol solution was added to 15.3 kg of water. The mixture was heated slowly to a temperature of about 80°C. Thereafter, 4.875 kg of ascorbic acid was added to, and dissolved in, the aqueous sorbitol solution at a temperature of about 80°C. Subsequently, 15.3 kg of gelatin was added to the sorbitol-ascorbic acid solution and the mixture which was thus obtained was stirred and heated at a temperature of about 80°C for a period of about 6 hours. At the end of the 6 hour period, the aqueous reaction mixture was cooled to room temperature.

In the next step of the process, 15.3 kg of the aqueous reaction mixture, produced as described in the preceding paragraph was charged into a suitable vessel, following which 1.562 kg of sodium citrate was added. In a separate vessel, a mixture of 0.302 kg of butylated hydroxy toluene, 0.09 kg of α-tocopherol, 2.941 kg of vitamin A palmitate Type P 1.8 and 2.2 kg coconut oil (Cobee 76) was prepared. This mixture was added to the aqueous gelatin reaction product-containing phase and emulsified therein, using an Eppenbach emulsifier. During the emulsification, 1.5 kg of corn syrup (80%) solids were added to the system. The emulsion was converted into beadlets by passing the emulsion through a revolving spray head and introducing the emulsion droplets into an agitated cloud or suspension in air of a collecting powder.

Example 2: In this example, 10.38 kg of a 70% aqueous sorbitol solution was added to 15.3 kg of water. The mixture was heated slowly to a temperature of about 80°C. Thereafter, 4.875 kg of ascorbic acid was added to, and dissolved in, the aqueous sorbitol solution at a temperature of about 80°C. Subsequently, 15.3 kg of gelatin was added to the sorbitol-ascorbic acid solution and the mixture which was thus obtained was stirred and heated at a temperature of about 80°C for a period of about 6 hours. At the end of the 6 hour period, the aqueous reaction mixture was cooled to room temperature.

In the next step of the process, 14.3 kg of aqueous reaction mixture, produced as described in the preceding paragraph was charged into a suitable vessel, following which 1.562 kg of sodium citrate was added. In a separate vessel, a mixture of 0.302 kg of butylated hydroxy toluene, 0.09 kg of α-tocopherol, 2.941 kg of vitamin A palmitate Type P 1.8 and 0.013 kg of vitamin D_2 (calciferol, potency 40 M I.U. per gram) and 2.2 kg coconut oil (Cobee 76) was prepared. This mixture was added to the aqueous gelatin reaction product-containing phase and emulsified therein, using an Eppenbach emulsifier. During the emulsification, 1.5 kg of corn syrup (80%) solids were added to the system. The emulsion, produced as described in the preceding paragraph, was converted into beadlets by the procedure described in Example 1.

Fat-Protein Emulsion

A. Rosenberg; U.S. Patent 2,973,266; February 28, 1961 reported that dry, free-flowing, beadlets involving encapsulation of vitamin A (one of the more labile members of the fat-soluble vitamin groups) in fats, in waxes, in gelatin or in pectin lack adequate stability when stored in a mineral mixture at 45°C in a moist environment, viz, about 85% relative humidity. The following mineral mixture has been used in the studies.

	Grams
Manganese sulfate, $MnSO_4 \cdot H_2O$	74
Potassium iodide, KI	6
Ferrous sulfate, $FeSO_4 \cdot 4H_2O$	73
Copper sulfate, $CuSO_4 \cdot 5H_2O$	11
Zinc sulfate, $ZnSO_4 \cdot 7H_2O$	4
Cobalt sulfate, $CoSO_4 \cdot H_2O$	0.6
Calcium carbonate, qs 20 lb	

When this mixture is supplemented with a few percent of the products to provide mixtures containing about 600 USP units of vitamin A per gram, the retention of the vitamin under the conditions of test is poor. The mixtures show a loss of from 50 to 100% of the vitamin A at the half-way stage of the storage period, i.e., 10 days at 45°C, the equivalent of 3 months at room temperature. The destruction of the vitamin D also under those conditions, although not as complete as vitamin A, is, however, severe. The mineral mixture fortified with these compositions to an initial potency of 100 USP units of vitamin D per gram show a destruction of that vitamin of from 30 to 60% at the end of ten days' storage at 45°C.

It is to be noted that mineral supplements contain, the so-called trace minerals, iron and copper, notorious for their influence in promoting the rapid oxidation of both fats and vitamin A. Even if iron and copper salts are not deliberately included as components of some mineral supplements, the other mineral salts furnish, as contaminants, substantial quantities of these essential elements. It was found that it was possible more effectively to stabilize the fat-soluble vitamins, by the creation of dry, discrete spheres or beads of small size, i.e., of about 2 mm or less in diameter, in which the vitamin is a component of a central core that is encased within an outer protective shell. The central core is a solid fat, the melting point of which, can be substantially below high melting point types.

The use of lower melting point fats makes it possible to omit emulsifying agents such as lecithin; and also to assure the physiological availability of the vitamin. The outer protective shell is a water-resistant heat-denatured proteinaceous material. In general, the spheres or beads are manufactured by dissolving a vitamin bearing oil or fat-soluble vitamins in the basic molten fat; adding (1) this vitamin-containing fat to (2) an aqueous solution of the proteinaceous material; heating the mixture of (1) plus (2) to a temperature to effect incipient denaturation of the protein; homogenizing the mixture to form an oil-in-water emulsion and spray-drying the emulsion, and at the same time irreversibly denaturing the protein in the outer shell. The resulting product is a mass of tiny spheres containing the fat-soluble vitamins, each sphere being formed of fatty vitamin-containing cores coated by, or enclosed within, a dry, water-resistant proteinaceous film. The dry vitamin spheres are free flowing, uniform and similar in structure, and will not oil off under regular conditions of storage for indefinite periods.

The vitamin bearing oils are fish liver oils, containing vitamin A and/or vitamin D, the synthetic vitamins in an edible oil solvent or in crystalline form, carotene in an edible oil vehicle, vitamin E in an edible oil solvent, and vitamin K in an edible oil vehicle or in crystalline form. When oils were used to make the vitamin spheres the following order of preference was noted in so far as vitamin stability, is concerned: hydrogenated soybean oil of 50°C melting point, hydrogenated soybean oil of 43°C melting point, hydrogenated cottonseed oil, limpid corn oil, hydrogenated peanut oil of 45°C melting point, limpid peanut oil, hydrogenated babassu oil, and lastly, hydrogenated coconut oil. However, by supplementing some of these basic fats with antioxidants which are soluble in oils, such as butylated hydroxy-anisole to the extent of 0.02% of the fat used, the poorer fats are made equal to the best fats as a stabilizing vehicle for vitamin A. Polyphenols are very effective in stabilizing fats, they do not remain in the basic fat component of the preparations but migrate to, and are found in, the outer proteinaceous shell. In the presence of iron salts in mineral feed mixtures, they impart an objectionable blue discoloration to the vitamin spheres. Likewise, acid stabilizers, such as citric acid, dissolved or dispersed in the basic fat component migrate to, and are found in, the outer proteinaceous shell. However, advantage has been taken of this apparent disadvantage of citric acid. In such a system there is realized the maximal value of the citric acid by metal sequestration in protecting the inner fat-soluble vitamin against the pro-oxidant effect of the trace minerals migrating into the beadlet in the presence of moisture.

When the basic fat components are supplemented with the phenolic type antioxidants as butylated hydroxy-anisole in 0.02% concentration, the order of preference for the basic fat components in so far as vitamin A stability is concerned, is directly related to preference based on the active oxygen method (AOM) values. The proteinaceous material used should contain on the dry basis at least 20% of protein that is heat coagulable and completely denatured in aqueous solution at a temperature of 65° to 85°C when the pH of the solution is between 4.0 and 5.0. Proteinaceous material that has already been denatured by heat-processing so that the material is no longer soluble in water, cannot be used. Solvent-extracted, but not heat-processed soy flour, is a highly satisfactory material to form the outer protective shell that encases the vitamin-fat cores of the spheres.

The proteinaceous material in water solution, in concentrations expressed on a protein basis, from 5 to 8%, is brought to a temperature of about 10°C above the melting point of the vitamin oil-basic fat blend. The fat-soluble vitamin-basic fat blend is prepared separately by dissolving the vitamin oils or synthetic vitamins in the basic fat component held at a temperature of about 10°C above the melting point of the fat. The aqueous and fat solutions are mixed, preferably under nitrogen, in a tank equipped with a motor-driven propeller agitator to form a coarse oil-in-water emulsion. During the period of agitation of about 10 to 40 minutes, the mixture is heated to a temperature of from 65° to 85°C, 40 minutes at 65°C.

To establish the time-temperature relationship for effecting incipient denaturation of the proteins small portions, each of about 15 ml, of the coarse oil-in-water emulsion prior to the heating are transferred to 100 ml test tubes and the latter are then immersed in a series of water baths held at 65°, 70°, 80° and 85°C. The contents of the test tubes are stirred and examined after 10 minute intervals of time. The maximal heating conditions which still provide a flowable emulsion, in contrast to a nonflowable emulsion or one containing coagulated protein curd in suspension in the case of the next higher member in the series, are selected in heating the bulk emulsion. Thus, if after 30 minutes of heating at 75°C, the test aliquot of emulsion remains flowable, but sets to a nonflowable mass after 30 minutes at 80°C, or after 40 minutes at 75°C, then the bulk emulsion is processed for 30 minutes at 75°C.

The aqueous-fat mixture with a total solids content of about 30 to 40%, water being added if necessary, is cooled to about 55° to 60°C (to stop further protein denaturation), is then homogenized at a pressure of 2,000 to 4,000 lb/in^2. This action leads to the formation of a fine oil-in-water emulsion, i.e., very small fat globules dispersed in the continuous proteinaceous aqueous phase. The emulsion is then passed through a high pressure pump necessary to force the emulsion at a pressure of 2,500 to 4,000 lb/in^2 through the nozzles of the spray dryer.

Spraying systems' core-type nozzles with removable core and orifice inserts give excellent results. The optimal nozzle opening in the spray-drying operation is about 0.030 inch. The temperature of the incoming hot air ranges between 150° to 160°C, and the outgoing air between 80° to 85°C. During the spray-drying opration, the moisture content of the spheres with their shells of denatured protein is reduced to a maximum of 2%. An automatic screw-type unloading unit continuously removes the product from the dryer.

Example: In one tank 97.0 parts of coconut oil, hydrogenated to an iodine number of 1.0 and having a melting point of 37°C, is heated to about 47°C. To this are added 0.02 part of butylated hydroxyanisole, 0.025 part of crystalline vitamin D_3 of a potency of 40,000,000 AOAC units per gram, and 3.0 parts of vitamin A palmitate in corn oil having a potency of 1,100,000 USP units per gram. The mixture is stirred under nitrogen for a period of about 15 minutes to effect a clear solution.

In another tank 31 parts of solvent-extracted nonheat processed soy flour containing about 50% protein (nitrogen x 6.25) are suspended in 300 parts of warm water at about 45 to 50°C and stirred for a period of about 15 minutes until the soy flour is dissolved (or finely dispersed). The contents of the two tanks are mixed under nitrogen to form a coarse oil-in-water emulsion; heated by steam coils to a temperature of about 75°C for a period of 30 minutes; and then cooled to about 55° to 60°C to stop further protein denaturation. The aqueous-fat mixture is homogenized at a pressure of 3,000 lb/in^2 to yield a finely dispersed oil-in-water emulsion.

The emulsion containing about 32 to 33% of solids is spray dried as described above to produce dry, discrete beads or spheres (referred to as the final product). This final product has the physical structure of a vitamin-fat core encased within a protective spherical coating of the water-resistant heat-denatured protein film, the diameter of the spheres measuring less than 2 mm and the spheres containing less than 1% moisture. The ratio of the fatty-vitamin-containing core to the protective coating is as 75:25. Assays conducted on the final product as thus manufactured indicated a vitamin A content of 10,800,000 USP units per pound (a loss of only 4% as a result of the process) and 3,500,000 AOAC units of vitamin D_3 per pound (no measurable loss as a result of the process.)

When one part of this product is added to 39 parts of the mineral supplement above described, and this mixture of minerals and the vitamin product is stored in a moist environment of about 85% relative humidity at 45°C for a period of 21 days, fully 85% of the vitamin A and all of the vitamin D_3 are retained, the vitamin A dropping from an initial value of 593 units per gram to 504 units per gram while the vitamin D_3 remained constant at about 187 units per gram. No measurable losses of the vitamins were noted after the 21 day holding test at 45°C. In the above and following examples, AOAC units of vitamin D_3 are synonymous with internal chick units.

ENCAPSULATION WITH CEREAL FLOURS

S. Katzen; U.S. Patent 3,786,123; January 15, 1974 uses a food system consisting of an encapsulating agent and various nutrients. The encapsulation allows the nutrients to be kept in a stabilized state for a long period of time. The cereals or other vegetable encapsulating agents are gelatinized or polymerized under elevated pressure and at elevated temperatures. The process causes the nutrients to be encapsulated (coated or protected) by the gelatinized or polymerized encapsulating agent. When cooled and the pressure removed, the resultant product is dry and can be stored for extended periods of time.

The encapsulation prevents the nutrients from being oxidized, reduced, solubilized, sublimated or otherwise diminished in potency which would occur if left in an unprotected state. The product can be sized, if desired or necessary by any convenient means such as grinding. The rate of release of the encapsulated nutrients can be varied according to the solubility of the gelatinized or polymerized encapsulating agent. The time in the digestive tract may vary from about one hour to about eight hours, with four hours being normal. Digestion depends upon enzyme action as well as acid hydrolysis. Digestion begins in the mouth by action of salinary amylase (ptyalin) with an optimum pH of 5.5 to 6.5. The stomach pH is 0.85 to 1.00. There are no carbohydrate enzymes in the stomach, but the small intestine contains many enzymes in addition to pancreatic amylase. The high protein vegetable agent can only contain up to 40% by weight of nonprotein material such as starch or nonprotein nitrogenous material (for example, ammonium phosphate). Preferably, 100 parts by weight of the encapsulating agent are mixed with about 40 to 50 parts by weight of the nutrients. That weight range for nutrients includes such additives as surfactants, preservatives, etc. The encapsulating agent is normally gelatinized or polymerized at a temperature at about 350°F. The temperature utilized must be such that the potency of the nutrients and/or additives is not seriously impaired. The pressure utilized will be above atmospheric and will vary with the type of encapsulating agent, nutrients and additives and the operating temperature, but will generally be about 1,000 psi.

The product is formed in an extruder or expander. If the encapsulating agent is made viscid for the mixing and encapsulating steps by the addition of water, or like material, the end product is made dry by the flashing off (evaporation) of the entrapped water as steam when the product exits from the extruder or expander. The resulting expanded product has air spaces which make the product brittle and easily ground. (The end product can be further dried before or after the encapsulating step, if necessary.)

The resultant product can have any shape and is often an irregular shape due to its method of manufacture. Complete encapsulation of the nutrients is not based on the use of a carrier as a viscate state when the two are mixed. Thus it is seen that the nutrients may be in a dry particulate form for use. To assure encapsulation the nutrients should have a particle size of about 20 to 200 mesh. If desired, before encapsulation, several types of nutrients may be coparticulated so that each type of nutrient used is contained in each encapsulated unit.

Example 1: 1 kg of vitamin A palmitate, having a rating of 1.7 million IU/g, and was thoroughly mixed with 4 kg of corn flour. The encapsulating agent was the corn flour and the nutrient was vitamin A palmitate. The admixture was extruded in an extruder (Wenger), which was operated at a pressure of 1,000 psi and has a temperature of 240°F. The temperature used was sufficient to gelatinize the corn flour but yet was not of such a level as to decompose, deactivate or otherwise impair or destroy the potency or effectiveness of the vitamin A palmitate.

The glatinized corn flour completely enclosed individual or groups of particles of vitamin A palmitate. The extruded material was dried, cooled and reduced in size by grinding to particles of an average diameter of about 0.1 mm. The resultant product contained 340 million IU of vitamin A per kg, which is above the conventional market standard of 325,000 IU/g. The resultant product which contained encapsulated dry, stabilized vitamin was stored for two months and was then consumed by a man and a rat. The gelatinized encompassing corn flour was solubilized in the digestive tract, the vitamin A material was released, thereby making vitamin A activity available to the organism.

Example 2: Example 1 was repeated except that a mixture of 3.0 kg of corn flour and 2.0 kg of flash desolventized cottonseed flour was used as the encapsulating agent. As the material was extruded, the gelatinized encapsulating agent expanded causing a flash evaporation of steam still trapped therein. When the expanded product was cooled, it contained air spaces. The bulk of the nutrient was encapsulated by the gelatinized corn flour and the polymerized cottonseed flour, even though air spaces were present. The product was easily ground, after being thoroughly dried in a conventional oven, to reduce its particle size to about 0.1 mm. The dry product was brittle and easily ground. The material was stored for three months and the encapsulated nutrient at that time showed no loss of vitamin A activity during the period.

Example 3: 1 kg of a xanthophyll oil containing 100 grams of xanthophyll was mixed with 4 kg of soybean flour. The material was extruded at a pressure of 1,000 psi and at 230°F. The extruded material was ground until it had an average particle size of 0.1 mm. The resultant product contained 20 grams of encapsulated dry stabilized xanthophyll per kg of the product. The resultant product was stored for three months without any noticeable loss of the potency of the nutrient.

Example 4: Example 1 was repeated except that the mixture also contained 100 grams of an amino acid mixture. The composition of the amino acid admixture was: 1-arginine

HCl, 8.0 g; 1-histidine HCl·H_2O, 4.0 g; dl-isoleucine, 10.8 g; l-leucine, 15.4 g; l-lysine HCl, 12.3 g; dl-methionine, 6.1 g; dl-phenylalanine, 6.9 g; dl-threonine, 10.8 g; dl-tryptophan, 1.8 g; dl-valine, 13.9 g; and glycine, 10.0 g. The resultant product was fed to chicks without showing the amino acid toxicity that normally accompanies rapid absorption of free amino acids.

ANTIOXIDANTS FOR VITAMIN A

Two antioxidant synergists are used to stabilize Vitamin A as described by *H. Klaui and W. Schlegel; U.S. Patent 3,637,772; January 25, 1972; assigned to Hoffmann-La Roche Inc.* The system consists of (a) an antioxidant, (b) colamine and/or a higher fatty acid salt thereof and (c) a higher fatty acid ester of ascorbic acid. The antioxidant in this system is α- or γ-tocopherol.

Colamine can be used as the base or as a salt with a higher fatty acid having 10 to 20 carbon atoms. The palmitate and the stearate are the preferred salts. The ascorbic acid esters are esters of higher fatty acids having from 10 to 20 carbon atoms, preferably palmitic or stearic acid. For every part by weight of antioxidant, the amount of colamine or colamine salt can be from about 1 to 10 parts by weight and the amount of ascorbic acid ester can suitably be from about 1 to 10 parts by weight preferred.

Example 1: 1.5 grams of vitamin A palmitate, 1.5 grams of sunflower oil, 50 mg of α-tocopherol, 250 mg of ascorbyl palmitate, 250 mg of colamine palmitate, 200 mg of polyoxyethylene (2) oleyl ether (HLB value 4.9) and 0.3 mg of copper oleate are heated to 60° and homogeneously mixed. This mixture is stored in air at 45° and the vitamin A retention is measured and compared to compositions without colamine and compositions without ascorbyl palmitate. After a storage time of 600 hours, the vitamin A retention in the composition prepared according to this example is 96%. In the composition without colamine, the vitamin A retention is 0% and in compositions without ascorbyl palmitate the vitamin A retention is about 10%. In these examples temperature is in °C.

Example 2: 1.5 grams of vitamin A palmitate, 1.5 grams of sunflower oil, 50 mg of γ-tocopherol, 250 mg of ascrobyl stearate and 250 mg of colamine palmitate are heated to 60° and homogeneously mixed. This mixture is stored in air at 45° and the vitamin A content determined after 600 hours and compared to compositions without colamine palmitate and compositions without ascorbyl stearate. The vitamin A retention in the composition prepared according to this example is 95% while in the composition without colamine palmitate, vitamin A retention is 0% and in the composition without ascorbyl stearate the vitamin A retention is about 10%.

Example 3: 1.5 grams of vitamin A palmitate, 1.5 grams of sunflower oil, 50 mg of butylated hydroxyanisole, 250 mg of ascorbyl palmitate, 50 mg of colamine, 200 mg of polyethyleneglycol (2) oleyl ether and 0.3 mg of copper oleate are heated to 60° and homogeneously mixed. This mixture is stored in air at 45°. The vitamin A content is determined after 400 hours and compared to compositions without colamine and compositions without ascorbyl palmitate. The vitamin A retention in the composition prepared according to this example is 94% while in compositions without colamine, the vitamin A retention is 0% and in the composition without ascorbyl palmitate the vitamin A retention is about 10%.

LIQUID IRON FORTIFYING SOLUTION

The Food and Nutrition Board of the National Academy of Science has set the recommended dietary allowance of iron at 18 mg/day, an amount so great that it cannot be supplied by existing foods without increased iron fortification. For instance, in bread the present standards require 13.0 to 16.5 mg iron per pound of flour, and it has been proposed by the baking industry that this should be increased to 50 to 60 mg iron per pound of flour. The Commissioner of the Food and Drug Administration has proposed 40 mg iron per pound of flour. Among the best sources of iron for food uses are salts such as ferrous sulfate, ferric ammonium citrate, ferrous fumarate, ferric choline citrate, and ferrous gluconate. Commercial preparations of these compounds include the relatively insoluble $FeSO_4 \cdot 1.2H_2O$ (32.1% Fe). When the ferrous sulfate compound is combined with wheat flour, off flavors develop during storage. Therefore, it is usually added during the baking process at the dough mixing stage in tablet form. The tablets, added at the rate of one per 100 pounds of flour, are broken up by hand which causes some doubt as to the completeness of distribution. This method is not easily adaptable to automation.

For baking and other food processing, a liquid iron composition would have great advantages. However, present commercial liquid preparations are not particularly suitable for food uses. One, for example, combines a nontoxic ferrous salt with a polyhydric alcohol, 1-ascorbic acid, and sometimes ethyl alcohol to maintain the iron salt in an aqueous solution. These relatively expensive compositions are pharmaceutical preparations for treating hypochromic anemias and are not intended for general food use. They are usually a dark amber color which indicates that part of the soluble iron has reacted with dissociated hydroxy ions from the water. The solutions must be stored in dark amber bottles in a cool place to prevent further degradation.

Water solutions of iron salts such as ferric ammonium citrate, ferric choline citrate, or ferrous gluconate are fairly stable but are quite highly colored. Water solutions of ferrous sulfate, on the other hand, are substantially lighter in color but much less stable. Iron rapidly comes out of these ferrous sulfate solutions as an insoluble red precipitate. An easy-to-use liquid preparation can be used according to *G.N. Bookwalter; U.S. Patent 3,809,773; May 7, 1974; assigned to U.S. Secretary of Agriculture.* The process requires first the preparation of aseptic aqueous solutions of ferrous or ferric salts and completely blending the resulting solution with a high conversion corn syrup optionally containing up to 40% sucrose based on total weight of solids under aseptic conditions. The ferrous or ferric salts, syrup, and water are blended so that the resulting compositions contain up to about 1% by weight ferrous or ferric ions, from about 70 to 80% by weight total solids, the remainder being water.

The iron salts must be water-soluble and nontoxic at addition levels and these include ferric ammonium citrate, ferric choline citrate, ferrous gluconate, and ferrous sulfate. Ferrous sulfate heptahydrate ($FeSO_4 \cdot 7H_2O$) is preferred primarily because it provides compositions which range from colorless (compositions containing up to 0.1% ferrous ions) to very slightly blue (compositions containing 0.1 to 1.0% ferrous ions) while other ferrous or ferric salts produce compositions which range from medium to dark amber. These amber colors are easily detected in food products which are not themselves highly colored. Ferrous sulfate is preferred also because it is the least expensive source of iron for physiological purposes.

High conversion corn syrup is defined as a syrup made by the partial acid hydrolysis of corn starch and having a dextrose equivalent (DE) of from about 60 to about 68 and a solids content of about 80% by weight. An extra high conversion corn syrup having a DE of 70 was also used to prepare the compositions and is considered equivalent to the high conversion syrup described above. Simply mixing together the desired amounts of iron salts, high conversion corn syrup, and water does not produce a stable liquid iron-fortified composition. Prepared in this manner, compositions quickly darken due to precipitated iron. However, iron will not precipitate when prepared according to the following procedure. All the process steps and storage should be conducted under aseptic conditions to insure biological stability of the products. The preferred manner of achieving this stability is described for each step.

The first step of the process is the preparation of an aqueous solution of ferrous or ferric salts. It is preferred that the water be boiled before dissolving the iron salts to provide an aseptic solution. When ferrous salts are used, boiled deaerated water cooled to about 70° to 90°F eliminates clouding of the resulting solution, probably caused by oxidation of the ferrous ion. Clear solutions containing up to 26% by weight $FeSO_4 \cdot 7H_2O$ can be prepared in this manner. The concentration of iron in these solutions is varied according to the amounts of iron and water desired in the final composition. Final compositions containing more than 30% water by weight tend to be biologically unstable, and at this water content any dissolved ferrous or ferric ions over about 1% by weight of the final composition will tend to precipitate. The concentration of ferrous or ferric ions in the examples has been calculated from the total amount of iron salt in solution. It is possible that all dissolved iron is not in free ionic form, but could be bound in some manner to the anion or to other compounds present in the composition. However, this is not important for nutritional purposes. Concentrations of iron described in terms of weight percent of ferrous or ferric ions will be understood to include all iron contained in the added salt.

A small amount of acid can be added to the aqueous iron solution, to bring the acidity of the final composition down to about pH 2. In some cases where the composition is used or stored under drastic conditions, such as high temperatures (e.g., 120°F), stability is decreased. Adjusting the pH downward to about 2 will maintain stability. Without addition of acid a composition containing 1% ferrous ions will have a pH of about 3.5.

After the aqueous iron solution has been prepared, it should immediately be mixed with the syrup. This can be as long as one hour. The temperature range for the syrup during the mixing step is from about 65° to about 300°F. Viscosity noticeably increases at temperatures below 65°F, and the syrup begins to discolor at temperatures of about 300°F. To maintain aseptic conditions, it is preferred that the syrup to which the iron solution is added be at a temperature of from about 180° to about 200°F.

The syrup can be all high conversion corn syrup or it can contain up to 40% sucrose, based on total solids weight. Sucrose solutions can be prepared which have up to 67% solids and 33% water. When ferrous sulfate is dissolved in this solution, a black precipitate forms which changes to red on standing. A similar precipitate forms when ferrous sulfate is dissolved in high conversion corn syrup containing less than 70% solids. It is, therefore, believed that the precipitation of the ferrous salts which occurs in blends containing less than 70% solids is due to excessive amounts of water. Stable liquid iron-fortified compositions can be prepared that contain 70% by weight solids consisting of 40% sucrose and 60% high conversion corn syrup solids, and that also contain up to 1% ferrous ions. Dextrose can also be added in the same manner as sucrose but in smaller amounts since the maximum amount of solids in a dextrose solution is only 50% by weight. While the lower limit of solids content in the liquid iron-fortified composition is preferably 70% by weight, stable compositions can be prepared which contain as high as 80% by weight solids.

To insure a final composition that maintains biological stability under storage and use conditions, it should be transferred after preparation, at a temperature of at least 180° to 200°F into sterilized containers. In the laboratory these containers consisted of clear glass pint bottles which were tightly screw-capped and inverted. Compositions containing ferrous sulfate are colorless even when the most concentrated ferrous sulfate solutions were used. The slightly blue color of the iron solutions is not detected in the final product. All final products are stable during storage and in food processing at temperatures of 0° to 100°F in darkness or direct sunlight. They may be frozen for extended periods of time without loss of stability or functional properties. Good flavors, physical stabilities, and biological stabilities were maintained after storage for 1 year at room temperature. A slight yellowing occurred to the same extent in both the compositions containing ferrous sulfate and samples of nonformulated high conversion corn syrup, but for all purposes for which these compositions are intended they are considered to be essentially colorless.

Examples 1 through 4: Aqueous iron solutions were prepared by mechanically mixing for about 10 minutes a ferrous or ferric salt in distilled water which had previously boiled (to deaerate and make aseptic) and cooled to between 70° and 90°F. At this point a mineral acid was added to some of the solutions to lower the pH of the final composition.

The iron solutions were then added by means of a funnel having its exit end protruding beneath the surface of the liquid to a syrup which had been previously heated to between 180° and 200°F. During the addition and for about 10 to 30 seconds after all the iron solution had been added, the syrups were mechanically stirred to insure complete mixing. The compositions, still at a temperature of 180° to 200°F, were transferred to clear glass pint bottles which were then screw-capped and inverted to eliminate fermentation during storage. Amounts and kinds of starting materials and final products for several composition preparations are described in the following table.

	Examples			
	1	2	3	4
Starting material, g.:				
High conversion corn syrup DE 65, 82% solids by weight	3,619	3,570	3,938	3,938
Sucrose	524	523	0	0
$FeSO_4 \cdot 7H_2O$ solution:				
Salt	3.5	164.5	230	207
Deaerated distilled water	100	400	380	360
Water	412	115	0	0
Acid (1 N HCl)	0	0	0	20
Final composition:				
Total solids content, wt. percent	75.5	74.9	75.4	75.1
Ferrous ion, wt. percent [1]	0.015	0.69	1.01	0.92
Water, wt. percent	Remainder			
pH	5.0	3.1	3.6	2.3

[1] Calculated from total amount of iron salt added.

All compositions were colorless and were stable at least 1 year at room temperature (70°F). At storage temperatures of 120°F, Examples 1 through 3 showed a tendency to discolor indicating some unstability. Example 4, on the other hand, showed complete stability at that temperature.

Example 5: An iron-fortifying composition was prepared as in Examples 1 through 4 with 3,690 grams of regular corn syrup (DE 42, 82% by weight solids), 524 grams sucrose, 341 grams water, and a ferric ammonium citrate solution containing 4.1 grams iron salt and 100 grams deaerated distilled water. The product was amber in color, contained 0.015% by weight ferric ion, 76.4% by weight total solids, and had a pH of 5.8.

AMINO ACID FORTIFICATION

Amino acids are essential for normal growth and well-being in animals, including man. Since they cannot be synthesized by the body in adequate quantity, they must be provided in free or combined form in the diet to meet the requirements of cellular protein synthesis and to fulfull other metabolic roles. The essential amino acids include arginine, histidine, isoleucine, leucine, lysine, methionine, phenylalanine, threonine, tryptophan and valine, although in man the first two are regarded as nonessential. It is known that a part of the methionine requirement can be replaced by cystine, and a part of the phenylalanine requirement by tyrosine. This information is obviously very useful in evaluating potential dietary sources of protein. Thus, for example, gelatin, which, contains no tryptophan, is clearly inadequate as a sole source of dietary protein.

For optimum nutrition, however, it is desirable that the amino acid requirements of the body be placed on a more quantitative basis, so that the essential amino acids can be provided in the daily diet in the proper proportions. If a poor protein is ingested, large excesses of some amino acids will have to be metabolized and excreted, whereas others will be in such short supply that the body's amino acid reserve will be partially consumed. Many attempts have been made to establish approximate quantitative requirements by elaborate experimentation. Unfortunately, however, it is found that these requirements seem to vary with species, and, to a lesser extent, with age and sex. Furthermore, there is some variation in amino acid requirement from individual to individual. Conclusions drawn from an emprical approach, observing the effect of varying one amino acid while the others are held constant are necessarily restricted in their applicability.

It has been discovered that a close interrelationship exists between dietary amino acid requirements and the amino acid content of the blood plasma. Although the amino acid content of the plasma is found to change shortly after eating, determination made upon fasting, e.g., 18 hr after eating, are remarkably reproducible. Important advantages are achieved by supplying animals, including man, a diet whose relative proprotions of nutritionally available amino acids conform substantially to the respective proportions of these acids as found in that animal's fasting blood plasma. By nutritionally available amino acid is meant amino acid present in the diet in forms utilizable by the body, i.e., free amino acids, their salts, and digestible protein. A method proposed by *C.I. Jarowski; U.S. Patent 3,080,234; March 5, 1963; assigned to Chas. Pfizer & Co., Inc.* permits more efficient utilization of proteins.

In diet supplementation it may not always be economically practical to bring the relative proportions of all the essential amino acids into complete conformity with the fasting plasma proportions. Important advantages may still be achieved if the essential amino acids present in most limiting proportion in the diet are so adjusted. The first limiting amino acid is that one which is present in the basic diet in the smallest proportion relative to the amino acid profile. The diet is supplemented with respect to the first limiting amino acid to the extent that its total proportion in the dietary protein will balance with the proportion of the second limiting amino acid. When this has been achieved, supplementation with both the first and second limiting amino acids follows, in sufficient quantity to achieve balance with the third limiting amino acid. This process may be continued even further if desired or economically practical, until all the essential amino acids are in proper balance.

Other applications of the process are possible. For example, two or more complementary protein sources may be blended in such proportions as to minimize their respective amino acid deficiencies and to reduce if not eliminate the amino acid supplementation required. Such protein sources may include, for example, corn meal, cottonseed meal, soybean meal, peanut meal, sesame meal, sunflower meal, alfalfa, casein, milk, wheat gluten and gelatin.

Furthermore, concentrated foods can be prepared containing mixtures of the amino acids in such proportions as to conform substantially to their respective proportions in the fasting plasma, together with adequate concentrations of vitamins, minerals and carbohydrate to satisfy established requirements. Such diets may be administered with water during periods of illness, recuperation, or stress, or where weight or space considerations make conventional food temporarily impractical. Sterile solutions suitable for intravenous feeding may be formulated in similar manner, by combining pure amino acids, by fortifying protein hydrolysates with amino acids, or by combining various protein hydrolysates in a physiologically acceptable aqueous medium and in suitable proportions to conform substantially to the amino acid profile determined in the plasma upon fasting. Such solutions may, if desired, also contain carbohydrate, e.g., dextrose or fructose, and salts, e.g., sodium chloride, and are suitable for intravenous feeding where the patient is temporarily unable to ingest or digest food so as to maintain a favorable nitrogen balance.

Example: Diet A – The following liquid diet, a 50% w/v aqueous solution found to be highly efficient in studies is prepared, using amino acids in the form of their L-isomers:

Arginine HCl	9.4 g
Histidine HCl·H_2O	5.7 g
Isoleucine	8.8 g
Leucine	14.0 g
Lysine HCl	13.0 g
Methionine	6.3 g
Phenylalanine	6.3 g
Threonine	8.8 g
Tryptophan	2.8 g
Valine	9.8 g
Tyrosine ethyl HCl	16.8 g
Alanine	6.4 g
Aspartic acid	13.5 g
Sodium glutamate	51.2 g
Glycine	4.1 g
Proline	25.4 g
Serine	13.2 g

(continued)

Cysteine ethyl HCl	1.1 g
NaOH	4.1 g
Dextrose	637.1 g
NaCl	4.8 g
Mineral solution A	100 ml
Mineral solution B	50 ml
Choline chloride	2.5 g
Ca fructose 1,6-diphosphate	50.5 g
Vitamin mixture (H_2O-soluble)	50 g
Polysorbate 80*	6 g
Vitamin solution (fat-soluble)	1 ml
Ethyl linoleate	4 g
Water to make 2 liters	

*Polyoxyethylene ether of sorbitan mono-oleate, non-ionic surface active agent.

Mineral solution A in the above diet has the following composition:

	Grams
Glucono delta lactone	26.1
Magnesium oxide	0.73
KOH	6.17
Ferrous gluconate	1.72
Water to make 100 ml	

The composition of mineral solution B is as follows:

	Mg
$Mn(C_2H_3O_2)_2 \cdot 4H_2O$	260
$Cu(C_2H_3O_2)_2 \cdot H_2O$	15
$Co(C_2H_3O_2)_2 \cdot 4H_2O$	9
Zinc benzoate	22
$(NH_4)_6Mo_7O_{24} \cdot 4H_2O$	6
KI	30
Water to make 50 ml	

The mixture of water-soluble vitamins in the diet consists of the following ingredients per 100 grams:

Thianine HCl	20.1 mg
Riboflavin	15.1 mg
Pyridoxine HCl	12.6 mg
Niacin	75.1 mg
Inositol	500.1 mg
Calcium pentathenate	100.1 mg
Biotin	0.67 mg
Folic acid	1.15 mg
Ascorbic acid	1000 mg
Vitamin B_{12}	0.2 mg
p-Aminobenzoic acid	600 mg
Glucose	97.67 g

The solution of fat-soluble vitamins in the diet formulation contains the following ingredients per milliliter:

	Mg
Vitamin A acetate	10
Vitamin K	4.2
Vitamin D	7
Vitamin E	50
In absolute ethanol	

If the relative proportions of essential amino acids in this diet are compared with their respective proportions in the plasma of fasting rats, it is noted that the diet is most deficient with respect to lysine, followed by tryptophan, arginine, threonine, and histidine; i.e., the ratios of the concentrations of these acids in the diet to their concentrations in the plasma are lowest.

Diet B – A second diet is formulated, in all respects identical with Diet A except that the concentrations of the five limiting amino acids are raised to the following values (per two liters of solution):

	Grams
Arginine HCl	19.7
Histidine HCl·H_2O	8.0
Lysine HCl	46.8
Threonine	15.8
Tryptophan	7.3

Since this change represents an increase of 57.9 grams, the dextrose content of Diet B is reduced by the same amount, from 637.1 to 579.2 g per 2 liters, so that the two diets are substantially isocaloric, and so that the 50% w/v concentration is adhered to. Diet B was 15% more efficient than Diet A in this experiment. Since Diet B contains 32.5 grams of amino acid nitrogen per 2 liters, as compared with 25.2 grams for Diet A, it is questioned whether the increased efficiency may be attributed to this higher nitrogen content. The amino acid content of Diet A is therefore increased at the expense of the dextrose to a value equivalent to 32.5 grams of nitrogen per 2 liters, without, however, altering the relative proportions of amino acids. When this diet is tested it fails to provide any increased efficiency over that of Diet A, demonstrating that the higher efficiency of Diet B results from matching the proportions of the six limiting amino acids to their proportions in the plasma.

BLAND AMINO ACID MIXES

Compositions containing one or more pure amino acids have been administered parenterally, rectally and orally to provide low residue nitrogen nutritional requirements while simultaneously resulting in decreased digestive activity. Such properties are partially useful in treating patients whose normal digestive processes are impaired as, for example, patients in catabolic disease states such as duodenal fistula, the short gut syndrome, pancreatitis, and ulcerative colitis. The compositions are also useful in providing nitrogen nutrition to pre- and post-operative patients when it is desired to limit their digestive activities.

A drawback associated with the oral administration of such compositions is that amino acids exhibit a particularly intense obnoxious taste and can, in fact, be orally ingested only with great psychological stress. Prior attempts at solving this problem have involved adding to these compositions various flavoring materials in large quantities whereby they exhibit an intense flavor note of their own sufficient to overwhelm that of the amino acids. In the main, only a few flavors have produced an intensely flavored composition which is less objectionable than the unflavored amino acid compositions and, in these few cases, the intense substituted flavor note has, without exception, been found to be organoleptically oppressive. This problem is particularly acute when a patient must subsist for an extended length of time on such dietary compositions. Further, the heavily flavored compositions have caused secondary dietary problems such as diarrhea and nausea and, still further, because these flavoring materials are costly, the necessity of using them in large quantities has greatly increased the expense of dietary regimens for supplying nitrogen nutrition in this manner.

K. Mohammed; U.S. Patent 3,821,432; June 28, 1974; assigned to Johnson & Johnson developed a process to reduce the intense and obnoxious amino acid flavor without utilizing great quantities of a substitute flavoring material. It has been found that oil of ginger will organoleptically interact with the amino acids in the finished product to greatly reduce the obnoxious taste intensity of the amino acids without itself producing a strong objectionable ginger taste. Unlike past attempts to mask the unwanted amino acid taste by adding large quantities of flavoring material which will superimpose a more acceptable but organoleptically oppressive flavor, the ginger acts to neutralize the taste of the amino acids and thereby render the resulting product relatively bland without contributing a high intensity flavor note of its own.

It is preferred that the ginger component be provided in the form of pure oil of ginger but other ginger forms which contain substantial quantities of this oil, such as the commonly available oleoresin of ginger, may be used. While both the oil and the oleoresin are available in many physical forms, it is preferred that one of the dry, free-flowing forms be utilized thereby facilitating the blending of the ginger into the dry particulate amino acids and other dietary components. The amino acids and ginger components interact organoleptically in the finished product to reduce the extremely unpleasant flavor intensity of the amino acids. This result occurs at such low proportions of the ginger component that the presence of the prescribed quantities of ginger in the finished beverage or pudding is generally unevidenced, except by the desired reduction in the unwanted amino acid flavor note.

No organoleptical interaction has been detected between ginger and other nutritional components such as, for example, vitamins, minerals or carbohydrates. While the interaction between the amino acids and ginger occurs over a substantial range of relative proportions of oil of ginger to amino acids, it has further been determined that the most effective suppression of unwanted flavor notes occurs at ratios within the specific range of about 0.04 to 1.2 mg of oil of ginger per gram of amino acids. Within this range the combined intensity of the unwanted flavor notes of both components is minimized.

Example 1: Four sample solutions are prepared using tap water and containing the components and proportions tabulated below:

Sample	Oil of Ginger (mg/1000 ml water on a carrier-free basis)	Component Amino Acids (gm/1000 ml water)	Ratio (mg oil of ginger/gm amino acids)
1	0.0	0.0	–
2	4.24	0.0	–
3	0.0	25.5	0.0
4	4.24	25.5	0.166

The ginger component is a spray-dried form of oil of ginger on an inert carrier wherein 20% by weight is the oil. The amino acid component is a mixture of individual amino acids and alkyl esters, hydrochlorides and hydrates of individual amino acids which, when provided in sufficient quantities to a human, will supply his complete nitrogen nutritional requirements. The mixture has a ratio of dispensable to indispensable amino acids of 0.667.

A panel of qualified judges, in a blind test, were asked to appraise these samples organoleptically with respect to both the intensity of the ginger flavor note and the intensity of the amino acid flavor note; 90% of the judges correctly responded to sample 1 which contained only water by reporting no ginger or amino acid flavor note at all. 60% of the judges reported that sample 2, containing only ginger and water, exhibited only a slight ginger flavor note. 30% reported this sample as having a moderate ginger flavor intensity. Only 10% reported a distinct ginger flavor note. The uncertainty in detecting a ginger flavor note in this sample is further indicated by the fact that 30% of the judges reported that the sample exhibited a slight amino acid flavor note, whereas in fact no amino acid was present.

In contrast with the judges' difficulty in identifying a distinct ginger note in sample 2, sample 3, containing only amino acids, received a response of at least a strong amino acid flavor intensity from 90% of the judges. 60% characterized the amino acid intensity as extreme. When presented with sample 4, the sample containing both the amino acids and the ginger, 90% of the judges reported a distinct decrease in the intensity of the amino acid flavor note relative to that of sample 3, To illustrate that the organoleptical interaction described above appears to be unique to amino acids, the following examples are given.

Example 2: A set of three sample solutions was prepared, the first sample containing 0.76 g of a vitamin blend per 1,000 ml of water. The second sample contained the same concentration of the vitamin blend and, in addition, contained 1.15 mg/ml of water of oil of ginger. The third sample contained the same concentration of the vitamin blend and an oil of ginger concentration of 4.24 mg of oil of ginger per 1,000 ml of water. The three samples were submitted to the taste panel and, again, the judges were asked to evaluate each sample in order of decreasing level of preference on a scale of 1 to 3. 60% of the judges assigned to the solution containing no ginger a high preference rating of 1. 40% of the judges assigned this rating to the solution containing the low quantity of ginger, and 40% assigned the high preference rating to the solution containing the high quantity of ginger. Again, these results indicate no meaningful preference for ginger-containing vitamin solutions versus nonginger-containing vitamin solutions.

Example 3: Four beverages are prepared, each containing the blend of amino acids described above in Example 1 and, in addition, the oil of ginger used in Example 1, in quantities tabulated in the table below. A panel of competent judges was asked to evaluate the samples of the beverages based on the relative intensities of both the ginger and the amino acid flavor notes, the evaluation to be based on a scale ranging from extreme to moderate to none. The results of this panel evaluation are also tabulated below.

		% of Judges Rating Intensity - - - Greater than Moderate - - - - -	
Sample No.	Oil of Ginger Conc. (mg ginger/g amino acid)	Amino Acid Flavor Note	Ginger Flavor Note
1	0.0	60.0	0
2	0.332	10.0	0
3	0.830	0.0	0
4	1.33	0.0	80

It is apparent from the above table that the majority of the judges could detect a more-than-moderate amino acid intensity of Sample 1, containing no ginger. In the subsequent samples of increasing ginger concentration, all but a small minority of the judges evaluated the taste intensity of the amino acid flavor note as no more than moderate, the judges lacking unanimity only in rating the sample containing the least amount of ginger, and in that case only 10% of the judges rated the amino acid intensity as greater than moderate.

In contrast to the evaluation of these samples on the basis of amino acid intensity, when the panel rated the same samples on the basis of ginger intensity it was only Sample 4, containing a substantial quantity of ginger, which failed to get a unanimous rating of moderate or less. Accordingly, at the preferred range of about 0.04 to about 1.0 milligram of oil of ginger per gram of amino acids, a relatively bland solution is obtained wherein essentially all the judges agreed that the intensity of both the amino acid and the ginger flavor notes were moderate or less.

HIGH ENERGY FOOD SUPPLEMENTS

Linear polyalcohols which contain 5 to 15 carbon atoms are used by *J.W. Frankenfeld and S.A Miller; U.S. Patent 3,791,241; February 12, 1974; assigned to Esso Research and Engineering Company* in foods as sources of metabolic energy. It is reported that if particular polyols are used as additives in foods an excellent, high quality, high energy product results. These aliphatic polyols, such as aliphatic 1,3-diols and 1,3,5,x-polyols of five carbon atoms or more are nontoxic, energy-dense compounds which are very excellent when utilized for the production of energy. These compounds also have additional advantages over traditional caloric sources, such as fats and carbohydrates, and may be used to replace fats and carbohydrates in human and animal diets.

These materials are synthesized easily, they have desirable properties, and are readily metabolized. These materials have several advantages over usual energy sources in that they are higher in caloric density than carbohydrates and proteins and are more stable than fats and thus may be used as replacements for natural foods. These materials are colorless, virtually odorless and tasteless and are easily formulated into human or animal foods or feeds. Since they are produced by chemical means from readily obtainable starting materials, the 1,3-diols and 1,3,5,x-polyols have marketing advantages in that they are available at stable prices and in consistent quantities. They also may be produced readily in localities where arable land for grazing or cultivation is scarce. A typical compounded food formula that may be supplemented with the 1,3 and 1,3,5,x-polyols is given in Table 1.

TABLE 1: LIQUID FORMULA DIET FOR HUMAN FEEDING

	Weight %
Soy Protein	7.59
Skimmed milk	9.40
Dextrose-maltose	0.86
Corn oil	3.85
Lemon juice (or other flavoring material)	1.00
Water	77.30
Salt	qs
Vitamins and minerals	qs

The usefulness of these polyols is demonstrated by the typical growth curves shown in Figure 9.1. These curves were obtained by feeding test groups, of 5 to 10 rats each, various amounts of several high energy supplements. The basal diets in each case contained sufficient protein, salts, vitamins and minerals to support normal growth. However, the basal diets were deficient in energy (calories). This deficit was, in part, overcome by adding varying amounts of several energy sources. The greater the amount of energy, the better the growth of the animals. This is evidenced by the consistent, positive slope of the lines in Figure 9.1.

The curves in Figure 9.1 were obtained by plotting the average change in body weight of the test animals against the amount of high energy supplement added to the basal diet. Curves **1** and **2** were obtained with glucose and sucrose, two typical, natural foods of the carbohydrate type. Curve **5** was obtained with lard, a typical, natural fat. Curves **3** and **4** represent two of the compounds, 1,3-hexanediol and 1,3-octanediol. The growth obtained with lard (curve **5**) is considerably better than that from the carbohydrates (curves **1** and **2**) which is to be expected since lard provides twice the amount of calories per unit weight than does glucose or sucrose.

The data clearly indicate the value of the 1,3-diols as dietary energy sources. These compounds are superior to the carbohydrates and are virtually as good as lard. In addition, they possess all the advantages over natural foods. The feeding studies were conducted ad libitum, that is, the animals were given free choice as to food intake. The results shown in Figure 9.1 indicate that there is no palatability problem; the animals readily accepted the new diets. Thus, the data in Table 1 shows the diols are potentially high in energy (calories) and possess certain desirable physical properties.

The 1,3-diols and 1,3,5,x-polyols are nontoxic and, by comparison with other diols it was found that the 1,3- and 1,3,5,x-configuration uniquely reduces the toxicity of polyalcohols. The 1,3-diols and 1,3,5,x-polyols are metabolized well while other diols, such as the 1,5-

hexanediol, are not. This further implies that the diols are readily absorbed and are palatable to the animals.

FIGURE 9.1: HIGH ENERGY FOOD SUPPLEMENTS

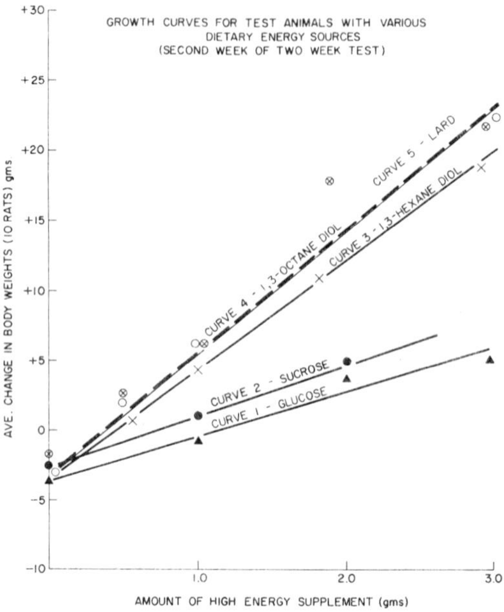

DIETS CONTAIN 18% CASEIN AS A PROTEIN SOURCE PLUS SUFFICIENT VITAMINS AND MINERALS TO MAINTAIN GOOD GROWTH

Legend for Figure 9.1

Curve	Symbol	Represents
1	Δ	Basal diet plus various amounts of glucose, a typical carbohydrate
2	•	Basal diet plus various amounts of sucrose, a typical carbohydrate
3	×	Basal diet plus various amounts of 1,3-hexanediol, one of the test compounds
4	0	Basal diet plus various amounts of 1,3-hexanediol, one of the test compounds
5	⊗	Basal diet plus various amounts of lard, a typical fat

Source: J.W. Frankenfeld and S.A. Miller; U.S. Patent 3,791,241; February 12, 1974

A typical high energy bread formula is shown in Table 2. The formula is baked in the normal manner using the sponge method of baking.

TABLE 2: HIGH ENERGY BREAD FORMULA CONTAINING 1,3-DIOLS

	Weight, grams
Flour	700.00
Water	430.5
Yeast	14.0

(continued)

TABLE 2: (continued)

	Weight, grams
Yeast food	3.5
Nonfat milk solids	21.0
Sugar	35.0
1,3-Diol	5.0
Carp	12.0
Salt	14.0

XYLITOL SUBSTITUTION

Advantage is taken of xylitol in the process of *O.N. Miller; U.S. Patent 3,717,711; February 20, 1973; assigned to Hoffmann-La Roche Inc.* as a replacement of regular carbohydrates to inhibit lipogenic enzyme activities with resultant inhibition of fatty acid synthesis. Xylitol is a five carbon sugar alcohol. It occurs as an intermediate in carbohydrate metabolism (member of the glucuronic acid xylulose cycle) as well as naturally, for example, in certain mushrooms.

The inhibition of fatty acid synthesis in biological systems by the use of xylitol is believed to arise from an inhibition of induction by xylitol of fatty acid synthesizing enzyme systems. Thus, the use of xylitol in place of rapidly absorbed carbohydrates which enhance induction of fatty acid synthesizing enzyme systems results in a reduction in fatty acid synthesis. Fatty acids are synthesized, for example, wherein the cleavage of citrate is catalyzed by citrate cleavage enzyme according to the stoichiometry:

$$\text{citrate} + \text{CoA} + \text{ATP} \longrightarrow \text{acetyl-CoA} + \text{oxaloacetate} + \text{ADP} + \text{P}_1$$

Actyl coenzyme A is then carboxylated by acetyl coenzyme A carboxylase to malonyl coenzyme A. The latter serves as the substrate for the fatty acid synthatase enzyme complex, thus producing fatty acids. In the conversion of carbohydrate and various amino acids to fat by nonruminant mammals, citrate is a major source of the acetyl group of acetyl coenzyme A which is utilized for synthesis of fatty acids. Citrate is formed in the mitochondria by the citrate synthatase reaction. It is then metabolized via the citrate acid cycle. Under conditions when energy intake exceeds energy demand, some citrate is diverted to extra-mitochondrial space of the cell where it is used for fatty acid synthesis, i.e., energy storage.

If xylitol is present in the diet in place of some or all of the rapidly absorbed carbohydrates which enhance induction of fatty acid synthesizing enzymes, i.e., mono- and disaccharide carbohydrates such as glucose, mannitol, sorbitol, sucrose, and fructose, then the fatty acid synthesizing enzyme systems are adversely affected and synthesis of fatty acids is reduced.

Xylitol, is useful in the treatment of obesity and in the correction of lipid abnormalities. The compounds are used generally in compositions that are incorporated into various dietary preparations. For example, the compounds can be used in pediatric formulations which are fed to babies, it can be used in various food preparations such as candies, bread, baked goods, and pastries. It has generally been found that if the xylitol is present from about 5 to 25% by weight of the food consumed and replaces from 5 to 25% by weight of the rapidly absorbed monosaccharide or disaccharide carbohydrates in the diet, then the desired effect is achieved. The particular amount of xylitol used and carbohydrates replaced depends upon the requirements and dietary pattern of the particular patent.

It is particularly advantageous to utilize xylitol in place of the rapidly absorbed carbohydrates in the diet of infants because this reduces the formation of fatty acids which thus reduced the formation of adiposites in the infant. The adiposites are cells where excess lipids are stored and one theory is that a reduction in the formation of the adiposites reduces the tendency for obesity in later life. When using xylitol as a replacement for the rapidly absorbed carbohydrates, it is considered preferred to build up the tolerance for xylitol in the mammal utilizing it. This is accomplished by gradually increasing the amount of xylitol in the diet until the amount required for maximum inhibition of the induction of the fatty acid synthesizing enzyme systems and efficient reduction of fatty acid synthesis is reached. This amount varies depending upon the condition and size of the patient as well as its sensitivity to xylitol. The reason for this regimen is that the administration of xylitol to those who have not been preconditioned to its use has a tendency to cause disturbances of the intestinal flora and results in diarrhea. However, the problem can be overcome by conditioning the patient to the administration of the compound.

Example 1: This example demonstrates the effect of xylitol on in vivo lypogenesis. (a) The test was carried out on Charles River female rats weighing 150 to 190 grams. The rats were divided into four groups and for 20 days meal-fed diets containing the following:

Group 1: 70% glucose
Group 2: 70% starch
Group 3: 20% glucose and 50% starch
Group 4: 20% xylitol and 50% starch

Only those of group 4 were adapted. The rate of lipogenesis was measured by determining the amount of ^{14}C-alanine incorporated per gram of tissue per 30 minutes. The effect on weight was also measured. Rats fed the diet containing xylitol exhibited only 61.5% the rate of lipogenesis when compared to the nonadapted animals on the 20% glucose containing diet.

(b) A similar test was run on another group of Charles River female rats weighing 150 to 190 grams. The only difference was that the rats were adapted to the carbohydrate added to their diet in the same manner as described for xylitol in part (a). Rats fed the xylitol containing diet exhibited only 50.6% the rate of lipogenesis in vivo of that of glucose-fed rats; 35.8% of that of fructose-fed rats and 41.7% of that of sucrose-fed rats.

Example 2: This example illustrates the effect of xylitol on lipogenesis as determined in vitro by lipogenic enzyme levels in rat liver. Charles River female rats weighing 150 to 190 grams were adapted to glucose, fructose, sucrose and xylitol as described in Example 1. The lipogenic enzyme activity was measured by determining the amount of ^{14}C citrate incorporated per gram of tissue per 30 minutes. The weight of the rats was also determined. Rats fed the xylitol diet exhibited an induction of lipogenic enzymes of only 28.0% that of glucose-fed rats, 30.8% that of fructose-fed rats and 28.9% that of sucrose-fed rats.

YEAST-STARCH TABLETS

Yeast-starch tablets useful for dietetic and therapeutic purposes are produced in the process of *H. Griffon and G. Tixier; U.S. Patent 3,378,377; April 16, 1968*. It is known that the yeast cells of ordinary fresh or untreated yeast are extremely sensitive to temperature, humidity and pressure conditions. In some instances it is desired to maximize the number of living yeast cells. However, under other circumstances it has been found best to kill as many yeast cells as possible. The yeast cells are deliberately subjected to conditions which kill most if not all of the yeast cells without, however, destroying their morphological and organoleptic properties.

It is found that dietetic yeast-starch tablets can be readily prepared by thoroughly mixing fresh untreated yeast of about 70% moisture content and powdered starch, dehydrating the mixture until the moisture is approximately 11% in order to place the mixture in equilibrium with the atmosphere (which corresponds to 9 to 13%) so that there is little or no tendency of the mixture to pick up moisture from the atmosphere or lose moisture to the atmosphere. The dehydration can be carried out in any suitable manner such as by forming the mixture into a relatively thin layer on drying plates or equipment and/or subjecting the mixture to a mild current of warm air. After the moisture content is reduced to approximately 11% the dehydrated mixture is then granulated, and the granules are finally subjected to sufficient compression on a standard tableting machine to form the tablets while at the same time killing at least a predominant proportion of the yeast cell. By using relatively high tableting pressures all or substantially all yeast cells can be killed. The effect of tableting pressure on the number of living yeast cells is well illustrated by the fact that when the compression of tableting amounts to 1,600 kg/cm^2 for a tablet of 6.3 mm the number of living yeast cells per gram is 1.8 billion. When the compression is increased to 2,600 kg/cm^2 the number of living yeast cells per gram is 260 million and pressures above 2,600 kg/cm^2 and for example, up to 4,000 kg/cm^2 kill 100% of the yeast cells.

The process is predicated on the discovery that yeast cells can be killed by a purely mechanical phenomenon, namely compression, so that while the yeast cells lose their vitality they retain their other properties, notably as to color, odor and taste, thus in contrast with other methods which involve relatively drastic heating and dehydrating conditions thereby modifying adversely the organoleptic properties and producing products which are brown in color and have the taste of peptone. The process is further predicated on the discovery that by subjecting the dehydrated yeast-starch granulation to sudden or violent instantaneous mechanical shock which is correlated with thermoshock, then ideal tablets result for these purposes. In carrying out the necessary compression with modern tableting machines, the tableting occurs in a period of time of the order of $^2/_{10}$ of a second. These machines give for example, 300 compressions to the minute of tablets having an average weight of 0.44 gram.

In order to carry this out the machine acts on successive quantities of material corresponding to the volumetric capacity of the admission chamber formed by the matrix and the punch beneath (a distance of 22 mm), and for a cylinder diameter of 11 mm. The proportion of living cells diminishes as a function of the compression and beyond the conditions above set forth living yeast cells no longer exist. By way of example a quantity of fresh untreated yeast having a moisture content of approximately 70% is thoroughly mixed with dry powdered starch in such proportions that the starch constitutes between 75 and 25% of the mixture prior to dehydration.

The yeast-starch mixture is then dehydrated in any suitable or known manner as by spreading it out in a thin layer on drying plates or equipment, and/or a mild current of warm dry air is passed over the mixture. Dehydration is continued until the moisture content of the mixture is approximately 11%. It has been found that this is very beneficial in that this moisture content contributes to the formation of a stable product which is not susceptible to deterioration or moisture content change because the moisture content is in equilibrium with the atmosphere, and consequently there is practically no tendency for the mixture to pick up moisture from the atmosphere or to lose moisture to the atmosphere. Prior to dehydration the relative proportions of fresh untreated yeast and starch are in the ratio of 100 grams of yeast to 60 grams of starch but variations from this ratio are permissible within the percentage ranges set forth above. The dehydrated mixture having a moisture content of approximately 11% is subjected to dry granulation in a manner well known in the pharmaceutical industry, and is then tableted on a standard tableting machine adjusted to produce tablets of the requisite size and thickness.

Compression in kg. per square centimeter	Thickness of the Compressed Tablets in millimeters	Number of Cells per gram (in millions)			
		Total Number of Cells	Number of Living Cells	Number of Dead Cells	Percentage of Living Cells
0		10,000	4,400	5,250	44.0
800	7.4	10,000	4,000	6,000	44.0
1,000	7.1	10,000	4,000	6,000	44.0
1,200	6.9	10,000	3,500	6,250	35.0
1,400	6.6	9,000	2,500	6,350	28.0
1,600	6.3	8,800	1,800	7,000	20.5
1,800	5.9	8,350	700	7,650	8.4
2,100	5.7	8,180	360	7,820	4.4
2,600	5.4	8,000	260	7,740	3.25
2,800	5.2	8,000	0	8,000	0
3,300	5.2	7,250	0	7,250	0
4,000	5.2	7,250	0	7,250	0

The data given in the above table show the effect of compression on the number of living cells. The above table well illustrates how the number of living cells per gram of yeast can be reduced to zero or any other desired value. The starch employed may be of any suitable nature such as corn starch, rice starch, potato starch, or cereal starch. The yeast which is used is preferably the ordinary fresh untreated commercial yeast having an initial moisture content of approximately 70%.

DIET SUPPLEMENTS FOR BODY ODOR

According to *C.W. Erwin; U.S. Patent 3,124,459; March 10, 1964* it is possible to prevent the formation of unpleasant odors generated by mammals by ingestion of water-soluble ferrous and cupric compounds. For most effective deodorizing effect, the ferrous iron and the copper are both present in two forms: as a water-soluble ionizable salt and as a water-soluble nonionizable complex, specifically, a chelate. The water-soluble ionizable salts may amount up to 90% of the total amount of iron and copper compounds, the remainder being the water-soluble nonionizable copper and iron complexes. The composition also includes a soluble edible phosphate.

Among the ionizable cupric copper and ferrous iron salts that may be used are the sulfates, chlorides and acetates. Among the water-soluble cupric copper and ferrous iron complexes that may be used are the gluconates, citrates, tartrates, the salts of ethylene diamine tetra-acetic acid, and the sodium or potassium iron and copper chlorophyllins.

There are various unpleasant body odors generated by mammals. These body odors include what may be called body odors proper which emanate from the skin and the hair and which are often due to perspiration (in those mammals which secrete perspiration); breath odors, which are due, in part, to materials adhering to the teeth, tongue and gums (which odors may be referred to as mouth odors) and, in part, to odors released in the lungs, as on the ingestion of odoriferous materials (such as onions and garlic) which are absorbed through the lining of the gastrointestinal tract and carried in the blood stream to the lungs; fecal and flatulence odors; and urine odors.

Some of these odors are referred to as systemic odors. This term designates odors due to odoriferous substances carried in the blood stream. These systemic odors appear in the body odor, particularly in the body odors of animals secreting perspiration; in the urine odor; and in the breath odor, in particular that part of the breath odor which may be referred to as lung odor, since it originates in the lung.

Other odoriferous substances, e.g., indole and skatole, are generated in the lower intestine by putrefaction of proteins and, by absorption through the intestinal wall, introduced into the blood stream. Other body odors are referred to as local body odors. These include the above-noted mouth odors due to putrefaction of food material lodged on the teeth, tongue and gums and also fecal and flatulence odors due to putrefactive decomposition of proteins in the lower intestine. The product described in this process, when incorporated with a food which is ingested by a mammal, eliminates, prevents or reduces both systemic and local body odors.

This food composition includes from 1 to 100 parts per million of ferrous iron and from 0.1 to 10 (preferably 5 or 6 or 7) parts per million of cupric copper. Preferably, the ratio of iron to copper is from 1:1 to 50:1 or 100:1. It was found that there is a synergistic effect between ferrous iron and cupric copper in these ratios, with respect to elimination, prevention and reduction of body odors. Included also are from 1 to 10 parts by weight, for each part by weight of water-soluble iron and copper compound, of an edible water-soluble phosphate, such as mono-, di-, or trisodium phosphate, or a sodium salt of a polyphosphoric acid, for instance, disodium dihydrogen pyrophosphate, sodium tripolyphosphate, tetrasodium pyrophosphate, sodium hexametaphosphate, the corresponding potassium and ammonium phosphates. The nature of the above-noted body odor inhibition is illustrated by the following experiment. Seventeen groups of three mice each were kept in separate cages and fed Rockland small animal food, ground, for one week. The cages were provided with fine screen floors so that the feces could not drop through the floor. A circle of filter paper was kept in a pan below the floor of each cage. The feces were removed each day and the paper circles were changed each day. After one week, the odor of the mice, the feces and the paper circles which trapped the urine was measured as follows.

The odor of the whole mouse was measured by placing each individual mouse in a plastic bag for three minutes at room temperature. The odor in the bag was then evaluated with an osmoscope graduated to distinguish six odor levels indicated as 1, 2, 3, 4, 5 and 6 in order of ascending odor levels. The fecal odor was determined by placing 25 feces pellets in a large test tube which was capped and allowed to stand in an incubator at 37°C, for 5 minutes. The odor level was then determined with an osmoscope. The urine odor was determined by removing the circle of paper from the pan below the cage and placing it in a closed plastic bag which was placed in the incubator for 5 minutes. The odor level was then determined with an osmoscope. The groups of mice were then fed with a diet of ground Rockland feed.

The Rockland rat or small animal food mentioned above is a standard food containing soybean oil, meal, cane molasses, fish meal, condensed buttermilk, corn gluten meal, irradiated brewers' type yeast, 4 oz/ton wheat germ oil, o.p. linseed oil meal, corn oil meal, ground oats, wheat bran, wheat flour midds, ground yellow corn, ground hulled barley, ground hulled oats, ground whole wheat, whole milk powder, alfalfa leaf meal, vitamin A oil, ½% steamed bone meal, 1% calcium carbonate from limestone, and 2% salt. This food contains at least 21% crude protein, at least 4% crude fat, not more than 6% crude fiber, and 5,000 USP units vitamin A per pound. Complete all around odor inhibition was found in groups which were fed with both ionizable and nonionizable ferrous iron and cupric copper compounds combined with various sodium phosphates.

Trisodium orthophosphate was included to show the effect of this specific phosphate. While not as effective as the other phosphates in urine deodorization, it does not significantly interfere with total deodorization and with fecal deodorization, as compared with the metal compounds used without any phosphates. There are other advantages connected with the use of soluble phosphates (including trisodium orthophosphate) which provide reasons for including them with deodorizing compositions even when, as in the case of trisodium orthophosphate, the deodorizing effect is not enhanced.

The combinations have been tested, with excellent results, on dogs, cats and humans. For humans, the supplement was mixed with oatmeal; for cats, with fish which was thereafter canned; and for dogs, with horse meat, which was also canned. The deodorizing food supplement when added to smelly goods, for instance, canned fish, or canned horse meat flavored with garlic, greatly reduces the smell of such food.

Ferrous and cupric gluconates, and other soluble chelates of these metals, and also water soluble ionizable ferrous and cupric salts, when mixed with foods in the amounts indicated may confer on these foods a metallic, astringent aftertaste. However, no such aftertaste can be noted when the above indicated amounts of edible soluble phosphates are also included with the foods. Sugar may be included with the deodorizing compositions in amounts ranging from the total weight of the other ingredients up to 10, 20,or 100 or more times the weight, to cancel the sour taste of some soluble phosphates.

SPECIAL PROCESSES AND PRODUCTS

HIGH-PROTEIN FOOD BAR

Aeration Process

The need for providing a shelf-stable, compact, nutritious food product for individuals such as campers, soldiers and space travelers has long been recognized. Food products that contain the maximum nutritional value especially in respect to protein and caloric content without concomitant, unnecessary bulk and weight (i.e., the minimum volume-to-weight ratio) are deemed essential for such a utility.

Solely on a volume-to-weight basis, the edible fats are most suitable in providing the necessary caloric requirements in such a food product. However, from a physical, chemical and bacteriological stability viewpoint, the utilization of fats presents many problems. A major problem with fats is its incompatibility with other solid food ingredients. Considerable difficulty has been encountered in preventing fat separation from the other solid constituents especially when the food products are subjected to abnormal conditions (e.g., temperature and humidity). Once separated from the other constituents, the fat readily degrades (e.g., oxidative degradation).

An additional requirement for such a food product is that it maintain its product identity until consumed. Palatability in respect to flavor, texture and the like should remain substantially unchanged notwithstanding exposure to abnormal conditions over prolonged periods of time. Also, it has been desired to provide a solid food product which contains a homogenous distribution of the caloric and proteinaceous requirements (preferably along with the other essential nutrients). Such a food product can be orally consumed in such a manner that a given amount of the product will provide ordinary and necessary nutritional food requirements for the consumer.

J.R. Durst; U.S. Patent 3,434,843; March 25, 1969; assigned to The Pillsbury Company gives a process to prepare compact food bars designed to provide a completely balanced diet. The bar is a solid, flexible, water-dispersible, unitary food piece consisting essentially of an external phase and an internal phase. The external phase contains a homogeneous dispersion of a hydrophilic film former and an edible humectant which has a boiling point and decomposition temperature in excess of 300°F. The internal phase contains edible fat globules encapsulated by the external phase. The food pieces are characterized by flexibility, impact strength against shattering, lack of hygroscopicity and a low permeability to gases and moisture.

The food pieces are stable against substantial reduction or increased flexibility when subjected to humidity variations of the surrounding atmosphere. At elevated temperatures (e.g., 300°F) the food pieces will ultimately degrade by charring rather than separation of the internal phase from the external phase. Similarly, the food pieces are not deformed when subjected to elevated temperatures (e.g., 100° to 200°F). Since the fat is totally encapsulated and protected by the external phase, the pieces are not subject to rancidification (e.g., oxidative degradation).

Examples of suitable edible film-forming substances for the external phase include egg albumin, human serum albumin, sodium proteinate of soybean, wheat germ, gelatin, sodium caseinate, nonfat dry milk solids, whole egg, pea flour, bean flour, corn germ, gelatinized starch, carboxymethylcellulose, agar agar, fish protein, gelsoy (soybean protein), whey, dextran, dextrins, bran protein, polyvinyl alcohol, polyvinylpyrrolidine, calcium alginate, potassium alginate, sodium alginate, carrageenan and salts thereof, hydroxypropyl cellulose, methylcellulose, methylethylcellulose, gum acacia, locust bean gum, gum ghatti, guar gum, gum karaya, oat gum and gum tragacanth.

In addition to the edible film-forming substances, the external phase contains water and an edible humectant. The edible humectants have an affinity for water and maintain a narrow range moisture-content fluctuation when exposed to a wide and varied range of humidity. In general, the humectants exhibit either solid or liquid properties (i.e., nongaseous) at temperatures in excess of 300°F at 760 mm pressure. Those humectants which are solid or liquid at 760 mm pressure at 350°F or above are most suitable. Suitable humectants which are normally liquid at room temperature but possess a definitive boiling point are glycerine and propylene glycol. These normally liquid humectants provide a permanent plasticizing effect.

Other materials which exhibit humectant properties are those which in the dry form decompose at elevated temperatures. Suitable solid humectants include the dry solids of invert sugar-containing materials such as molasses, honey, brown sugar, papylon, fruit juices, starch hydrolyzates (e.g., corn syrup), sorghum, etc., as well as the aqueous solutions thereof and invert sugar itself. Other solid humectants include such edible materials as calcium chloride, fructose, maltose and sorbitol.

In addition to the film formers and humectants, the external phase contains water in an amount ranging from at least 1 to about 30 parts by weight water for each 10 parts by weight hydrophilic film former therein. A substantial portion of the water may be provided in the external phase by the employment of humectants such as honey, molasses, starch hydrolysate syrups, cane juice and inverted fruit juices. The amount of water and edible humectant to hydrophilic film former in the external phase respectively ranges from about 5 to 15 and 2 to about 30 parts by weight for each 10 parts by weight hydrophilic film former.

It is essential that a stable dispersion containing the fat, film formers, water and humectant be provided. Stability of the dispersion (i.e., a stable dispersion) can be readily ascertained by mixing a drop of the mixture in 100 ml of hot water (130°F) and then observing whether or not fat separation results. A stable dispersion will not release the fat (i.e., fat lakes will not appear at the water surface).

A suitable method of preparing the food pieces is to first mix the edible fat and edible hydrophilic film former with a sufficient amount of water and edible humectant at a temperature sufficient to melt the fat and for a period of time sufficient to provide a stable dispersion consisting essentially of an external phase of hydrophilic film former, water, humectant and an internal phase of an edible oil and then remove a sufficient amount of water to set the dispersion. When this method is used, the fludity of the stable dispersion is generally too great to provide the food piece in the ultimate desired form. Consequently, it is desirable to either refrigerate, to set the dispersion or place the dispersion in a mold and then dry the dispersion to permanently set it in the form of a food piece. Another method of preparation is to mix the fat with a heat coagulable film former and a sufficient amount of water and edible humectant at a temperature sufficient to melt the fat but below the coagulation temperature of the film former for a period of time sufficient to provide a stable dispersion. The stable dispersion is then heated to a temperature and for a period of time sufficient to heat coagulate the film former. Suitable heat coagulable film formers for this method include animal albumins such as egg albumin. After heat coagulation, the stable dispersion may then be cut into the desired size and shape and dried.

A further alternative method is to mix the hydrophilic film former with a sufficient amount of water and humectant and then aerate the mixture. Suitable gases for aeration include edible gases such as air, nitrogen and carbon dioxide. These gases are incorporated into viscous mixtures of hydrophilic film former, water and humectants in an amount sufficient to increase the volume by at least about 100% (i.e., a two-fold increase) and generally greater than 200% (i.e., a three-fold increase). Suitable apparatuses for aeration include any high-speed food mixer such as a Waring Blender, or any standard household mixer.

Following aeration, edible fat is then slowly added to the aerated product while subjecting the product to high-speed mixing in a manner such that the ingested gas voids are filled (i.e., the gas is displaced) with the fat without disrupting the continuous external water, humectant and hydrophilic film former phase. In order to incorporate the fat into the aerated product, it is essential that the fat be added in a liquid state.

High-speed mixing conditions should be conducted in a manner such that the resultant added fat provides a stable dispersion. When fats which are normally solid at room temperature are used, the aerated product and fat are maintained at a temperature greater than the melting point of the fat. Other food additives such as sugar, ungelatinized starch, coloring and flavoring agents, etc., may be mixed and uniformly distributed throughout the stable dispersion after formation of the stable dispersion. By extruding the extrudable mass through a die orifice and drying the extrudate (e.g., in an air dryer), the food pieces can be produced in a continuous manner.

Drying the food pieces is conducted in a manner such that the resultant product has a moisture content of about 8 to 15% by weight. Based upon a food piece having a weight ranging from about 350 to 1,000 g, such food pieces are adapted to provide at least 2,800 kilocalories and at least 40 g of protein containing the essential amino acids (i.e., the minimum daily adult male protein and caloric requirements).

Illustrative proteinaceous film formers for this product include egg albumin, hydrolyzed soy protein, corn germ, gelatin, sodium caseinate, nonfat dry milk solids, wheat germ, fish protein, bran protein, etc. If nonproteinaceous film formers (e.g., carboxymethylcellulose, gelatinized starch, agar agar, dextran, polyvinyl alcohol, etc.) are used the proteinaceous requirements may be provided in the external phase by non-film-forming food components such as torula yeast and denaturized egg albumin. Essential amino acids may be used to provide the prerequisite protein requirements. Amino acid supplements may also be incorporated in the food pieces.

A major portion or all of the caloric requirements in the food pieces may be provided by the internal fat phase. Additional edible caloric ingredients such as torula yeast, dextran, gelatinized and ungelatinized starches, sugars or sugar-bearing substances may be incorporated into the external phase. Essential food nutrients can be provided as food constituents. Also, iodine compounds may be included. Other food piece constituents include the dietary and mineral nutrients.

Example 1: Food Rods – (A) Stable Dispersion Preparation: The following food ingredients in the designated amounts were used in making a stable dispersion.

Ingredients	Parts by Weight
Sucrose	36.0
Edible oil*	15.2
Dextrin	13.0

(continued)

Ingredients	Parts by Weight
Corn syrup solids**	10.5
Sodium caseinate	8.3
Glycerin	7.0
Gelatin (275 bloom)	4.8
Cocoa	4.5
Vanilla	0.5
Citric acid	0.1
Potassium sorbate	0.1

*Refined domestic vegetable oil having a Wiley melting point of 71° ± 3°F and a solid fat index as follows: 25 to 31% at 50°F; 7 to 13% at 60°F; 3% at 70°F; and 0.5% at 80°F (percent solids).

**Dextrose equivalent 26 to 30.

A stable dispersion was prepared by first heating the oil, in the amount designated, until translucent (at about 150°F). The oil was then placed into a Waring Blender with the designated amount of sodium caseinate and sucrose. The resultant product was then blended at a low speed to form a uniform slurry (the sucrose and caseinate were coated with the oil). An aqueous portion was simultaneously prepared by adding the gelatin to 42 parts by weight cold water (about 60°F). After allowing sufficient time for the gelatin to swell, the glycerin and citric acid were added thereto and the resultant aqueous portion was heated to 150°F and then rapidly poured into the blender containing the oil portion while maintaining vigorous agitation. The portions were then mixed at a high speed for one minute.

To insure a homogeneous mixture, the material sticking upon the inner wall of the blender was scraped into the mixed product and the blender was operated for an additional two minutes at a high speed. The dispersions thus formed consisted of fine oil droplets (i.e., the internal phase) dispersed throughout the plasticized hydrophilic film former (i.e., the external phase). Stability of the dispersions thus formed was tested by mixing a drop of the dispersion in 100 ml of hot water (130°F). No separation of the oil portion from the dispersion was observed thus indicating that finite oil droplets were encapsulated within the water, glycerin and plasticized hydrophilic film former portion.

After formation of the stable dispersion, another 20 parts by weight cold water (at 60°F) was added to the blender and mixed at a low speed for 2 min. To the stable dispersion there was added in the designated amounts the remaining ingredients (i.e., cocoa, dextrin, vanilla and potassium sorbate). The remaining ingredients were uniformly mixed by operating the blender at a high speed for 2 min. The resultant product, a thick syrupy liquid, was then deaerated in a vacuum filter flask to remove any entrapped air.

(B) Preparation of Food Rods: The product from (A) was formed into rods by forcing the product into polyethylene casings of a 0.35" diameter and 30" length. The casings were then heat-sealed on each end and the product was refrigerated at 35° to 39°F. After 1 hr of refrigeration the casings were removed from the product. The product which contained 38.2% by weight moisture was air dried, under ambient conditions (75°F and 35% relative humidity) for 3 days, to a moisture level of 10% by weight.

(C) Rod Characteristics: The dried rods from (B) possessed excellent chew, mouthfeel and flavor properties. The resultant flexible rods were of a uniform nutrient composition adapted to provide 9 g of protein and 406 kcal for each 100 g of product. Employing a standard test for stability of oils, general glass rods 2" in length, were coated with about a ¹⁄₂₀" layer of the stable dispersion prepared in accordance with Example 1(A).

The coated glass rods were placed in test tubes provided with a means of venting an air current. The coated glass rods were maintained at 97.8°C while subjected to a heated air current. After 8 days, the test was discontinued with no observable rancid odor, thus indicating the oils (i.e., internal phase) were protected from decomposition by the external phase.

Under identical conditions, the food rods were stored for 477 hours (test then discontinued) with no detectable rancid odor. The flexible rods were stabilized against loss or moisture gain even when exposed to varied temperature and humidity conditions. Specific density of the rods was 1.2 g per cubic centimeter.

Example 2: (A) Preparation of Stable Dispersion – A salami-flavored food rod was prepared from the following food ingredients.

Ingredients	Parts by Weight
Dehydrated potato flakes	28.6
Corn syrup solids*	28.5
Edible oil**	15.7
Sodium caseinate	11.9
Gelatinized tapioca starch	5.0
Glycerin	3.5
Salami-flavoring agent	2.5
Monosodium glutamate	2.0
Table salt	1.4
Garlic powder	0.2
Onion powder	0.2
Potassium sorbate	0.1
Citric acid	0.1
Red coloring agent (2.5% aqueous solution)	0.3

*Type employed in Example 1(A)
**Type employed in Example 1(A)

Into a Waring Blender, 25 parts by weight water, the glycerin, the potassium sorbate and 9 parts by weight of the sodium caseinate were added and aerated at a high speed for two minutes (volume increased 200%). Red coloring agent was then added to the aerated product. Hot edible oil (at 120°F) was slowly added to the aerated product while operating the blender at a high speed. After 90 sec, all of the oil was dispersed; however, the blender was operated for an additional two minutes. When subjected to the stable dispersion test there was no observable fat separation from the resultant product. The resultant stable dispersion having a temperature of 115°F was then transferred to a 2½-gal stainless steel Read mixer equipped with a variable speed controller, thermal jacket, sigma blades and vacuum controller.

The remaining ingredients including the remaining 2.9 parts by weight sodium caseinate were added to the Read mixer and the entire formulation was mixed for two minutes at 180 rpm. The resultant product was then extruded at 96°F through a ⅜" diameter circular polytetrafluoroethylene die orifice with a mechanical press supplying the extrusion pressure. The extrudate containing 20% by weight water was air dried at about 75°F and 35% relative humidity to a moisture level of 13.6%. The resultant rods possessed properties similar to that described in Example 1, but also possessed an enhanced chew and texture character. The specific density of the rods was 1.325 g/cm^3. The rods had flavor and texture similar to salami.

Example 3: An imitation cherry-flavored flexible rod from the following food ingredients was prepared.

Ingredients	Parts by Weight
Sodium caseinate	8.3
Gelatin (274 bloom)	4.8
Edible oil*	15.7
Powdered sucrose	38.1

(continued)

Ingredients	Parts by Weight
Glycerin	7.0
Citric acid	0.22
Potassium sorbate	0.12
Dextrin	15.0
Cornstarch	10.404
FD&C red #2	0.016
Cherry-flavoring agent	0.34
Water	12.36

*Similar to that of Example 1(A)

The rods were prepared in a manner similar to that disclosed in Example 2 except that the Read mixer thermal jacket was maintained at 150°F, mixing was for three minutes at 170 revolutions per minute and a vacuum of 24" gauge pressure. The resultant product having 12.36% by weight water was extruded at 130°F upon a polytetrafluoroethylene-coated conveyor belt and then air-dried at 72°F and a 35% relative humidity to a moisture level of 10.5% by weight (about 24 hr). It was observed that the product was easily extruded through the die orifice. The dried extrudate was resistant against impact shattering and flexible with an excellent flavor and eating character.

Protein-Containing Filling

The process of *J.E. Morgan; U.S. Patent 3,814,819; June 4, 1974; assigned to The Pillsbury Company* for the preparation of a high-protein food bar is based upon the use of a protein in a mixture containing shortening, cereal flour, and flavorings. The bar consists of at least two phases, one being plastic, i.e., creamy or pasty and another which is rigid, both having edible protein added to them. It was found that by distributing the protein used for fortification between two phases, that many of the deficiencies associated with the nutritional foods could be overcome. The bar is composed of at least three crisp wafers stacked one above the other with the creamy filling between them. Typically, each wafer consists of 10 parts of cereal flour (e.g., wheat flour), 6 or more parts of added protein, e.g., the calcium caseinate or a lactalbumin-caseinate coprecipitate, about 0.8 part oil as a release agent and a small amount of chemical leavening.

The wafer is dry, crisp, relatively thin and is fortified with added edible protein. It consists of 10 parts of wheat or other cereal grain flour, a minor amount of shortening, e.g., 2 to 5 parts to act as a separating agent from molds in which the wafers are baked. One suitable shortening consists of about 0.7 part oil such as cottonseed oil and about 0.25 part soy lecithin. A minor amount of chemical leavening is also used, e.g., about 0.5 when ammonium bicarbonate is used as the leavening agent.

Protein fortification is accomplished with any edible protein that does not impart an undesirable taste to the wafer. Examples are lactalbumin-casein coprecipitate, calcium caseinate, purified or refined grades of casein or mixtures thereof. Sodium caseinate and soy isolates, if used, should be used in conjunction with calcium ions, e.g., by adding about 0.5 part calcium lactate, calcium chloride or other calcium salt for each 10 parts of sodium caseinate. Of these, calcium caseinate is preferred because of cost and batter characteristics at relatively high concentrations.

There is no minimum amount of protein that can be added since any additional amount will have a small but measurable effect upon increasing the protein content of the wafer. It has been found, however, that to meet established nutritional standards, above 20% and preferably over 25% of dry composition should be added protein. When lactalbumin-casein coprecipitate or calcium caseinate is employed, they can be used in the amount of about 4 to 7 parts for each 10 parts of flour. To prepare the wafers a batter is made by adding the dry ingredients in a ratio of from about 150 to 250 parts water for each 100 parts dry ingredients with about 175 to 225 parts water being preferred. If soy protein or sodium caseinate is used, the calcium salt is added to the water before other dry ingredients.

The protein is then added slowly to the water with high shear mixing. The remaining ingredients are then added with continued mixing. The batter is then poured into wafer molds and baked at 450°F for three minutes. The filling is a water-free combination of about equal parts of shortening and edible protein in finely divided powdered form. To prepare the filling, the protein is added to the shortening and mixed vigorously until a stable cream (or dispersion) is formed.

The shortening is used in amounts up to about 40% of the filling. It can be any edible shortening which is plastic (not brittle) at room temperature. One suitable shortening is less than 50% solids and typically about 20 to 30% solids at room temperature. High melting-point shortenings will not provide the proper eating characteristics and will not provide the required creamy texture. Of the proteins that can be used in the filling, the preferred protein is a lactalbumin-casein coprecipitate, hereinafter referred to as coprecipitate.

Coprecipates can be prepared by precipitation from milk by the addition of calcium chloride as described in "The Manufacture of Calcium Coprecipitation," by Buchanan, Snow and Hayes, *The Australian Journal of Dairy Technology,* September 1965, page 139. The coprecipitate which results from such processes is relatively insoluble and is gritty when added to a food composition. It is preferred that the coprecipitates be resolubilized to a predetermined degree.

While the coprecipitates can be resolubilized in many different ways, for example, by exposure to strong alkali followed by ammonium hydroxide or by the addition of ammonium hydroxide to provide a pH of about 7.4 to 7.8, it is preferred to resolubilize the coprecipitate by the addition of sodium tripolyphosphate. Resolubilization of the protein with the addition of 1% sodium tripolyphosphate calculated on the dry weight of the precipitate has a higher degree of grittiness.

Good results are achieved with 2% sodium tripolyphosphate solution. When 6% of sodium tripolyphosphate is used, the resulting coprecipitate is much more soluble and if used in the wafer batter relatively large amounts of water are absorbed. The coprecipitate is resolubilized to such a degree that the water absorptivity is increased sufficiently to remove the gritty mouthfeel when eaten but without sufficient additional resolubilization to absorb large quantities of water. Thus, the resolubilization is stopped upon reaching a point just sufficient to remove the gritty mouthfeel. It is best that the coprecipitate be resolubilized, for example, by redispersing the curd in water containing 2% sodium tripolyphosphate calculated on the dry weight of the curd.

While calcium caseinate, sodium caseinate, soy proteins and other edible proteins can be used, the taste is not as good as with the milk coprecipitate. To improve the taste of the filling, a minor amount of sugar can be added. Confectioners' sugar is preferred. The filling can contain, for example, about 10 to 30% sugar and other flavor. When the bar composition is to contain added vitamins and minerals, they are incorporated into the filling which has been found to stabilize them probably by protecting them from moisture and oxygen.

To prepare the filling, all ingredients except the protein are added to the shortening and mixed to form a cream. The protein is then added with continued mixing to obtain a homogeneous creamy consistency. During this time, a certain amount of air may be incorporated into the composition which adds to its creaminess. The shortening is then spread and applied as a layer about $^1/_{10}$ of an inch thick between the wafers.

The bars are of such dimensions that one serving of two bars provides one-quarter of the recommended daily allowance of high-quality protein, vitamins, minerals and contains about 12 g of protein, about 12 g of fat, about 21 g of carbohydrate and about 250 calories. The product is acceptable in flavor and textural quality after storage at 70°F and 12 months as judged by standard taste panel tests. The shelf stability of the bars is exceptional. This stability is believed to be in large measure due to the exclusion of water from the overall formulation without sacrificing the creamy eating quality of the bar. In this way the eating quality, characterized by the sharp textural diversity between the crisp wafer and creamy filling, is maintained through extended periods of storage. Furthermore, the stability of the vitamins present in the bar is outstanding. This is apparently due to the fact that the vitamins are not exposed to water in the filling and additionally are shielded from ambient oxygen and moisture by the protective shortening matrix of the filling. By the term "high utilization value" is meant having a biological utilization value of at least 70 compared to egg as 100.

Example: A filling is made up having the following composition.

Plastic shortening (vegetable)	35
Milk protein (casein-lactalbumin coprecipitate)	34
Powdered sugar	20
Vitamin and flavor blend*	10
Color	0.6
Emulsifier (lecithin)	0.4

*Vitamin and flavor blend:

	Percent
Vitamin and mineral mixture	44
Powdered sugar	38
Caramel flavor	10
Salt	4
Vanilla flavor	2.5
Butterscotch flavor	1.5

The vitamin and flavor preblend is dry blended until homogeneous. The shortening, caramel color and emulsifier are placed in a horizontal dual sigma mixer and run at high speed until the texture is creamy and uniform; approximately three minutes. The vitamin flavor preblend is then added together with the powdered sugar and mixing is continued until a uniform texture is obtained; approximately three minutes on high speed. The milk protein is then added to the mixer and mixing is continued until a uniform color and texture is achieved; approximately another three minutes on high speed. Mixing temperature should be about 75°F. The wafers are prepared in accordance with the following formula.

	Percent
Wheat flour	62.25
Milk protein (casein-lactalbumin coprecipitate)	35
Cottonseed oil	2
Lecithin emulsifier	0.5
Ammonium bicarbonate	0.25

This composition is mixed with about 200 parts water for each 100 parts dry mix. The batter is then poured into wafer molds and baked for about three minutes at approximately 450°F. The filling is then spread to a thickness of about 1/10" and a complete bar is formed by providing three wafers measuring about 1½" x 3½" separated by two layers of filling. A small amount of flavored confectioners' coating is applied to the finished bar.

Nutritious Frosting

A unique and nutritious food product in the form of a cereal bar is described by *P.D. Halladay and J.W. Dougherty; U.S. Patent 3,821,443; June 28, 1974; assigned to General Foods Corporation.* Its prime components are a creamy nutritious frosting-like confectionery component and a farinaceous component containing cake or a cereal bound in a matrix containing fat and sugar. The filling composition is based on a mixture of an oil, sugar, and a mixture of soy protein and egg white solids and may also contain sugar, vitamins, flavorings, colorings, stabilizers and the like. The filling is prepared by adding the protein source material, sugar and other ingredients to the oil or liquid fat, and mixing these vigorously to form a creamy dispersion of the ingredients in the fat. Under vigorous agitation such as Hobart mixer, the confection exhibits a whipped character.

The oils used in preparing the confection of this process may be vegetable or animal oils exhibiting liquid properties at or below body temperature, that is, about 100°F. In addition, they should have a suitably bland flavor. Examples of suitable oils are vegetable oils such as coconut oil, cottonseed oil, and safflower oil. The oils may be partially or totally hydrogenated.

The protein sources used in the filling are a mixture of egg white solids (egg albumin) and soy protein. Any form of soy protein may be used, such as soy flour, which contains 30 to 70% protein, soy protein concentrates containing 70 to 90% protein or soy protein isolates containing at least 90% protein. The preferred protein source material is the soy isolate containing 90 to 100% protein. These materials may be used in the relative proportions of 30 to 70% (by weight) of the egg white solids and correspondingly 70 to 30% by weight soy protein. This combination imparts an unexpectedly high protein efficiency ratio to the filling composition, much higher than would be expected by calculation, as will be discussed.

The amount of the protein source combination may vary depending on the nutritional value desired. Generally speaking, these materials may be used at a level of about 10 to 40% by weight of the confection. Aside from the nutritional aspects, the egg white solids also impart whipping properties to the filling composition. This whipping property is due to the combination of the vegetable oil and the solids. The protein holds the aeration and makes possible the heating of the filling without excessive loss of aeration and the running of the filling. The filling also contains a sweetener such as sugar present to suit taste. The sugar should be very finely granulated (6X) or in liquid form. Coarsely granulated sugar would severely detract from the palatability and mouthfeel of the filling. Sugar may be present to suit taste and generally in the range of about 10 to 40% by weight of the filling composition.

Vitamins such as A, B_1, B_2, B_6, B_{12}, C, D, E and niacin may be added in appropriate amounts to supply any desired percentage of the minimum daily requirement. Minerals such as calcium, phosphorus, iodine, iron or magnesium may also be added. Where vitamins and minerals are both used in the formulation of a cereal bar containing a filling layer it is desirable to include the vitamins in one phase, say the filling, and minerals in the other phase, i.e., the cereal binder. This segregation of these ingredients prevents the possibility of their degradation caused by the reaction of the vitamin compounds with the mineral source materials.

The filling may also contain additives which impart improved heat stability to the filling. For example, the incorporation of a mixture of silicon dioxide and a polyhydric alcohol or polyhydric alcohol ester bridging agent into a low melting point oil increases the high temperature stability of the oil and prevents the oil from running or weeping out of the filling composition after prolonged periods of storage at room temperature or temperatures within the range of 70° to 110°F.

No water, milk or other aqueous ingredient is used in preparing the filling composition. The only moisture present in the filling is that which is present in the ingredients used in its preparation. Because of the very low moisture content, the filling exhibits shelf stability over a period of six months or more. In general, the moisture content of the filling is less than 5% by weight, preferably 2 to 3% by weight.

The filling composition is best prepared by mixing the dry components, i.e., the sugar, egg white solids, soy protein, and any other additives with the oil. The oil is heated above about 110°F at the time of mixing, as this facilitates the aeration of the oil. The mixture is then aerated by subjecting it to vigorous agitation using a high speed mixer or an impellor mixer until a filling having the desired light, fluffy consistency is obtained. In some instances, it has been found that the filling may have a tendency to become crumbly or mealy after being subjected to elevated temperatures for an extended period of time. The condition is believed to be the result of the absorption of the lower melting point fractions of the oil by the protein sources which causes the filling to lose its plasticity or creaminess. This condition may be overcome by thoroughly mixing the protein sources at ambient or elevated temperatures with a sufficient amount of oil to saturate the protein and form a paste, and then incorporating the oil saturated protein with the remaining oil and other ingredients and mixing as described above.

The filling may be used in combination with a cereal-bar-type product, cake-type product or other farinaceous component. A cereal bar may generally be prepared by combining commercially available cereal flakes with a liquid binder composition, thoroughly mixing the cereal with the binder and forming a sheet of the bound cereal. After permitting the cereal sheet to harden, the sheet may be coated with the filling, more of the bound cereal may be added on top of the filler to form a laminate of cereal, filling, and cereal.

Cereal binders may be made of sugar, polyhydric alcohol, oils, or combinations of these. The preferred binder contains mainly a vegetable oil such as corn oil or cottonseed oil. It is not necessary that the fat or oil used in the binder be liquid at body temperature; fats having a Wiley melting point of up to 120°F may be used. The binder may also contain sugar, salt, corn syrup solids and other ingredients such as minerals and vitamins. The binder is thus prepared by adding sugar, minerals and other ingredients to liquid oil and mixing thoroughly at a temperature above the melting point of the oils. The binder is then combined with cereal flakes in a ratio of about 20 to 80 parts of binder per correspondingly 80 to 20 parts of cereal.

Preferably, the binder and cereal are combined in approximately 50/50 ratios. Materials are mixed by combining the cereal and liquid binder and gently agitating in a mixer until the cereal flakes are completely and thoroughly dispersed with the binder. Mixing may be carried out using a Ribbon Mixer, Paterson Kelley V Blender, a rotating drum or other device which provides gentle mixing to prevent excessive breakage of the cereal flake. Where the binder contains an oil, it is preferred to mix the cereal and binder at temperatures above about 110°F or the melting point of the oil.

The cereal bar may be formed by placing the desired amount of the cereal/binder mixture into a mold having the desired dimensions and applying gentle and uniform surface pressure sufficient to compact the mixture without causing excessive fracturing of the cereal particles. Where the binder contains a fat or oil, the temperature of the mixture prior to the molding operation should be such that the fat or oil is liquid. The filling layer is then applied evenly over the lower layer of compressed cereal and the top layer of cereal/binder mixture applied and compressed in the same manner as was the lower layer.

Best results in terms of bar appearance are obtained if the lower layer of bound cereal is quick-cooled in order to solidify the fat present in the binder prior to the application of the filling, as this prevents substantial commingling of the filling and cereal layers. The filling may be applied at a temperature above the melting point of the fat or oil contained therein or it may be applied cold in the form of a solid sheet which has been precut to the appropriate dimensions.

Example 1: A nutritious aerated filling composition was prepared from the following ingredients.

	Parts
Hydrogenated vegetable oil (Durkee Kaola D)	12.00
Soy protein isolate (Supro 610, 95% protein)	4.50
Egg white solids (Henningsen's type P-20)	4.50
Sugar (6X)	6.40
Glycerine	0.13
Silicon dioxide (Cab-O-Sil)	0.51
Flavor	0.01
Preservatives (BHA + EDTA)	0.02
Total	28.07

The filling was prepared by adding the glycerine, silicon dioxide and preservatives to the oil at about 140° to 180°F and mixing for about two minutes in a Hobart Mixer. The sugar, soy protein, egg white solids and flavoring were then added and the composition was mixed at high speeds for several minutes until a viscous, aerated frosting-like filling was obtained. The filling composition was then poured onto a cool cookie sheet and cooled to about 40°F, well below the melting point of the oil. Rectangular sections of the filling having the approximate dimensions of 1⅝" x 3⁹⁄₁₆" x ³⁄₁₆" were cut for use in combination with the cereal layers of the bar. The cereal portion of the bar was prepared by mixing corn flakes sized through a U.S. No. 5 screen and retained on a U.S. No. 10 screen with an equal amount of binder. The binder was prepared by mixing the following ingredients at about 140°F.

	Parts
Hydrogenated vegetable oil	19.04
Glycerine	0.39
Silicon dioxide (Cab-O-Sil)	0.57
Sugar (6X)	9.52
Salt	0.12
Corn syrup	3.50
Minerals	1.90
Total	35.04

About 20 g of the cereal/binder mixture at a temperature of 140°F was placed in a wooden mold having an inner dimension of 1⅝" x 3⁹⁄₁₆", leveled, and gently compressed using a plunger to a thickness of about ¼". A 16-g sheet of the precut filling as prepared above was placed in the mold on top of the compressed layer of cereal and subsequently a second layer of the cereal/binder mixture applied on top of it and compressed as above. The assembly was quick-cooled to about 40°F and removed from the mold.

The finished bar has a cereal content of about 35.7%, a binder content of about 35.7% and a filling content of about 28.6% by weight. The overall moisture content of the bar is about 2.5%. The bar was packaged in a moisture-impermeable wrapping material and exhibited shelf stability for a minimum period of at least six months. The oil component of the bar was found not to run even after prolonged exposure to temperatures in excess of the melting point of the oil. As indicated above, the actual protein efficiency ratio (PER) for the filling composition is higher than the PER as theoretically calculated. PER may be calculated according to the following mathematical formula:

$$\text{PER} = 5.264\ (0.065\% \times \text{percent deficit limiting amino acid})$$

The percent deficit limiting amino acid is determined by analyzing each protein source to determine the amount of each of the essential amino acids present. These values are also available in the literature. Using whole egg as a standard, the percent deficit of limiting amino acids is obtained by comparing the combined amounts of amino acids in the given product with the combined amounts in whole egg.

The actual PER is determined by rat feeding studies conducted over a 28-day period. In this test the amount of protein eaten (fed at 10% in the test diet) is related to the amount of weight gain by the rats during the 28-day period. Thus, a PER of 4.00 means that for every gram of protein consumed during the 28 days, 4.00 g of weight were gained per day.

The calculated PER for the filling prepared according to this example is 3.81. The actual PER as determined by 28-day rat feeding studies was 4.36. The filling is thus specifically formulated using the combination of soy protein and egg white solids to give rise to PERs in excess of 4.00.

Example 2: A nutritious aerated filling composition was prepared from the ingredients shown on the following page.

	Parts
Hydrogenated safflower oil	45.5
Soy protein (Supro 610, 95% protein)	14.0
Egg white solids	14.0
6X Sugar (vitamin fortified, ⅓ MDR)	25.0
Flavorings	1.5
Total	100.00

The filling was prepared by combining the ingredients, mixing and whipping as in Example 1. A cereal binder was prepared by mixing the following ingredients.

	Parts
Hydrogenated vegetable oil	59.5
Silicon dioxide (Cab-O-Sil)	1.2
Glycerine	1.0
Sugar (6X)	29.6
Minerals	7.2
Flavorings	1.5
Total	100.0

Samples of three cereal bars based on corn flakes, puffed rice and wheat flakes were prepared by mixing the above binder and appropriate cereal and fabricating bars, by the method used in Example 1 such that the finished three-layered bars were of the following composition.

	Percent Corn Flakes	Percent Puffed Rice	Percent Wheat Flakes
Cereal	30.7	20.3	31.3
Binder	38.6	42.4	38.6
Filling	30.7	37.3	30.1
Total	100.0	100.0	100.0

The bars were packaged in moisture-impermeable wrapping material and exhibited no deterioration in texture or flavor after storage for a period of six months. Bakery products which are not inherently shelf-stable for periods of up to six months or more should be specifically formulated in order to impart extended stability under room temperature storage conditions. For example, the shelf life of conventional cake and pastry items having moisture contents of 15 to 30% may be greatly extended by inclusion in the formulation of an edible polyhydric alcohol such as glycerol, lower alkylene glycols, and sorbitol.

These additives serve to retard staling as well as to prevent bacteriological degradation. It is also desirable to include up to 2% by weight of an antimycotic such as potassium sorbate, benzoic acid, propionic acid, alkali or alkaline earth metal salts of these acids. After conventional baking of the cake, the protein nutritious confection may be applied to the cake as a frosting, filling or frosting and filling. The cake may then be packaged in a moisture-impermeable wrapping material.

HIGH-PROTEIN SNACK

The high-protein snack of *W.T. Atkinson; U.S. Patent 3,480,442; November 25, 1969; assigned to Archer-Daniels-Midland Company* is obtained by extruding a protein mix by a special process.

The preparation of a resilient, chewy, expanded protein product, defined as plexilamellar protein, is described which not only is hydratable but, further, on hydration, has the texture of cooked meat. In addition to being coherent and resilient, the plexilamellar protein is characterized by an open cell structure in which the cells have a greater length in the direction of extrusion than average width as measured in the transverse direction. In contrast to plexilamellar protein, the product of the process is characterized by crispness, friability, and a more random cellular structure.

The differences in the texture and physical properties of this protein snack and plexilamellar protein are believed to be the result of differences in the conditions at which each of these products is extruded. The process described for plexilamellar protein requires the mixture of protein and water to be masticated at the elevated temperatures and pressures until a plastic mass exhibiting plastic flow is formed which is simultaneously subjected to unidirectional flow and then extruded through an orifice.

In this process, however, the protein mix contains a lower concentration of water and is extruded at generally higher pressures, which is believed to prevent the formation of a plastic mass having plastic flow and its orientation when subjected to unidirectional flow, even though extrusion equipment containing mixing sections is used which, at a higher water content of the protein mix, would result in plexilamellar protein. On the other hand, it is to be recognized that the extrusion involved in this process does not require a masticating or mixing zone in the extruder and thus permits the use of less sophisticated, simpler extrusion devices.

It is believed, therefore, that the structural differences between these two products are based on the orientation of the protein molecules occurring in the process. Molecular orientation, if any, in this product is more random, whereas the orientation of plexilamellar protein is unidirectional, i.e., in the direction of extrusion.

Equipment which is capable of heating the protein mix to or maintaining such at the required temperatures, of exerting the pressures above indicated and of pushing the heated, pressurized protein mix through an orifice into the atmosphere can be used. The extruder need not necessarily be a screw extruder but ram extruders and other machines can be used. Because of the high degree of compaction, i.e., extrusion pressure and the low plasticity of the protein mix used, extruders normally employed in the extrusion of thermoplastic resins do not have sufficient power to continuously push the protein extrudate out of the extrusion orifice.

As a result of the pressure drop across the extrusion orifice, the water dissolved in the protein or finely distributed throughout the protein mix vaporizes and causes the extrudate to expand into the cellular product. The process is not limited to any particular type of protein. Any type of edible protein of vegetable, fish, or animal origin can be used. The term "protein derivative" as employed herein is intended to define an edible product having a protein content of at least 30% by weight.

The protein can be employed in substantially pure or water-soluble form or, as is preferred, in the form of flakes or flour, generically referred to as meal, in which the protein is in mixture with other ingredients. The preferred protein derivatives, from the standpoint of an optimum product, are solvent-extracted oil seeds such as peanuts, cotton seeds, sesame seeds, or soybeans.

The preferred oil seed meals are those which have a protein content of 40 to 70% by weight. If desired, such protein means can be diluted with nonproteinaceous fillers such as corn, sorghum, tapioca, rice, potato, wheat or bran. In general, the protein concentration of the protein derivative employed in the extrusion should not be decreased below 30% by weight in view of the decreased nutritional value and increased caloric content of such blend. The total concentration of water in the protein mix is, as indicated, from 12 to about 20% by weight of the mix. The water is necessary to obtain a coherent product and cause the expansion of the extrudate.

Not all of this water, however, is available for the expansion of the extrudate, some of the water being retained by the protein molecules as nonreleasable water. The extent of nonreleasable water depends on the specific nature of the protein involved as well as on the extrusion temperature, generally decreasing with increasing extrusion temperatures. The amount of nonreleasable water can be established by careful drying methods. Sufficient water is added above this level to permit the extrusion of a coherent extrudate. The water content of the mix should, however, not be increased to the extent that orientation causing plastic flow is induced, since such will result in the resilient, plexilamellar protein described above.

The upper water level will, therefore, depend to a certain degree on the type of equipment used. If the equipment contains no masticating, unidirectional flow-imparting zone, a higher water concentration above 20% can be used. The expansion of the extrudate normally occurs at atmospheric pressure and the minimum temperature to which the protein mix must be heated is the boiling point of water, i.e., 212°F. Application of a vacuum to the extrudate may, of course, allow the use of somewhat lower temperatures such as 200°F.

An increase in the extrusion temperature, other variables being maintained constant, will result in a finer celled structure. The upper limit of the extrusion temperature is dictated by the stability of the protein mix. Thus, the temperature should not be so high as to cause substantial degradation of the protein or any additive being present. Within the operable temperature range, temperatures are adjusted in order to result in the desired cellular structure in the extrudate. In general, extrusion temperatures, i.e., the temperature of the protein mix at the extrusion orifice, are maintained within a range of 240° to 350°F.

The extrusion pressure, or, more specifically, the pressure drop across the extrusion orifice, is at least 1,000 psi and generally 1,000 to 25,000 psi. As indicated, such high pressures are necessary to form a continuous extrudate from the protein mix which, because of its low water concentration, exhibits a low order of plastic flow. Hence, any pressure above 1,000 psi which is capable of continuously pushing the protein mix through the orifice can be used. A further increase in the pressure generally results in a higher extrusion rate.

The upper limit of extrusion pressure is generally caused by the limitations of the design of the equipment used, although the pressure should not be so high as to result in a discontinuous extrudate. The design of the extrusion orifice is a matter of choice and can vary from a slit or band orifice to a circular or square orifice. A die containing a multiplicity of orifices can be employed. In general, the resulting extrudate is cut into the desired size for the intended application.

In order to maintain the friability and crispness as well as the blandness of taste of the extrudate, the pH of the protein mix is adjusted to be within the range of 6.5 to 7.5. At pH levels below 5.5, the extrudate can lose its crispness and assume a sour taste. At a pH level above 8.5, a bitter-tasting extrudate is frequently obtained, making the product unsuitable for consumption. Protein derivatives obtained by the solvent extraction of oil seeds are generally too acidic for the blandness desirable in the unflavored product. Hence, a base such as sodium hydroxide is generally added to the protein mix in order to raise the pH to within the desired range.

If the extrudate does not have the desired crispness, edible metal salts such as sodium chloride, potassium chloride, or calcium chloride can be added to the protein mix. The addition of such salts, generally used in concentration of up to 3% based on the weight of the protein mix, tends to further reduce plastic flow in the protein mix, thereby increasing the friability of the extrudate.

In some instances it may be desirable to prepare a protein snack product which has a lower density than is obtained by expanding the product with steam. Lower density products can be readily prepared by adding to the protein mix auxiliary blowing agents such as sodium carbonate, which decompose during the heating of the protein mix to form inert gases, e.g., carbon dioxide, which causes a greater degree of expansion in the extrudate.

The protein extrudate is normally bland and of white to light tan color, depending on the protein derivative used to form the protein mix. Flavoring agents and food dyes are normally added to the protein mix to prepare a protein snack of particular taste and color.

Example: Into a ribbon blender maintained at 120°F was charged 5,675 g of Ardex 550 flakes, a commercially available soybean product containing 50% of soy protein and 9.5% of moisture, 20 g of imitation hickory-smoked flavor, 505 ml of water, and 1.8 g of Tween #40, a commercially available detergent used for the distribution of the flavor in the mix. The resulting mix was blended until uniform, and 25 g of sodium hydroxide flakes (97%) was added. Blending was continued for an additional 7 to 8 minutes.

To the resulting blend was then added a dye solution containing 1.1 g of FD&C Red #3 dye in 250 ml of water. Blending was continued until the dye was uniformly dispersed throughout the resulting protein mix. The protein mix was extruded, cut and cooled to room temperature by the following procedure. The protein mix was charged to a Kell Dot extruder, maintained at a temperature of 275° to 300°F, and extruded at that temperature through a die containing an orifice.

The extrudate expanded on emerging from the extrusion orifice and on cooling to room temperature resulted in a friable, open-celled structure which disintegrated readily when chewed. Similar results are obtained if instead of the soy flakes used in the foregoing example, other defatted vegetable proteins or animal or fish proteins are used.

FAT COATING OF PROTEIN CONCENTRATES

Protein sources such as casein, sodium caseinate, soy protein, fish protein, and other sources usually exhibit characteristic flavors which may contribute an undesirable off-taste to foods containing them. They may also linger in the mouth as a chalky-tasting precipitate. For this reason, many nutritious sources of protein have been left unexploited for human consumption.

According to *D.T. Rusch; U.S. Patent 3,793,464; February 19, 1974; assigned to ICI America Inc.,* protein concentrates can be more palatable by encapsulation or coating the protein material with an edible fat. The sequence of steps for combination of fat-water emulsions with proteinaceous materials is critical.

In conventional techniques the protein is combined with an aqueous system prior to the addition of fat which is subsequently emulsified. Such compositions are used in topping mixes, frozen desserts, ice cream mixes, whipped creams, high-protein drinks, and milk substitutes. The process differs in that fat and protein are mixed together prior to emulsification with the aqueous system.

Practically any source of protein can be used, such as sodium or calcium caseinate, casein, soy protein, fish protein, gluten, and other more recently discovered protein sources can be made palatable by encapsulation with lipids such as edible fats and oils. These can be saturated or unsaturated vegetable oils or animal fats. Among the most suitable edible lipids are cottonseed oil, safflower oil, corn oil, soybean oil, butterfat, coconut oil, peanut oil, lard, chicken fats, hydrogenated cottonseed oils, corn oils, soybean oils, peanut oils, olive oils, and coconut oils. These can be used in the form of solids and plastics.

The fat-encapsulated proteinaceous source can be dispersed in an aqueous medium using an emulsifying agent but in some instances the agent is not required. Examples of suitable emulsifying agents to be used alone or in combination are lecithin, glycerol esters, glycol esters, polyglycerol esters of fatty acid, as well as fatty acid esters of sorbitol, sorbitan, and mannitol and ethoxylated derivatives. The aqueous phase may consist of soluble additives such as sugars, glucose, sucrose, artificial sweeteners as well as flavoring, coloring agents, vitamins, and other beneficial food ingredients such as stabilizing, flavoring, and coloring agents and starch.

The composition may also contain gases such as air, carbon dioxide, or nitrous oxide to provide an effervescent beverage or a foam useful in artificial toppings and whipped cream substitutes. Liquid fat can be blended with solid proteinaceous material employing an emulsifying agent if needed. The type and level of emulsifier used is dependent upon the function desired in the finished product. In systems where water is a large portion of the product composition (such as in a beverage), care must be taken such that an excess of emulsifier does not contribute a detergent action which would strip the fat coating from the protein.

A surfactant level of less than 15% (based on fat) will normally be appropriate for most preparations. In cases where the fat has a melting point above 70°F, the fat may be placed in a mixing bowl, heated to above the melting point so that the protein can be mixed. The weight ratio of fat to protein is usually a ratio of 1:1 to 4:1. The emulsified fat-protein blend may then be mixed with the aqueous base. Usually the emulsion of the fat-aqueous base intermixture is conducted at a temperature above the melting point of the fat mix.

The conditions of the emulsifier and procedure are adjusted such that the materials form the desired oil-in-water or water-in-oil dispersion. The emulsion is easily stabilized by cooling as rapidly as possible to well below the congealing point and preferably to about 45°F. The faster the fat is congealed, the more stable the emulsion and the less chance the protein has of being leached into the aqueous phase and thereby losing its protective coating.

Example 1: 31.5 g of coconut oil (MP 98° to 101°F) is liquefied by heating to a temperature of 130°F and mixed with 3.5 g of a molten emulsifier blend comprising 85% sorbitan monostearate and 15% by weight polyoxyethylene (20) sorbitan monostearate emulsifier. To this is added 35 g of casein with vigorous agitation. This hot fat-emulsifier-casein blend is then added to 1,500 g of an aqueous base heated to 140°F and which contains 15% by weight sugar plus phosphoric acid and lactic acid such that the pH is adjusted to 2.8. The mix is then run through a homogenizer adjusted at 2,000 psi in the first stage and 500 psi in the second stage to form an oil-in-water emulsion. It is thereafter cooled to 45°F and stored. Excessive foaming does not occur in the homogenizer since the reactivity of the protein is diminished.

Example 2: An unacidified preparation similar to that of Example 1 having a chocolate flavor dissolved in the aqueous phase prior to homogenization is cooled and charged with carbon dioxide to provide a pleasant-tasting carbonated chocolate beverage.

Example 3: An emulsion concentrate is prepared by the process of Example 1 by blending 50 parts of soy protein with 80 parts of partially hydrogenated soybean oil using an emulsifier and thereafter blending the proteinaceous fat with 1,000 parts water. After homogenization and cooling, the concentrate may be added to any number of aqueous phases having different flavors and compositions.

Example 4: High-protein margarines or butter is made from fat and protein by dispersing protein in desired levels of fats. For example, 140 g of fat and 20 g of protein can be blended as described in the previous examples and thereafter combined with about 40 g of water to form a water-in-oil emulsion and then congealed by rapid cooling to 50° to 60°F and working in the normal fashion. The protein is easily added to the fat since it does not hydrate and subsequently lump.

Example 5: High-protein-content shortenings are prepared from lard, partially hydrogenated vegetable oils, or other shortenings having suitable melting points. The fat with or without an emulsifier is melted and admixed with protein. The shortening is then congealed and worked in the normal fashion to form a product thereafter useful in foods where fat is desired and protein supplementation is needed for cakes, coatings, toppings, whiteners, confections, icings, shortening in bread rolls, etc. The advantage offered in this type of shortening is that protein supplementation can occur without the deleterious effects of too much protein that is often present with uncoated protein sources.

SPECIAL PROTEIN FRACTIONS

R.A. Johnson and P.T. Anderson; U.S. Patent 3,397,991; August 20, 1968 found that the assay of essential amino acids of protein can be adjusted to any desired set of values by combining the protein from two or more different protein sources where each source has the appropriate amino acid content to achieve the desired balance in the end product. Further, the assay of the end combination product will be substantially uniform for any given sample of the product if the protein sources are simultaneously processed and intermixed during colloidal solubilization and recovery of their protein content. (Since the protein content does not ionize to form a true solution in the process but is present in the solvent as a colloid or dispersoid, the mechanism of causing the protein content to go into the solvent is referred to as colloidal solubilization.)

As is well understood, in order that man may survive he must be supplied with approximately 70 g of protein a day which includes eight amino acids that he cannot physiologically synthesize in his body. These eight amino acids are commonly referred to as indispensable amino acids and are tryptophan, lysine, leucine, isoleucine, methionine, phenylalanine, threonine and valine. The daily requirements for these essential amino acids of an average adult man in grams per day are as follows:

Tryptophan	0.25
Lysine	0.80
Leucine	1.10
Isoleucine	0.70
Methionine	0.22
Phenylalanine	1.10
Threonine	0.50
Valine	0.80

In addition to the fact that all eight of these are essential and cannot be synthesized by the body, if any one of these eight amino acids is not present at the time that the other seven are introduced into the body, man cannot utilize the other seven. Withholding of an indispensable amino acid even for a few hours has been observed to produce growth retardation or a negative nitrogen balance.

All proteins are not equally nutritious and do not all contain equal amounts of the essential amino acids. Animal proteins are generally superior to vegetable proteins. Thus, a major cause of the poor nutritional value of the vegetable proteins is a low content or unavailability of one or more of the indispensible amino acids. For example, vegetable proteins of the type here of most interest tend to be lacking in methionine and lysine.

In this special process to balance the lack of a particular essential amino acid a shift is made in the first limiting amino acid in the protein. The first limiting amino acid is that amino acid available in the least amount with reference to the requirements of the protein pattern being followed. In addition to making possible a change in the first limiting amino acid, the process can be used to shift the relative amount of the second or third or any other amino acid of the protein. These latter shifts may be desirable when attempting to obtain a particular balance of the amino acids.

In addition, when mixed sources of vegetable proteins are combined, alternate patterns of amino acid complements for dietary purposes may be sought to stimulate the amino acid content of other proteins such as found in egg albumin, the casein of milk, meat or other specially prescribed rations for the dietary needs of infant feeding, geriatrics and pregnancy. Sometimes a dry mixing of previously isolated proteins does not usually achieve a homogeneous blending of the materials because of the physical properties of dried protein. The low bulk density and electrostatic charges on the protein which may be imparted by the drying process prevents the achievement of the necessary uniformity in the mass of the combined materials.

For example, where two proteins are dry blended after having been separately isolated and prepared, the amino acid assay of a given sample of the mixture may vary by about 5 to 10% from other randomly selected samples of the mass. Use of such a material is therefore not feasible where a careful control over nutrition is desired. In this process, uniformity of the amino acid assay throughout its mass is assured.

Each source of protein is selected for its amino acid content and is combined in an amount so as to provide the desired assay of amino acids after solubilization. The aqueous solution of the solubilized protein is separated from the residue of the protein sources and the aqueous solvent is removed from the protein to yield the desired end product of uniform amino acid assay. In the preferred form of this process, the sources of the protein are vegetable meals since they represent a readily available inexpensive source. It should be understood that any other protein source could be used in combination with these vegetable sources or by themselves and the principles of the process remain unchanged.

Other phenomena are sometimes encountered which result in end products deficient in the desired amino acid content. Thus, in the extraction and isolation of proteins by chemical means from vegetable sources, it is not generally possible to completely predict the amino acid pattern in the final product with any precision based upon what may have been its starting amino acid pattern. This is because there are several classes of proteins in these various meal sources. In the simple proteins there are albumins, globulins, glutelins, protamines, albuminoids, prolamines and histones. In most economically feasible processes, only one or two of these main groups of proteins may be isolated or extracted thereby frequently altering the amino acid pattern in the isolate as compared with that in the protein prior to extraction.

This loss in extraction is of no great concern since the loss of protein upon extraction from one vegetable source can be complemented by the protein obtained from another source that has been selected with the deficiency of the first source in mind. Because of the inability to precisely predict what will be obtained from any given source of protein by the usual extraction processes, a certain amount of trial and error may be required before the precise end product desired is obtained by this process.

Vegetable means may be derived from the seeds of vegetables such as cotton, sesame, soy, rape, wheat bran, safflower, chick pea, linseed, peanut and coconut. The term vegetable meal refers generally to the by-product produced when such seeds are treated to remove the vegetable oil carried by the seeds. The remaining by-product of this operation is known as a vegetable meal or press cake.

Isolation of protein from the vegetable meal is executed for other reasons in addition to this one. Thus, the protein when consumed must be available and digestible by the human system. For this reason, it is important to isolate the protein from the extraneous, non-usable fibrous materials associated with the vegetable meals. In the isolated form the protein is also more easily utilizable in the preparation of food formulas.

The isolation or extraction process generally follows that which is used for colloidally solubilizing protein and removing it from vegetable meal. The slurry may be prepared so as to contain about 10 to 20% solids and in any event is made sufficiently liquid so as to render the slurry flowable for liquid transport. Colloidal solubilization of the protein in the meal is accomplished in this aqueous slurry.

The next step is a partial alkaline hydrolysis and sufficient alkali is added to adjust the pH of the slurry to about 10.5 although a pH range of about 9 to 12 may be used. This partial hydrolysis and colloidal solubilization of the protein is enhanced by heating the slurry between 150° and 200°F for a time necessary to achieve the required degree of hydrolysis. After the protein has been solubilized, the aqueous solution containing the protein may be physically separated from the residue of the protein source such as by filtration or centrifuging. The protein can then be recovered from solution by precipitation by adjusting the cation or hydrogen ion concentration of the solution to the appropriate point.

Generally, this is accomplished by acidifying the solution upon the addition of a suitable acid such as hydrochloric acid. The precipitated protein is suitably separated from the aqueous layer by a convenient processs which may take the form of filtration or centrifuging. The protein can then be dried in the isoelectric state if desired by any convenient technique. The yield of the protein isolate can be improved by the combination of other techniques with the alkaline hydrolysis step, for example, by the use of proteolytic enzymes.

The following procedure was followed in obtaining product (A). 37.5 lb of solvent-extracted soy meal (protein content, 44.6%) and 37.5 lb of solvent-extracted sesame meal (protein content, 46.8%) were slurried with 81 gal of water with good agitation. The resultant slurry had a volume of 95 gal. 4,750 ml of a sodium hydroxide solution (approximately 16%) were added to the slurry to give an initial pH of 10.5. The slurry was heated to 166°F and held for 30 min with good agitation. The pH had lowered to 9.7. 1,300 ml of the same sodium hydroxide solution were added to the slurry thus raising the pH to 10.3 and insuring complete reaction of the hydrogen peroxide. 2,000 ml of 35% hydrogen peroxide were added to the slurry.

The slurry temperature was raised to 193°F by steam injection (approximately 3 min come-up time) and held at the temperature for 25 min until the hydrogen peroxide reaction was complete. The slurry was then cooled to 156°F. It was passed through a pulper to remove the coarse solids and then passed through a desludging centrifuge to remove the rest of the solids from the iiquid solution. The coarse and fine solids from the pulper and centrifuge were composited. A volume of water approximately equal to the volume of the liquid solution was added to the composited solids, the resultant slurry heated to 140°F and again run through the pulper and centrifuge to separate the solids from the liquid solution. This recycle is a washing procedure.

The two liquid portions were combined and the temperature adjusted to 80°F. The protein was precipitated from the solution by adding a hydrochloric acid solution (approximately 10% HCl) rapidly with good agitation until a pH of 4.0 was reached. The precipitated protein was allowed to settle and 65 gal of the whey were decanted from the mixture. 86 gal of acidulated water (adjusted to pH 4.5 with the HCl solution) were added to the precipitated curd, the mixture well mixed and the curd allowed to settle. 75 gal of liquid were then removed from the mixture by decantation. 77 gal of acidulated water were added to the precipitated curd, again remixed and the curd allowed to settle.

89 gal of liquid were then removed from the mixture by decantation. 90 gal of acidulated water were added to the precipitated curd, well mixed and the curd again allowed to settle. 98 gal of liquid were removed by decantation. The precipitated protein was then passed over a vacuum drum filter to remove excess free water and the curd was washed again on the drum with fresh water. The curd was then made into a dispersoid by raising the pH to 7 by the addition of a sodium hydroxide solution (10% sodium hydroxide), the product heated to 130°F and spray-dried with influent air at about 400° to 450°F. (The effluent air temperature in the spray dryer was approximately 180° to 190°F.)

All fractions separated during the process were analyzed for total solids (or moisture) and protein. The spray-dried protein isolate was analyzed for the eight indispensable amino acids (tryptophan, lysine, leucine, isoleucine, methionine, phenylalanine, theonine and valine). The spray-dried protein isolate had a protein content of 85.5%. It represented 34.2% of the solids in the original material and 52.8% of the protein in the original material. The dried sludge solids from the alkaline extraction step had a protein content of 27.0%. It represented 34.7% of the solids in the original material and 25.1% of the protein in the original material.

Approximately 31.1% of the solids and 22.1% of the protein in the original material were present in the acidified whey and washes in the form of soluble material. Products B, C and D were prepared in substantially the same manner as Product A except that in the case of Products C and D the pH was adjusted to 10.5 before being put in the pulper. Product B was made from 75 lb of solvent-extracted soy meal (protein content 44.6%) by itself without combination with another source of protein. Product C was made from 75 lb of solvent-extracted sesame meal (46.8% protein) by itself without combination with another source of protein, and Product D was made from 41.25 lb of solvent-extracted soy meal (protein content 43.8%) in combination with 33.75 lb of solvent-extracted sesame meal (protein content 46.8%). The analyses of the amino acid content of the four products are shown in the table.

Amino Acid	Product A (mg/g)	Product B (mg/g)	Product C (mg/g)	Product D (mg/g)
Tryptophan	10.1	10.4	13.1	12.0
Phenylalanine	49.5	43.2	39.7	41.9
Isoleucine	42.8	36.1	41.0	43.7
Lysine	30.4	45.6	20.0	31.3
Valine	58.2	42.1	88.4	49.8
Threonine	41.2	30.4	27.8	28.3
Methionine	15.4	8.8	26.0	11.0
Leucine	64.8	60.1	54.0	59.5

The first limiting amino acid in the 100% soy isolate of Product B has been successfully shifted from phenylalanine to lysine in Product A. Therefore, the daily intake of protein to acquire the necessary phenylalanine requirement has been reduced from 20.6 g in Product B to 19.0 g in Product A, or a decrease of 8%. Of equal or even greater significance in comparing Products A and B, is the quantity of protein required for daily methionine needs which is so important to growth. This has been reduced from 20.2 g in Product B to 12.2 g in Product A. This amounts to a decrease of 40% in the amount of protein necessary to acquire man's daily maintenance needs of methionine.

NUTRITIOUS SOY BASE

To reduce the weight of obese persons, a limited caloric intake is nearly always essential. However, it is necessary to provide not only for the limitation of caloric intake but also for adequate nutrition to maintain health. For example, it is well-known that individuals consuming a limited or narrow diet may develop vitamin deficiencies. Furthermore, if only carbohydrates are ingested for an extended period, even if in minimal amounts, protein deficiency will ensue. Lack of unsaturated fats in the diet may result in the dermatological changes or dry scaling of the skin that follows fat deprivation in experimental animals.

A dietary nutritional product for weight-reducing purposes must have a high satiety value to eliminate as much as possible the temptation to overeat. Likewise it is necessary that a dietary product be palatable. For example, it is well-known that soybean flour or meal provides a high-protein food; however, products containing more than about 25% soybean flour are unpalatable.

A vitamin and mineral-fortified dietary nutritional product developed by *K.S. Kemmerer; U.S. Patent 3,097,947; July 16, 1963; assigned to Mead Johnson & Company* contains from 58 to 75% nonfat milk solids and from 1.5 to 2.5% milk fat, from 9 to 13% soy flour and from 5 to 9% sugar.

The compositions may be further characterized as containing from 25 to 35% protein, preferably from 29 to 33% protein, from 5 to 13% fat, preferably from 8 to 10% fat, and from 40 to 60% carbohydrate, all on a dry weight basis. The fat component of these compositions is divided in proportions of from 50 to 70% unsaturated fats and from 30 to 50% saturated fat. The soy (soybean) flour, as indicated, is one that has been precooked or treated to improve the digestibility and palatability and to provide for ready dispersibility. When low fat, or defatted soy flours are used other sources of edible fats or oils are added to adjust the composition of the product to within the specified range of fat content. Generally the products may be prepared in dry form which may be subsequently mixed with water to make a liquid dispersion. The products may be flavored with a variety of well-known flavors, such as chocolate, butterscotch or vanilla. It is also possible to utilize the nutritional compositions in the preparation of puddings or pudding mixes by the addition of gelatin or gelatinizing agents to the dry components of the composition which may then be gelatinized to pudding in the conventional manner.

The total caloric content of the powder-form composition ranges from about 700 to 1,200 calories per 227 g (½ lb) and preferably from about 800 to 1,000 calories on the same basis. The liquid composition contains from about 175 to 300 calories in each eight-fluid-ounce portion. The sodium content of the compositions is from about 0.3 to 0.5% on a dry weight basis. The products are vitamin and mineral-fortified to the extent that ½ lb of the dried material provides adequate vitamins and minerals to satisfy the daily minimum requirements recommended by the Food and Nutrition Board of the National Research Council.

The process to pretreat the soy flour involves heating an aqueous suspension of the soy flour at slightly alkaline pH values until undesirable flavor bodies are removed and the physical structure of the flour is altered sufficiently to provide a product of improved nutritional quality and suspendibility and resuspendibility characteristics. It has been found that soy flour of usual commercial grade, including the full-fat, low-fat, and defatted varieties, is markedly improved with regard to the above-indicated characteristics when subjected to this process.

For instance, it has been found that a 5% suspension of commercial soy flour in water will undergo complete phase separation in 2 to 3 minutes after agitating the flour with the water despite its fine particle size. The particle size of commercial soy flour is ordinarily such as to permit passage through a 100-mesh sieve. Untreated commercial grade soy flour frequently has a beany flavor and a mealy or sandy mouthfeel. Treated soy flour appears to lose these characteristics.

Aqueous suspensions of treated soy flour have been found to be entirely compatible with skim milk, permitting heat sterilization to be employed in the preparation of liquid nutritional products containing these ingredients. Curdling of the milk proteins has presented a long-standing problem in the preparation of liquid nutritional products containing both soy flour and milk solids as principal protein ingredients. This curdling effect on heat-sterilizing liquid products containing untreated or improperly treated soy flour and skim milk involves coagulation of the proteins resulting in their precipitation as aggregates varying from very fine particles having a grainy nature to marble-sized curds.

The soy flour precooking process is accomplished essentially as follows. First, an aqueous slurry is prepared by mixing one part by weight of the soy flour with from 3 to 7.5 parts by weight of water. Precise measurement of the water within this range is not entirely necessary, but it is preferred in the interest of economy and convenient operation on a large scale. If less than 3 parts by weight of water are used, a very viscous adhesive mass is obtained which is difficult to process, while the use of more than 7.5 parts by weight of water is uneconomical since distillation of excess water at a later stage in the process is usually required.

After formation of a homogeneous slurry, the alkalinity of the mass is adjusted to within the range pH 7.5 to 10. It is preferred to use a volatile base, such as ammonia, ammonium hydroxide, or a lower aliphatic amine having a boiling point of up to about 80°C, for pH adjustment, since it is readily removed by evaporation when no longer needed at a subsequent stage. Examples of operable amines include methylamine, ethylamine, ethanolamine, diethylamine, propylamine, trimethylamine, etc. Sodium hydroxide and other alkali metal, or alkaline earth metal oxides and hydroxides are also operable. However, since one of the objects is to prepare a composition of low sodium content, use of sodium hydroxide is undesirable for pH adjustment.

The alkaline suspension is then heated with agitation for a graded time interval at from 190° to 240°F. At 240°F, the desired alteration of the physical character of the flour takes place almost instantaneously. Heating periods varying from two or three seconds up to four minutes have been found to be operable at 240°F. At 190°F, ten to twenty minutes are required to bring about the desired modification of the flour. For intermediate temperatures within this range, time intervals gauged proportionately between two seconds and twenty minutes are employed. If treatment is extended beyond the above limits of time and temperature, undesirable flavors are developed making the flour unsatisfactory for use.

Coupled with this heat treatment is a distillation or evaporation feature. This results in removal of undesirable flavor bodies from the uncooked flour. Although it is possible to satisfactorily improve the flavor of the soy flour by vacuum distillation of aqueous slurries at temperatures below the 190°F heat treatment limit specified, the desirable alteration of the physical structure of the flour does not then take place. A convenient gauge for adjusting distillation or evaporation conditions to a particular apparatus is to condense the vapors evolved and to measure the amount of condensate.

It has been found that when an amount of condensate is removed corresponding to about 1% the weight of the water charged to preparation of the original soy flour suspension, that sufficient evaporation has occurred to result in removal of the undesired flavors. Once the distillation or evaporation conditions have been empirically determined in this fashion, of course, it is not necessary to condense the vapors in subsequent routine operation.

The process may be conducted in any convenient equipment, so long as the above conditions are met. It has been found most convenient, however to use a continuous process in which the aqueous slurry is rapidly heated, for instance by direct steam injection, to a temperature of 190° to 240°F, to then pass the flow of liquid through an insulated or heated tube of sufficient length to maintain this temperature for approximately ten seconds and to then allow the liquid flow to pass into an evacuated tank in which the vacuum is adjusted so as to result in nearly instantaneous evaporation of the desired quantity of water. In operating on a heat cycle of 230°F with a 24" vacuum, sufficient water to accomplish the above result is evaporated in the resulting flashing process to reduce the temperature of the liquid to within the range 120° to 135°F.

Alternatively in batch-type operation, the aqueous suspension may be agitated and heated in a tank by any desirable means at a temperature of from 190° up to the boiling point of the suspension at atmospheric pressure if an efficient condenser or ventilation system is provided so as to result in evaporation of an amount of water corresponding to at least 1% by weight of the water originally charged within the heating period prescribed. The mixture is then passed to further processing.

It has been found convenient to include various other of the ingredients in the slightly alkaline aqueous slurry prior to the heat treatment step. For example, the major fatty glyceride ingredients and at least some of the milk solids are incorporated at this stage. Emulsion stabilizers, mineral additives, etc., may also be added as components of the soy flour base prepared by the process. Additional fatty glyceride ingredients amounting up to about 40 to 50% by weight of the process mixture may be incorporated prior to the aqueous heat treatment step without interfering with alteration of the soy flour in the desired fashion.

The further process steps required are determined by the nature of the desired finished composition. In the preparation of solid nutritional compositions, the soy flour suspension containing additional ingredients as mentioned, is simply spray-dried. If the capacity of the dryer is sufficient, the hot liquid from the heat treatment process described above may be passed directly to the dryer without intermediate cooling. However, it is important that the extent of heat treatment be limited as described so as not to result in overcooking the flour resulting in the development of undesirable flavors. Therefore, if the mixture cannot be immediately dried, it should be cooled to a temperature of 160°F or less. For storage for short periods of time, 160°F is a convenient temperature, since a minimum of coolant is required, and sterility is readily maintained. In the preparation of the liquid nutritional products, the liquid soy suspension after heat treatment and evaporation may be used directly after appropriate pH adjustment. Skim milk is frequently an important ingredient of such compositions. In the preparation of liquid nutritional products containing soy flour and skim milk or milk solids as significant protein ingredients, the consistency of the final product is affected by the pH. Values less than about pH 6 are unsuitable because the milk proteins coagulate and precipitate similar to the phenomenon observed when fresh milk sours. At higher pH values the product becomes more viscous until a point is reached at about pH 9.0 where a wholly undesirable mass is obtained.

This property of the liquid compositions appears to be related to the soy flour content. For compositions having a nonfat milk solids to soy flour ratio of about 9:1 slightly alkaline pH values are preferred. For compositions having a higher proportion of soy flour, lower pH values are required to achieve a desirable viscosity. One of the major advantages of the process for the treatment of soy flour is that heat sterilization may be used in the preparation of liquid products containing it in combination with skim milk. It has been observed that there is little or no tendency for liquid products containing from 2:1 to 9:1 parts by weight of nonfat milk solids, supplied principally by skim milk, per part by weight of treated soy flour to curdle when heat sterilized. When untreated soy flour and raw or heat stabilized skim milk are mixed and heat-sterilized, curdling invariably occurs, resulting in a product which is thoroughly unappetizing.

The so-called heat-stabilized skim milk is skim milk which has been heated within a period of about one minute to a temperature of 220° to 260°F and then maintained at this temperature for a period of about seven minutes at the 220°F level or one to two minutes at the 260°F level. Holding periods for other temperatures within the range are interpolated between these two figures. Heat stabilization has the effect of reducing the propensity of the skim milk to curdle when it is concentrated or sterilized by further treatment.

Example 1: (A) Commercial full-fat soy flour (22.5% fat), 49 lb, is added to 20 gal of water at 120°F with agitation to provide a homogeneous suspension. Sufficient ammonium hydroxide is added to the resulting suspension with stirring to raise the pH up to 8.2. The following ingredients are then added with agitation to provide again a homogeneous suspension: 0.157 g of potassium iodide, 11.5 lb of corn oil, 2.73 lb of coconut oil, and 57.5 g of carrageenin (Kraystay S).

The carrageenin is an emulsion stabilizer and is dispersed in the corn oil prior to adding to the mixture. The amounts of corn oil and coconut oil added are selected to bring the total fat content of the mixture (dry basis) up to about 39.9%. The mixture is then heated by steam injection to maintain it at 190° to 195°F for 20 min in an open, well-ventilated tank permitting efficient removal of the vapors. The slurry is then cooled to 120° to 140°F prior to further processing. The cooled effluent is then homogenized under pressure at 2,500 psig. The homogenized product is spray-dried to provide a homogeneous powdered soy base having the following composition.

	Percent
Full fat soy flour (22.5% fat)	77.3
Corn oil	18.2
Coconut oil	4.3
Carrageenin	0.2
Potassium iodide	5.5×10^{-6}

(B) The soy base is then used in the preparation of various nutritional compositions in powder form by dry blending. An illustration follows.

Nonfat milk solids (instant)	161.7 lb
Soy flour base (as above)	46.5 lb
Skim milk powder	32.7 lb

(continued)

Whole milk powder	22.3 lb
Sucrose	17.6 lb
Starch	12.1 lb
Yeast	3.00 lb
Vanilla flavor (solid)	2.64 lb
Vitamin concentrate	476.00 g
Mineral concentrate	204.00 g

After blending the above, the product is passed through an 8-mesh screen and packaged for sale to the consumer. In use, the dry product is reconstituted by mixing with water in a ratio of from about 28 to 36 fluid ounces of water per one-half pound of the dry composition, preferably about 1 quart per 1 to 2 lb of the dry composition. The mixture may be blended with a mechanical stirrer or by hand.

Example 2: (A) Preparation of Liquid Soy Base – Into a mixing vessel equipped with heater and agitator, containing 88 gal of water heated to 150°F, are added the following ingredients, with agitation.

Full fat soy flour (22.5% fat)	100 lb
Corn oil	23.55 lb
Coconut oil	5.56 lb
Potassium iodide	0.32 g
Carrageenin (Kraystay S)	117 g

The carrageenin is dispersed in the oil prior to adding the ingredients to the mixture. Concentrated (28%) aqueous ammonium hydroxide is added to raise the pH of the mixture to 8.2. After all of the ingredients are thoroughly mixed, the contents of the vessels are passed through a confined stream steam injection heater and heated by direct steam injection to 230°F. The holding time in the steam injection heater is from about 8 to 12 sec.

The contents of the steam injection heater are then passed into a vacuum chamber maintained at about 24" of vacuum, where flash evaporation occurs, reducing the temperature to 120° to 135°F. The product is cooled to about 45°F and passed to a holding tank. The pH of the flour slurry is 7.5 to 7.7 at this point. It is adjusted to pH 7 by the addition of hydrochloric acid. The resulting neutralized suspension has a solids content of 27.5%. It is used in the preparation of liquid nutritional compositions of this process as follows.

(B) Production of Liquid Nutrient Product – The following ingredients are blended in a mixing vessel equipped with an agitator and heater.

Liquid skim milk concentrate (22% nonfat solids)	4,808.4 lb
Liquid soy base (Ex. 2A) (27.5% total solids)	844.64 lb
Cream (40% butterfat)	75.17 lb
Calcium cyclamate	3.52 lb
Vanilla flavor	22.22 lb
Sucrose	119.31 lb
Carrageenin (Kraystay S)	600.0 g
Mineral concentrate	178.32 g
Vitamin concentrate	75.00 g

The liquid skim milk is stabilized prior to mixing by heating in a steam injection heater to 250°F, held for 4 min and flashed to a holding chamber and permitted to cool to 150°F. The product is then concentrated to 22% solids by heating. The product preferably should have a solids content of from 20 to 25%.

The ingredients listed above are thoroughly blended at 150°F and then homogenized by passing through a pressure homogenizer at 3,000 psig. The homogenized product is cooled to 50°F in a holding tank. Dry vitamin mix of the following composition is added and incorporated by stirring.

Dry Vitamin Mix

	Grams
Sodium ascorbate	334.0
Thiamine hydrochloride	6.0
Riboflavin	9.0
Niacinamide	45.0
Pyridoxine hydrochloride	6.0
Vitamin B_{12}	0.006
Calcium pantothenate	30.0

The solids content of the product is adjusted to 22% solids and placed in 8-oz-content steel cans and sealed. The cans are sterilized for 6 min at 250°F. The product may be chilled prior to consumption if desired, but when canned and sterilized as indicated it may be stored at room temperature. The production of puddings based upon the dry nutritional product is described.

For example, the dry nutritional composition specifically illustrated in Example 1(B) may be admixed with gelatinizing agents by conventional procedures to produce a pudding mix which may be readily converted to puddings or utilized in the production of ice cream or frozen dessert. For example, alkali metal salts of alginic acid, such as sodium alginate, may be used in amounts of from 1.5 to 7% based on the dry solids weight of the nutrient composition with optimum amounts of about 3.5%. On the same basis carrageenin may be used in amounts of from 0.5 to 3.5%, rennet from 0.1 to 0.5%, gelatin from 3 to 12%, pectin (low methoxyl) from 2.5 to 10%, starch or pregelatinized starch from 7 to 26%.

The gelatinizing agents other than alginic acids, carrageenin, and pregelatinized starch require heat in the preparation of the pudding. The alginates, carrageenins, and rennet have the feature that they have little or no caloric value in the amounts used, and thus do not appreciably change the caloric contents of the nutritional composition used in the preparation of the puddings. In the use of alginates and carrageenins as gelatinizing agents it is usual to add small amounts of edible salts to aid in sequestering of the calcium of the milk and/or participate in the setting reaction. Such salts are, for alginates, disodium hydrogen phosphate, tetrasodium pyrophosphate, and hexametaphosphates; and for carrageenins, salts such as potassium citrate, potassium gluconate and sodium carbonate.

YEAST AND HVP COMBINATION

The flavor of yeast can be masked by addition of hydrolyzed vegetable protein (HVP). The details of the process are given by *I.E. Witwicka, J.S. Pavuk and K.M. Gaver; U.S. Patent 2,999,753; September 12, 1961; assigned to The Ogilvie Flour Mills Company, Limited, Canada.*

Shown in Figure 10.1 is a flow sheet of a process where hydrolyzed gluten (containing salt) is obtained in any convenient practical way. Several ways are indicated in the dotted box in the upper left hand part of the figure. In the process it is best to first remove humin, inorganic salt, impurities, etc., then the various amino acids separated, such as leucine, isoleucine, tyrosine, etc. Then it is possible to acidify to the isoelectric point of glutamic acid and separate out the glutamic acid. The glutamic acid is crystallized out and separated. The end liquor left at this point has a pH of about 3.0 to 3.2. To obtain hydrolyzed vegetable protein the pH is increased to at least 5 as by the addition of sodium hydroxide and then it is treated to reduce the ammonia content.

It is best to adjust the end liquor to a pH of 10.7 to 10.8 by the addition of caustic soda (sodium hydroxide), then to warm the adjusted end liquor to temperatures ranging from 100° to 175°F under vacuum for between about 8 to 20 hr to reduce the ammonia content to the 0.01 to 0.05% range or below, and then to neutralize to a pH of from 5.5 to 5.6 with hydrochloric acid to obtain HVP. In lieu of alkalizing, heating, and neutralizing, it is possible to neutralize the end liquor from its pH of 3.0 to 3.2 to a pH between 5.0 to 7.0 (e.g., 5.5) by the addition of sodium hydroxide and then warm to a temperature of 100° to 175°F for from 8 to 20 hr under vacuum to eliminate the ammonia.

The adjustment of the pH to an approximately neutral pH and the elimination of ammonia improves the taste of the food product and facilitates the drying of the subsequently formed autolyzed yeast-end liquor mixture. It should be noted, however, that it is best to use the neutralized end liquor after removal of ammonia in the making of the improved food products without drying such liquor. This neutralized and pretreated glutamic acid end liquor is called hydrolyzed vegetable protein (HVP).

Hydrated yeast is also obtained in any practical convenient manner. The hydrolyzed vegetable protein is then mixed with the hydrated yeast. The autolysis of the yeast is induced by the salt in the hydrolyzed vegetable protein. After autolysis, the autolyzate is dried in any convenient manner. It is preferred to dry by vacuum evaporation or by spray drying.

FIGURE 10.1: FLOWSHEET FOR PREPARATION OF YEAST-HVP COMBINATION

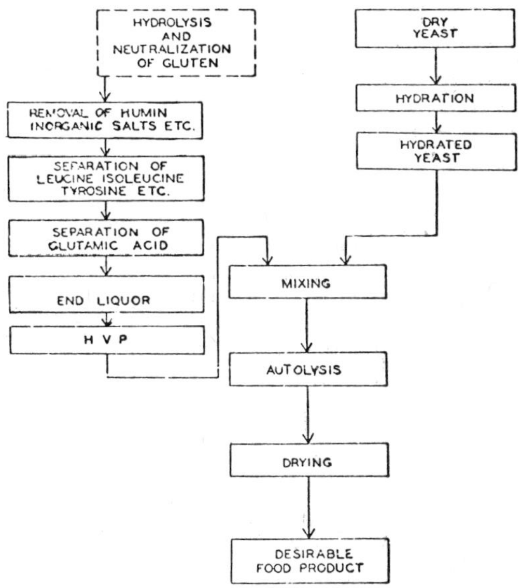

Source: I.E. Witwicka, J.S. Pavuk and K.M. Gaver; U.S. Patent 2,999,753; Sept. 12, 1961

Example: 1,000 grams of compressed yeast (27% solids) was mixed with 1,000 ml of processed HVP (1,250 g) and autolysis allowed to proceed with agitation at room temperature for 4 hr. Half of the autolyzate was vacuum-evaporated to a paste containing approximately 75% solids, and the other half was spray-dried to a tan powder. Analysis of the powder is shown on the following page.

Solids	95 %
Protein	51.25 %
Total nitrogen	9.18 %
Amino nitrogen	8.35 %
Salt	31.25 %
Vitamins	
Thiamine	16 mcg/g
Riboflavin	20 mcg/g
Niacin	147 mcg/g
Pantothenic acid	41 mcg/g
Pyridoxine	10 mcg/g
Folic acid	7 mcg/g
Biotin	1 mcg/g
Choline	760 mcg/g
Inositol	495 mcg/g
Amino acids	
Alanine	2.2 %
Arginine	2.5 %
Aspartic acid	2.3 %
Cystine	1.0 %
Glutamic acid	6.3 %
Glycine	1.8 %
Histidine	1.4 %
Isoleucine	2.0 %
Leucine	2.9 %
Lysine	2.4 %
Methionine	1.3 %
Phenylalanine	2.5 %
Proline	6.3 %
Serine	2.2 %
Threonine	0.9 %
Tryptophan	0.2 %
Tyrosine	1.0 %
Valine	2.4 %

BLAND PROTEIN FROM MEAT

Food can be fortified with bland protein materials obtained from meat according to *J.W. Staackmann and H.P. Furgal; U.S. Patent 3,062,655; November 6, 1962; assigned to Armour and Company*. Meat is first rendered substantially tasteless and colorless by water extraction and then alkali-treated in a manner which softens and ruptures the small thread-like fibrils without degrading the resulting insoluble protein. Rupture and softening of the meat fibrils, defibrillation, yields a meat preparation which has no typical meat texture and is susceptible to many subsequent operations, such as spray-drying, where fibrils cause interference. These fibrils, which give meat its typical fibrous texture cannot be effectively ruptured by use of heat or ordinary mechanical processing.

All types of meats may be used for raw material. Muscle meats are preferably used in that some animal organs possess marked flavors and odors which are more difficult to remove by water extraction. The first step in the process is the preparation of finely divided raw meat particles. Any common method of comminution may be used. Ground meat mixed with water is blended by means of equipment such as a Mincemaster, silent cutter, or a sausage mill.

The finely divided meat slurry is then mechanically separated as by centrifugation. The resulting supernatant contains fat and the soluble flavor components of the meat. The washing and blending steps are repeated to ensure that the meat is substantially free of color and flavor. After washing, the water-insoluble meat protein is mixed with fresh water and adjusted to pH 8.0 or higher. This alkaline treatment of the washed meat causes swelling and softening of the meat fibrils. Any alkaline level in excess of pH 8.0 is operable.

Although strongly alkaline conditions, as those above pH 12.0, may be used, upper levels require careful control of time and temperature conditions to avoid protein hydrolysis. NH_4OH is the alkali of choice due to its volatility. After or during the softening and rupturing treatment, the softened protein will usually be subjected to heat treatment, and during this heat treatment the NH_4OH because of its volatility is largely evaporated from the defibrillated product. Therefore there is little likelihood that the alkalizing agent will contribute flavor to the finished product.

However, other reagents such as complex phosphates may be used in this process. If it is deemed desirable to use high alkalinities, as those in excess of about pH 10.0 and complex phosphates are initially used, another type or secondary alkalizing agent such as sodium bicarbonate, sodium hydroxide or ammonium hydroxide is used to prevent formation of soapy flavors created by the use of high levels of some complex phosphates.

While the washed protein is maintained at a pH of from 8.0 to 14.0, it is subjected to heat treatment and/or a holding period sufficient to soften and rupture the meat fibrils. It is best to use a temperature of from about 140° to 212°F when the water-extracted meat protein is from pH 8.0 to 10.0. However, when facilities for very rapid continuous heating are available, as steam injection flash heaters, temperatures in the vicinity of 300°F may be used. When these high temperatures are used the heating period will be instantaneous. If very high alkaline pH's are used, temperatures much lower than 140°F can be satisfactorily used for defibrillation.

It should be stressed that alkali treatment conditions—pH, time, and temperature—must not be great enough to produce hydrolysis of the protein. Defibrillation is a function of the degree of alkalinity, time, and temperature: when higher pH levels are used, lower temperatures and shorter times suffice; when lower temperatures are used, longer times are required for softening and defibrillation. Thus at pH 8.0 to 10.0, and at temperatures from 212° to 300°F defibrillation is accomplished in a few seconds. At a pH of 8.0 to about 10.0, and at a temperature of 140°F a time interval of about 10 min is required for defibrillation.

The softened protein is again comminuted after the alkali defibrillation, but with proper alkali treatment a second comminution is not essential. Comminution may be desirable in that the softened fibrils, some of which might have escaped rupture during the alkali treatment, are easily broken apart by ordinary mechanical methods after this treatment.

After defibrillation, the meat protein should be stabilized if it is not to be consumed soon. Stabilization includes any of the common methods of food preservation such as heating, freezing and dehydration. The best method of stabilization is dehydration which may be accomplished by spray drying. Alkaline processing destroys fibrils which make spray drying of ordinary meat preparations impractical due to the tendency of the intact threadlike fibrils to clog minute orifices of spray nozzles and cores. Thus, the alkaline treatment has the added advantage of facilitating subsequent drying operations.

After it has been stabilized the bland meat product may be incorporated into many types of food products to enhance their protein value, but without altering their typical flavor and texture characteristics. Meat preparations, grain and cereal products, dairy products, candies and other confections, fruit drinks and juices, sauces, soups, gravies and prepared vegetables are some of the foods which may be fortified. In addition, special dietetic foods such as hypoallergenic preparations, infant formulas and geriatric foods may be improved by fortification with water-extracted fiberless meat protein.

Example 1: 600 lb of beef chucks were ground and mixed with 1,200 lb of water. The ground meat and water were blended using a Mincemaster. The 1,800 lb of blend was then pumped into a solid bowl centrifuge which had been filled with cold water. After centrifugation, the supernatant containing fat and soluble flavor components was removed, and 530 lb of residue remaining in the centrifuge was mixed with 1,200 lb of water. The diluted residue after mixing was again centrifuged.

The supernatant of both runs was collected for beef extract manufacture. The residue mixed with water was alkalized with 28% NH_4OH to a pH of slightly above 8.0. At this point the mixture became very viscous. The alkaline mixture was then reblended by mincing. The prepared blend was cooked by means of a flash cooker at 212°F, remixed and cooled to 90°F. After cooling it was packed into 100-lb molds and frozen. The finished product was white, smooth and tasteless. At 212°F the product was very fluid; at 100°F it was slightly viscous. Upon analysis this high-moisture protein product had the following analysis: moisture 95.3%, protein 4.13%, fat 0.3%, and ash 0.07%.

Example 2: Bland and defibrillated meat protein prepared as in Example 1 was spray-dried to a smooth soft powder. A rat feeding study was performed to biologically evaluate the protein quality of the bland meat powder. The method which results in a value termed protein retention efficiency, briefly consists of feeding a group of rats on a diet containing 10% of the test protein, and a group of litter-mates on a nonprotein diet.

After 10 days' feeding the algebraic difference between the gains in weight of the two groups, divided by the weight of the protein eaten, is defined to be the new protein ratio. This ratio multiplied by 16 is called the protein retention efficiency. Diets were formulated to be isocaloric and to supply 10% protein from the various materials being tested.

Material	Number of Animals	Average Initial Body Weight	Average Final Body Weight	Average Total Food Consumed per Animal	Protein Retention Efficiency
Protein-free control	5	62	51	45	–
Bland meat protein	5	62	110	121	78.0
Whole meat protein	5	62	105	121	71.4
Cottonseed protein	5	62	78	99	43.5
½ Bland protein	5	62	101	114	70.2
½ Cottonseed meal wheat gluten	5	62	66	81	31.5
½ Bland protein	5	60	105	119	75.2
½ Wheat gluten lactalbumin (standard)	5	62	95	97	72.5

The bland meat protein was found to be of higher quality protein than whole meat protein. A synergistic upgrading of cottonseed protein and wheat gluten quality resulted from the blending of these proteins with a 50% level of total water-insoluble meat protein. Therefore, the process, while imparting new and desirable properties to meat protein enhances the nutritional quality of meat protein. The addition of water-extracted and alkali-defibrillated meat protein to selected food material, such as cottonseed protein and wheat gluten, causes a remarkable improvement in protein nutritional quality.

Example 3: Water-extracted and alkali-defibrillated meat protein in wet form, 80% moisture, was incorporated in vanilla ice cream mix. The basic ice cream mix formula was as follows.

	Percent
Cream	32.5
Bland meat protein (80% moisture)	27.3
Water	21
Cane sugar	13.3
Nonfat dry milk	5.5
Gelatin	0.3
Emulsifier	0.1
	100

After all ingredients were blended, the mix was pasteurized at 150°F for 30 min. After overnight cooling to 35°F, vanilla extract was added just prior to freezing. The flavored mix was frozen in a commercial batch freezer to 90% overrun. After hardening at 0°F the product was compared with standard ice cream made according to the above formulation but with nonfat dry milk replacing the bland meat protein to adjust the standard mix to a protein level comparable to that of the meat protein mix. Taste panelists showed no preference for either the control or the meat-protein ice cream. Water-washed and alkali-defibrillated bland meat protein may successfully be added to a standard commercial ice cream mix.

Example 4: Water-washed and alkali-defibrillated bland meat-protein spray-dried powder was used in the formulation of a typical graham cracker crust such as that used in cream pies. The formula of the crust was as follows.

	Grams
Graham cracker crumbs	100
Bland meat-protein powder (76% protein)	26
Sugar	36
Melted butter	74

All dry ingredients were blended. The melted butter was blended into the dry ingredients, and mix pressed into the bottom and sides of the pie pan. Baking for 8 min at 350°F resulted in a typical graham cracker crust which could not be distinguished from a typical composition crust. Calculated protein content of this crust was 11.6% protein.

Example 5: Bland defibrillated meat-protein powder, spray-dried, may be successfully used as an enrichment for wheat flour and baked products. A whole wheat apple sauce cake incorporating the following ingredients was baked.

Cane sugar	96 g
Shortening	50 g
Whole wheat flour	50 g
Raisins	60 g
Bland meat-protein powder (76% protein)	20 g
Vital wheat gluten	20 g
Water	30 g
Salt	¼ tsp
Baking soda	½ tsp
Baking powder	½ tsp
Cinnamon	½ tsp
Cloves	¼ tsp
Apple sauce	125 g

Cream the shortening and sugar. Sift the remaining dry ingredients together and add to the cream mixture alternately with apple sauce and water. Add raisins. Divide into 10 muffin cups and bake at 350°F for 20 to 25 min. Typical moist apple sauce muffins containing 11.1% protein result from the preparation of this recipe.

DIETARY WAFER

Many of the dietary foods of the dehydrated type are characterized by a tendency to stick to the mouth, particularly to the roof of the mouth while eating. This condition represents an undesirable inconvenience to those persons who are obliged to conscientiously follow a diet program based on such dietary foods in order to control the intake of calories or cholesterol.

The process of *C.H. Barr, Sr., C.H. Barr, Jr. and J.W. Barr; U.S. Patent 3,360,374; Dec. 26, 1967; assigned to Courtland Laboratories* is specific for low-calorie dehydrated dietary food products which tend to be dry in texture and also stick to the roof of the mouth.

Briefly the product consists of a mixture of methylcellulose and a foodstuff which is substantially free from amino acid, the product containing amino acid in an amount sufficient to substantially prevent the product from sticking to the mouth. The foodstuff may be cooked before it is freeze-dried or may be cooked after freeze-drying and before eating in those cases where cooking is necessary or desirable.

Example: A mixture of 33⅓% by volume of cooked mashed carrots, 33⅓% by volume of methylcellulose and 33⅓% by volume of glycine was prepared. The mixture was then formed into a disk approximately 4" in diameter. The disk was first frozen and then freeze-dried by subjecting it to a vacuum of about 100 microns (McLeod Gauge). Radiated heat at a temperature of about 100°F was applied to the disk to cause the frozen liquids to sublime. This product was then removed from the vacuum.

It was found that the product, if hermetically sealed, can be stored indefinitely without regard to the ambient temperature. Before eating, the disk can be soaked in water so that the methylcellulose can absorb water. This imparts weight to the disk and enhances the transfer of flavor to the methylcellulose. Alternatively, the product can be eaten in dried form in the same manner as a cracker. The product takes on water in the stomach and swelling occurs as a result of the absorption of water, discouraging hunger. The glycine present in the dietary food product was found to prevent sticking to the roof of the mouth.

The product provides a convenient method for the control of the water content present in the diet. The dietary food normally contains on the order of 1% or less by weight of water. Accordingly, the desired control of water intake may be simply accomplished by using the prescribed measured amount of water to reconstitute the dietary product or by taking the prescribed amount of water in conjunction with the dietary product when it is consumed in the dried state. Similarly, the product allows control of the sodium chloride intake. The salt content of the dietary product can be adjusted to any desired level by leaching out any excess salt, or by the addition of salt as required to achieve the desired level. Alternatively, the salt intake can be adjusted by the use of salt water of known concentration to reconstitute the dietary product.

The dietary product is of further benefit in controlling the cholesterol content of the diet. This may be accomplished by adjusting the fat content of the dietary product to any desired predetermined level. The dietary product is often made up in large quantities on the order of 1,000 lb or more. In processing this amount of material it is relatively simple to perform a fat content analysis which enables the seller of the dietary product to indicate the exact fat content of the product on the package. Thus, the person on a low cholesterol diet may accurately control the fat intake by the consumption of indicated amounts of the dietary product.

The amount of methylcellulose used in the product may be varied within any limit depending upon the amount of caloric intake desired. It has been found that the use of methylcellulose greatly facilitates the freeze-drying procedure. Although the precise reason for this advantage is not known with certainty, it would seem that the methylcellulose functions both to absorb some of the juices in the undried foodstuff thereby allowing greater distribution of the moisture, and acts as a spacer which operates to increase the surface thereby allowing sublimation to occur at a more rapid rate.

Typical of the amino acids which may be incorporated in the dietary food product to prevent sticking to the roof of the mouth include glycine, alanine, valine, leucine, isoleucine, aspartic acid, glutamic acid, serine, proline, hydroxy proline, cystine, methionine, phenylalanine, tyrosine, tryptophan, arginine, lysine, histidine, or any other of the consumable amino acids. The amino acid may be added in dried form to the other ingredients after they have been freeze-dried. The amino acid may also be added before freeze-drying in either dried or solution form.

HIGH-PROTEIN PEANUT BUTTER

Soybean Protein and Added Fat

The nutritional quality of peanut butter is improved by addition of soybean protein. The process is described by *R.T. Darragh and J.A. Succo; U.S. Patent 3,580,729; May 25, 1971; assigned to The Procter & Gamble Company.* Many supplementary materials have been added to peanut butter ingredients to make peanut sandwich spreads. These supplementary materials include dextrose up to about 10% by weight concentration, partially or completely hydrogenated fats up to about 15% by weight concentration, and also combinations of sugar and fat additives.

These added materials have served either to make the product less sticky in eating quality or to make the product stable against oil separation, or to make the product more flavorful. Although these additives serve useful purposes, they also reduce the protein nutritive value of the resulting product since they do not contain protein and dilute the protein present in conventional peanut butter.

Animal proteins have been added, for example, lactalbumin, or nonfat dry milk solids, to peanut butter ingredients to improve the protein nutritive value of a peanut spread. It was thought that addition of these animal proteins to conventional peanut butter ingredients together with a fat, imparts to the resulting spread an eating quality equivalent to conventional peanut butter.

In this process vegetable protein with added fat is used to prepare a protein-supplement peanut spread. The steps are: (1) forming a homogeneous mixture by admixing spread ingredients comprising conventional peanut butter ingredients, soy flour, and a liquid vegetable oil in particular proportions; (2) heat-treating the homogeneous mixture under particular conditions; and (3) chilling the heat-treated mixture with agitation.

Turning first of all to the ingredients utilized in the homogeneous mixture-forming step, the homogeneous mixture, that is, the spread ingredients, necessarily contain by weight from about 42 to 48% peanut paste, from 1 to 4% solid fat stabilizer materials, from 17 to 22% soy flour, and from 27 to 32% liquid vegetable oil. The combination of peanut paste and soy flour should amount to at least about 63% by weight of the homogeneous mixture. The peanut paste and the solid fat stabilizer materials are conventional peanut butter ingredients.

In addition to these various ingredients which are required, certain optional ingredients which are sometimes added to conventional peanut butters can also be utilized; these normally amount to less than about 5% by weight of the total ingredients. The peanut paste ingredient is conveniently obtained by conventional methods by roasting and blanching raw peanuts and then grinding them. The resulting peanut paste is a mixture of peanut particles and oil which has been released from the cellular structure of the nuts during the grinding operation. If this product is allowed to stand for a period of time, the oil will separate from the product and will form a separate layer on top and a rigid crumbly mass underneath.

The solid fat stabilizer materials are utilized so that layer separation in the peanut paste will not occur. In other words, it stabilizes the peanut paste against separation of oil and solid components. The solid fat stabilizer materials can include nearly any edible glyceride fat solid. Hydrogenated natural fats such as peanut oil, corn oil, cottonseed oil, linseed oil, palm oil, whale oil, and other marine oils, alone or in combination are very suitable to provide stabilizers.

One preferred type of stabilizer is rapeseed oil having an iodine value not greater than about 10. Another preferred stabilizer is cottonseed stearin hydrogenated to an iodine value of less than about 12. The optional materials referred to above can be various flavor-providing materials including salt, sucrose, and honey. For example, the homogeneous mixture formed in the first step described above can contain additionally, by weight, from about 1 to 1.5% salt, from about 1 to 2% sucrose, and from 0.5 to 1.5% honey. It is preferred that when these flavoring components are utilized that they be balanced with an equivalent addition by weight of fats to provide good spread consistency. The solid fat stabilizer materials count toward this equivalent addition by weight of fats. If these flavor agents are utilized in an amount exceeding the above-described stabilizer component, an amount of vegetable oil, for example, peanut oil, or slightly hydrogenated soybean oil, is often added in an amount equal to the difference in weight between the weight of these flavor components and the stabilizer components.

The soy flour ingredient is available commercially. It normally contains from about 55 to 70% by weight protein. It is prepared, for example, by grinding soybean meal to a particle size ordinarily ranging from about 10 to 100 microns. Preferably the resulting ground meal is defatted, for example, by extraction with hexane. More preferably, the defatted ground meal is also extracted with methanol in order to reduce a bitter beany flavor normally associated with soy flour previous to this extraction. The soy flour preferred for use is a methanol-extracted defatted soy flour.

As indicated above, the total amount of peanut paste and soy flour, that is, the combination of these two ingredients, should amount to at least about 63% by weight of the homogeneous mixture of ingredients processed to give the final product; this level of peanut paste and soy flour is essential to assure the provision of a final spread with a protein nutritional value at least equivalent to that of conventional peanut butter.

The liquid vegetable oil ingredient is ordinarily utilized in a weight ratio with the soybean flour ranging from about 1.25:1 to 1.75:1. The use of oil and soy flour in these ratios provides an ultimate product of good consistency, that is, consistency equivalent to that of conventional peanut butter. The liquid vegetable oil can be a nonhydrogenated oil, a hydrogenated oil, or a combination of the two. Preferably for reasons of health, an oil having a high level of polyunsaturated acid moieties, for example, more than about 40% by weight of its acid moieties expressed as acids, is utilized.

Oils containing high levels of linolenic acid, for example, oils having acid moieties expressed as acids containing more than about 3% by weight of linolenic acid moieties expressed as acids, ordinarily are selectively hydrogenated by conventional methods to reduce the level of linolenic acids to less than about the 3% level; this is because oils wherein this level of linolenic acid moieties is exceeded can be subject to rancidity. Partially hydrogenated oils can be winterized to assure continued liquidity.

Suitable liquid oils without high levels of polyunsaturated acids are derived from coconut oil, olive oil, palm oil, palm kernel oil, peanut oil, and rapeseed oil. Suitable liquid vegetable oils with high levels of polyunsaturated acids are derived from corn oil, cottonseed oil, safflower oil, sesame seed oil, soybean oil, and sunflower seed oil. A preferred liquid vegetable oil ingredient is partially hydrogenated soybean oil which has been winterized, for example, soybean oil hydrogenated to an iodine value of 107 and winterized.

The homogeneous mixture of the above-described spread ingredients is formed by mixing the various ingredients in conventional mixing equipment. The various ingredients can be mixed in any order, for example, the soy flour and liquid vegetable oil ingredient can be mixed with already prepared peanut butter. Or, the conventional peanut butter ingredients can be individually mixed with the soy flour and liquid vegetable oil ingredient. Preferably, the soy flour is first slurried in the liquid vegetable oil ingredient, and this slurry is mixed with the rest of the ingredients.

The formed homogeneous mixture is then heat-treated by maintaining it at a temperature ranging from about 350° to 450°F, for a period ranging from about 2 to 4 min. This heat treatment can be carried out in any suitable equipment, for example, in a scraped wall heat exchanger, or in a jacketed stainless steel vessel equipped with an agitator and having heated oil circulating through the jacket. This heat-treating step is essential to provide to the ultimate product a color and flavor similar to that of conventional peanut butter and the combination of time and temperature utilized in this step should be sufficient to achieve this result. This heat-treated homogeneous mixture is then chilled with agitation to a temperature ranging from about 35° to 85°F. This chilling ordinarily is carried out over a time period ranging from about 20 seconds to 5 minutes, more preferably not to exceed 2 min. This chilling step is essential to bring to an end the heat-treating step and to provide an ultimate product having the consistency similar to that of conventional peanut butter. Moreover, it provides an ultimate peanut product with good nonstickiness properties, that is, a peanut spread which does not stick to the roof of the mouth and resists to some degree sticking to the peanut butter container.

Example: A homogeneous mixture of ingredients is prepared comprising the following percentages of ingredients.

Ingredients	Percentage by Weight
Peanut paste	44.7
Methanol-extracted, defatted soy flour	19.0
Soybean oil hydrogenated to an iodine value of 107 and winterized	28.45
Cottonseed stearin hydrogenated to an iodine value of 8	3.45
Salt	1.20
Sucrose	1.70
Honey	1.00
Peanut oil	0.5

The soy flour is analyzed to contain 63% by weight protein. The combination of the peanut paste and soy flour ingredients amounts to 63.7% by weight of the homogeneous mixture. The weight ratio of liquid vegetable oil, that is, winterized hydrogenated soybean oil, to soy flour is about 1.5:1.

The ingredients are mixed to form a homogeneous mixture as follows. The soy flour ingredient is slurried in the liquid vegetable oil ingredient by addition of the soy flour to the liquid vegetable oil with stirring in a stainless steel vessel. This slurry is then mixed with the rest of the spread ingredients, that is, with the conventional peanut butter ingredients, in a stainless steel vessel. The mixture is then heat-treated; in this operation it is pumped with agitation through a heat exchanger wherein it is maintained at a temperature of 400°F for 3 min until it attains the color and taste of conventional peanut butter.

This heat-treated mixture is then rapidly chilled to a temperature of 60°F in a scraped wall heat exchanger, the residence time in the scraped wall heat exchanger being 0.5 min. The spread so formed has a consistency, color, and flavor similar to that of conventional peanut butter and a nutritional value better than that of conventional peanut butter.

In another case, peanut spread is made as described above except that a conventional temperature of 150°F is utilized during the heat-treating step instead of the 400°F temperature utilized above. Even when this temperature is applied for as much as one hour, the heat-treated mixture and the ultimate product is gray and lacks the peanut flavor normally associated with peanut butter. Peanut spread is made as above except that the liquid vegetable oil ingredient is omitted. The homogeneous mixture has a consistency of fudge and cannot be processed further.

Consistency, color, flavor and nutritional results similar to that achieved above are achieved when other solid fat stabilizer materials, for example, rapeseed oil hydrogenated to an iodine value of 8, or a mixture containing 50% by weight of substantially completely hydrogenated soybean oil and 50% by weight of substantially completely hydrogenated rapeseed oil (that is, each hydrogenated to an iodine value of about 8) are substituted for the cottonseed stearin utilized above.

Skim Milk Powder and Fat

The process for making peanut butter includes roasting shelled peanuts, cooling and blanching the peanuts, removing the nibs, grinding the peanuts that passed inspection, mixing in the sugar and salt flavorings, and then regrinding. In making the stabilized peanut butters now on the market, a hydrogenated fat is added prior to the last grinding operation. About 92.5% of the product (90 to 95%) is ground roasted peanuts. The protein content of conventional peanut butter is about 29% (27 to 32%; N x 6.25) and the oil content is about 50% (48 to 52%).

The peanut butter consists essentially of ground roasted peanuts and sugar (dextrose and/or sucrose) and salt flavorings. Because this product exhibits an oil layer separating on top of the product it has become regular practice to add a relatively high-melting-fat component to the hot peanut butter (about 170°F) sometime prior to the filling of the product into jars. This high-melting-fat component usually has a melting point in excess of 110°F but less than 160°F and may be: a partially hydrogenated fat, a completely hydrogenated fat, mono- and diglyceride esters of the saturated fatty acids.

These high-melting-fat components, when added in small amounts (1 to 3% of the peanut butter) may be introduced as a supplement to the ground roasted peanuts or when added in larger amounts (5 to 10% of the peanut butter) may be introduced after an equivalent amount of the liquid peanut oil in the ground roasted peanuts had been removed. The added hard fat sets up as a continuous or semicontinuous stearin structure within the final peanut butter during the cooling of the product and in so doing prevents oil from separating from the peanut butter.

The sugar and salt flavorings are added in peanut butter manufacture in total amount, usually less than 5%, and this addition is balanced with an equivalent addition by weight of liquid nonhydrogenated vegetable oil, or the firming-up agent previously mentioned. When the latter approach is used, less of the native peanut oil has to be removed to compensate for the subsequent addition of the hard fat component. During the roasting of the peanuts, moisture content is reduced so that the final product will contain less than 4% moisture.

Conventional peanut butter is sticky in the mouth and because of this is characterized as a product of poor eating quality. This defect is due to the fact that the peanut solids, particularly the protein and polysaccharide components are in essentially the dehydrated state in peanut butter. These components have a high affinity for moisture, and when the saliva in the mouth comes in contact with the peanut butter a thickened, sticky, gummy mass forms.

The second major criticism of peanut butter is that the nutritive value of the original peanuts has been greatly lowered as a result of the roasting operations. The nutritional value of the protein is lowered and the magnitude of this change is related to the degree of the roasting of the peanuts. There is a marked destruction (sometimes almost complete) of thiamine (vitamin B_1) and significant destructions of other vitamins, namely 76% loss of pantothenic acid and up to 24% loss of riboflavin.

Attempts have been made by others to make peanut butter a more acceptable product by introducing supplementary materials which do not contribute to stickiness of the product in the mouth. For this purpose, simple sugars have been added, such as dextrose in about 5% concentration, or partially hydrogenated fats in about 15% concentration, or combinations of sugar and fat additives.

These added materials do indeed reduce the stickiness of peanut butter in the mouth but unfortunately also reduce still further the nutritive value of the peanut butter. For example, the product containing both added fat and simple sugars may contain as little as 23% protein (N x 6.25) whereas conventional peanut butter contains about 29% protein. The reason for this difference is that the supplemented peanut butter contains only about

75% ground roasted peanuts in contrast to the more than 90% in most peanut butters, and the sole source of protein in the product is the ground roasted peanuts. Others have added plant materials of more than 6% moisture content and rich in enzyme (proteolytic) content and have depended upon time and enzymic hydrolyses of the protein present in the expectation of obtaining better mouthing quality.

D. Melnick; U.S. Patent 3,216,830; November 9, 1965; assigned to Corn Products Co. states that the peanut is high in protein content, about 31% (N x 6.25), dry basis, and is an excellent source of the vitamins of the B complex. About 50% of the roasted peanut consists of an edible oil rich in polyunsaturated fatty acids (about 30% of the oil being the essential fatty acid, linoleic acid) and the ratio of the polyunsaturated to the saturated fatty acids is well above 1.0 and generally above 1.3.

In addition, the peanut is a good source of four of the essential minerals, phosphorus, iron, copper and manganese. As mentioned earlier, protein value is impaired and certain of the B vitamins are greatly reduced in concentration as a result of the roasting process in converting the raw peanut to its edible form in peanut butter. Certain essential nutrients are absent or present in limited amounts in the raw and roasted peanut.

A peanut spread of improved nutritional quality can be prepared by the use of protein supplements, such as skim milk powder containing about 35% milk proteins, lactalbumin powder containing about 80% protein, casein as sodium caseinate powder of about 90% protein content, egg albumin powder of about 85% protein content, dehydrated whole fish powder containing about 80% protein and mixtures of these protein materials.

The protein materials are in powder form sufficiently reduced in particle size to pass through a 200-mesh screen. These protein materials are added in an amount of about 5 to 20 parts by weight to 100 parts of the basic peanut butter, the protein in the protein material providing an absolute quantity of about 4 to 10%, preferably an absolute quantity of about 4 to 7.5% of protein to the total protein content.

The protein concentration (N x 6.25) in the total supplement, added to the conventional peanut butter, varies from 28 to 45%, and hence is about 100 to 150% of that of the basic peanut butter. The vegetable oil concentration in the total supplement is at least equal to the animal protein content, varying from about 30 to 52%, and hence is about 60 to 100% of that of the basic peanut butter. The ratio of the animal protein to oil in the supplement varies from about 0.55 to 1.0, and hence is about 100 to 180% of that of the basic peanut butter.

The vegetable oil in the supplement added to the conventional peanut butter may be liquid, nonhydrogenated vegetable oil or hydrogenated vegetable oils, or combinations of the two. When the supplementary oil additive includes a hydrogenated component (essential when the basic peanut butter lacks such a component), this component exhibits a melting point of from 100° to 160°F. The hard fat in this supplement serves to stabilize the final peanut spread against oil separation.

Because of the demonstrated need to add extra vegetable oils to balance off in part the nonfat components provided in significant amounts in the total supplement, it becomes unnecessary to remove any of the peanut oil from the ground roasted peanuts for subsequent replacement purposes. The vegetable oil component of the supplement can provide in itself adequate amounts of the hydrogenated component for stabilization purposes.

The added vegetable oil, in nonhydrogenated and/or hydrogenated form, is any vegetable oil such as peanut, cottonseed, corn and soybean. When it is desired to increase the vitamin E (alpha-tocopherol) level of the peanut spread, the preferred supplementary vegetable oil is cottonseed oil. Carbohydrate material, when added as such in the supplement, is in the form of a water-soluble sugar, added to the peanut butter containing the usual sugar and salt flavorings. The improved peanut spread contains less than 4% moisture and consists of from 10 to 55 parts of supplementary material added to 100 parts of conventional peanut butter, the ratio of animal protein to oil in the supplementary material varying from about 0.55 to about 1.0. The peanut spread contains from 60 to 85% of ground, blanched, roasted peanuts. There is present in this spread an absolute quantity of about 4 to 10% of high-quality animal protein, from 13 to 35% of the total protein present. Total protein (N x 6.25) content varies from about 27 to 32%. The peanut spread contains from 43 to 51% of vegetable oil. Thus, the peanut spread has a protein content which is not less than that of conventional peanut butter and an oil content which is not more than that of conventional peanut butter.

Additional nutrients can be added into the supplementary material in order to make the peanut spread in itself a nearly perfect food. Since peanut butter is a bread spread and in this capacity substitutes for butter and margarine, vitamins A and D have been added to peanut spreads in comparable concentrations. Vitamin E and the essential fatty acids (linoleic acid) have been provided by the native peanut and the cottonseed oil added by way of the supplement; ratio of linoleic to saturated fatty acids in the peanut spreads well above 1.0 and generally above 1.3.

Since raw peanuts contain about 1.0 mg of thiamine per 100 g and this vitamin is largely destroyed in the roasting process, sufficient thiamine has been added to bring the vitamin B_1 content to 1.0 mg/100 g (also the minimal adult daily requirement for thiamine). The other B vitamins, vitamin B_2 (riboflavin), vitamin B_6 (pyridoxine), vitamin B_{12} (cyanocobalamin), and calcium pantothenate have been added in quantities as required to effect nutritional balance with respect to daily requirements. No supplementation with niacin has been required since the peanut is one of the richest sources of this essential heat-stable nutrient.

With respect to the other vitamins of the B-complex, biotin, folacin, choline, and inositol, those which as yet have not been shown to be required in human nutrition, no additions have been made. Vitamin C (ascorbic acid) has been added to this product in order to round out the nutritional picture, both qualitatively and quantitatively, with respect to vitamins furnished. Peanut butter is deficient in calcium but rich in phosphorus, an unsatisfactory situation from the nutritional standpoint since a poor calcium:phosphorus ratio results.

A calcium salt containing no phosphorus (calcium carbonate) has been added in order to make the calcium:phosphorus ratio optimal, i.e., equal to 1.0, essentially the same as that found in milk. With respect to the minerals, iron, copper and manganese, required for blood (hemoglobin) formation, peanut butter per se rates as a fairly good source. Since the salts of these minerals are known to be prooxidant catalysts, it was not advisable to supplement the special peanut spread with them. Iodine may be supplied, if desired, in the form of iodized salt added during formulation.

The improved peanut spreads were rated for organoleptic qualities versus conventional peanut butters; i.e., color, flavor, body and eating quality. Because of the criticality and importance of the last measurement, an objective procedure was desired for that purpose. A Farinograph, usually used to measure flour dough consistency, was employed. The force required to rotate the Farinograph's blades, operating at constant speed, through a dough held at constant temperature is continuously recorded versus time to obtain a chart called a Farinogram.

In the Farinograph bowl at 27°C are added 300 grams (about 90 grams protein) peanut spread and 125 milliliters distilled water, both at 27°C. A 4,040-gram stainless steel block, fitting just within the bowl's walls is placed on the mixture. The block's top surface is flush with the top of the bowl, its bottom clearing the mixing paddles by about 5 millimeters. The Farinogram is taken when the mixer is operating at 30 revolutions per minute, at maximum sensitivity setting with the damper open. The area in square inches under the Farinogram curve from the point of uniform absorption of the water, usually after the first 2 to 2.5 min until the end of the 15-min mixing period, is a measure of the stickiness or eating quality of the peanut spread. The decrease in area between the new spreads and conventional peanut butter is the measure of the improvement in eating quality; it is recorded as a plus figure. An increase in area between other spreads and conventional peanut butter is the measure of the impairment in eating quality (greater stickiness); it is recorded as a minus figure. Good correlation exists between the above interpretations of the Farinograms and eating quality, when the latter is subjectively evaluated by trained scorers.

In this study, protein nutritional value was established by controlled rat feeding studies. All diets provided weaning male albino rats for a period of 28 days, 10% protein (N x 6.25) in an otherwise nutritionally adequate diet. Body weight and food intake were recorded weekly. In some tests blood hemoglobin levels were determined at the end of the four-week test period. The efficiency of food utilization was calculated from the gram gain in body weight per 100-g food eaten; and protein efficiency ratio was calculated from the gram gain in body weight per gram protein eaten.

Overall protein value was obtained by multiplying the protein efficiency ratio by the percent of protein in the food being tested. A modification of above procedure was used in determining the value of the proteins in the peanut spread in supplementing the proteins of standard white bread. The protein value of the latter product is known to be low; not only is protein content low (about 9%), but so also is protein quality. The protein efficiency ratio of the protein in standard white bread has been reported to average 0.74, ranging from 0.56 to 1.01.

Example 1: This is an example of a peanut spread of the special type. The added vitamins and minerals are of no import in the bioassays for protein quality since the basal protein-free diet furnished all essential nutrients in ample quantity. Only in the assay of the peanut-spread sandwich as the sole food consumed are the supplementary vitamins and minerals of importance. The total quantity of vitamins and minerals is so small, less than 1% of the product, that these exert no influence on the eating quality of the peanut spread. The following is the composition of the example.

Ingredients	Parts by Weight
Peanuts, roasted, blanched and ground	74.60
Blend of vegetable oils	9.50
Nonfat dry milk solids (200 mesh; 35.6% protein)	10.00
Lactalbumin product (200 mesh; 78.0% protein)	2.66
Flour salt	1.40
Dextrose powder	1.12
Sucrose (confectioners', 6X)	0.13
Calcium carbonate powder, precipitated USP	0.55
Vitamin A palmitate concentrate	0.0076
Vitamin D_2 (calciferol) concentrate	0.0014
Thiamine mononitrate (vitamin B_1)	0.0010
Riboflavin (vitamin B_2)	0.0011
Pyridoxine hydrochloride (vitamin B_6)	0.0010
Calcium d-pantothenate	0.0025
Cyanocobalamin (vitamin B_{12}) concentrate	0.0030
Ascorbic acid (vitamin C)	0.0300

The above composition, containing less than 4% moisture, is equivalent to the addition of 24.1 parts of supplementary material to 100 parts of average conventional peanut butter; the animal protein material (the nonfat skim milk solids and the lactalbumin product) is present in the amount of 15.7 parts by weight to 100 parts of the basic peanut butter. The peanuts are roasted in an atmosphere of 400°F to a slight over-roast in color and flavor and to a moisture content of 1.5%. After having been cooled by air blown over them for about 10 min, the peanuts at about 100°F are blanched for removal of the skins and nibs.

Defective nuts are eliminated by hand picking. The peanuts are subjected to an initial grinding. To the peanut paste now at a temperature of about 140°F are added the blend of vegetable oils at 150°F, and containing the just added vitamins A and D. After mixing the above for a period of about 10 min, a uniform blend of the remaining dry ingredients is gradually added with mixing. The total mix is now blended for a period of about 20 minutes while maintaining product temperature at about 140°F.

The mix is given a second and final grind reducing the peanut particle size to such a degree that 95 to 97% will pass through a 200-mesh screen. The peanut spread at about 165°F is transferred to a deaeration kettle and held under vacuum with mild agitation for about one hour. Product temperature is then lowered to about 120°F and the spread quickly chilled through a heat exchanger (a votator heat exchanger) to a temperature of about 85°F. The peanut spread is transferred to a second deaeration kettle and mildly agitated for a period of 10 min and then packed in glass jars and capped. The product is stored at room temperature (about 75°F).

Comparison of the product of Example 1 was made with an average conventional peanut butter (control a) and with a commercially available peanut spread (control b) containing partially hydrogenated fat in about 13% concentration and sugars in about 5% concentration. The latter product was designed simply to be an improvement on the eating quality of peanut butter. In color all three products were of good quality.

When evaluated over a temperature range of 45° to 95°F, the product of Example 1 was found to exhibit the least change in its desirable body, while the conventional peanut butter showed the narrowest plasticity range under the conditions of test. In flavor scoring of fresh and aged product by an expert panel, the peanut spread of Example 1 was found to be comparable to that of the conventional peanut butter and significantly more desirable (more peanut flavor) than the prior art peanut spread. In eating quality, the prior art peanut spread was less sticky in the mouth; indeed, to such a degree that peanut flavor was actually masked. The product of Example 1 was definitely less sticky in the mouth than the conventional peanut butter, and this was noted without impairment in intensity of peanut flavor.

Example 2: Into 100 parts of a conventional high-roast peanut butter (control c), described above, and containing a hydrogenated fat stabilizer, was gradually stirred a mixture of 6.4 parts of a lactalbumin product (200 mesh, 78.0% protein) and 6.4 parts of peanut oil. The mixture, at about 80°F, was gently stirred under vacuum for about 5 min to obtain a uniform product. It was then packed in glass jars. This example was compared with a product made with the same ingredients, but omitting the extra peanut oil, and with the original peanut butter similarly stirred. All three products contained less than 4% moisture.

The product of Example 2 was made by the addition of 12.8 parts by weight of supplementary material to 100 parts of the peanut butter, the animal protein material (the lactalbumin product) being present in the amount of 6.4 parts to 100 parts of the basic peanut butter. In color, flavor and body, all three products were good. The lactalbumin-supplemented product without the added peanut oil was definitely poorer in eating quality, poorer than the basic peanut butter, while the product of Example 2 was definitely superior in this respect to the basic peanut butter.

PROTEIN-FORTIFIED VEGETABLE PRODUCTS

Fabricated synthetic foods are increasing in importance and interest, but they have not yet made a significant contribution to the protein consumption level. Protein fortification has made some contribution to the protein levels in the diet. Bread is commonly fortified with whey solids and skim milk solids. Some breakfast cereals have added secondary protein.

W. Heusdens and B.F. Guevara; U.S. Patent 3,692,531; September 19, 1972; assigned to Ralston Purina Company have developed a process for fortifying food products with protein from secondary sources and a unique protein-fortified product. The process involves fortifying carbohydrate materials such as a substantially gelatinized starch with a substantially undenatured protein material. The carbohydrate material can be from a variety of sources. Starch-containing materials such as grain and grain flours can be used. Other sources which may be used are potatoes, tapioca, cellulosic vegetables such as carrots or beets, and fruit such as bananas, apples, or pears. Green bananas are particularly useful due to their high starch content. The source of carbohydrate used will depend on the geographical locality and will be determined by the dominant or most economical source.

Vegetable protein and protein isolates such as soy protein, sesame protein, or other vegetable sources may be used. Microbial protein extracts from yeasts may be used or fish protein concentrates. The food products are formed by mixing the protein material with the carbohydrate and working the material into a homogeneous mixture. The carbohydrate may be in a cooked or an uncooked state. The protein which is mixed with the carbohydrate material should not be heat-gelled, however. After the carbohydrate material and the protein material are mixed, they are formed into a desired shape, for example, a snack chip, and given a heat treatment to set the protein.

The heat treatment gives the material desirable taste and mouthfeel characteristics by fully gelatinizing the starch source, for example, by adding fat to the material or by cooking a crust on the surface of the material. In addition, it was found that the combination of cooking and the high initial content of the protein source which has not been heat-gelled adds unexpected desirable characteristics to the cooked product in addition to the nutritional benefit from protein fortification. The protein material adds to the stability of the product on storage. These products have an increased shelf life; they remain crisp and palatable for periods of time which are much longer than products which do not contain the protein-fortifying materials.

The exact mechanism by which the increased shelf life is obtained is not clear. However, it is believed that the protein material becomes resistant to moisture on being heat-set and actually inhibits absorption of moisture by the food material. By forming the food product at temperatures below the point where the protein material will set or gel, below 100°C, a product is formed under conditions which allow the protein-fortified material to be readily worked. The formed product can then be given a cooking treatment to gel the protein-carbohydrate mixture. The cooking treatment toughens the protein and inhibits absorption of moisture by the product. The cooking step also adds crispness to the final product and reduces the moisture content. The final moisture content of the cooked product will be about 6% by weight.

The cooking step can be performed in a variety of ways: by conventional ovens, microwave ovens, or by frying. Deep frying is particularly useful; it cooks the material quickly, reduces the moisture content, and adds fat to the material. The added fat improves the flavor of the cooked material. The product with protein added has the additional advantage of absorbing less fat than starch. As a result, the final cooked product has a lower fat content than an all-starch product. The mouthfeel and taste of the product is improved, because the fat content is low enough to prevent the product from feeling greasy in the mouth.

Food materials can be produced by this process which are designed to be frozen and later reconstituted by heat, such as frozen pie shells or filled puff pastry. The materials will reconstitute rapidly to form a palatable cooked product. For example, puff products produced by this method will reconstitute in about 3 min. This is much quicker than the 20 min required by a conventional product.

Example 1: Ground, frozen stew-cut potatoes were cooked in a Brabender extruder cooker in two zones. The first zone was at a temperature of 100°C and the second zone was at 80°C. The cooked potatoes were extruded from the second zone at 80°C and a moisture content of about 54% by weight. 100 parts by weight of isolated soy protein and 200 parts by weight of water were added to 500 parts of the cooked potatoes and mixed in a

Barbender extruder fitted with a low pressure screw. The mixed material was extruded cold (temperature between 80° and 100°C) through a one-inch ribbon die. The ribbon was cut into 1½" pieces and deep-fried in an oil bath at 170°C. The fried chips had a very crisp mouthfeel and a highly pleasing taste. The chips had a final content of 6% water, 20% fat, 20% protein, and 45% carbohydrate.

Example 2: A mixed material prepared as described in Example 1 was formed into a puff shell product by cold extruding the material through a 1" diameter tube die. The extruded tube was filled with a cheese filling, cut into 2" lengths, and the lengths were closed at both ends. The filled pieces were deep-fried in an oil bath at 170°C. The cooked pieces had a pleasant taste and mouthfeel. The puff shells were very crisp. The cooked pieces were frozen and were later prepared for consumption by heating the frozen pieces in a 205°C oven for 3 min.

Example 3: 300 g of all-purpose wheat flour was extruded through a Brabender extruder at a moisture content of 36% at 140°C. 70 g of isolated soy protein and 200 g of canned pear (moisture content of 85 to 90% by weight) were ground and mixed in with the gelatinized flour. The mix was cold extruded into ¼" rods. The rods were cut into 1½" pieces and deep-fried. The deep-fried pieces had a crisp, chewy texture and a pleasant fruit flavor. The fried pieces were judged to be an exceptionally palatable snack food.

HIGH-PROTEIN PUDDING

Puddings have been made with starch as the medium for obtaining the desirable thickening and gel-like consistency. These puddings contain a maximum of approximately 1% protein and, therefore, were not particularly beneficial in aiding in the protein content of the diet. The procedure followed in preparing puddings consisted of heating the starch-containing pudding material to a predetermined level, and thereafter subjecting the mixture to a cooling action for gel formation and then placing the slightly gelled mixture into cans for completion of the cooling of the mixture. This final cooling step served to set up the final gel structure to produce the desired consistency of the starch-type pudding.

Thus, in order to obtain the desired consistency and texture of these types of pudding products, extreme care had to be utilized in the cooling operation since the texture was finally determined after the material had been placed in cans. If care was not exercised in this operation, it was possible for successive runs of pudding to have wide variances in the consistency and texture. A high-protein pudding gives a tough gel structure which is somewhat rubbery. It was believed necessary to maintain the protein content of starch-type puddings at a low value so that the use of pudding as a protein supplement for the diet was nonexistent.

J.T. Duren; U.S. Patent 3,689,288; September 5, 1972; assigned to Ralston Purina Co. prepared a high-protein pudding which may be utilized as a protein supplement for dietary purposes. The high-protein pudding can be produced by forming a mixture of protein-containing material such as vegetable protein isolate, concentrate, or curd with other pudding ingredients such as sugar, nonfat dry milk, and flavoring materials. This mixture of protein-containing materials and flavoring material is then formed into an aqueous mixture by the addition of water. The protein content should be in the range of between approximately 3 to 15% by weight of the pudding mixture.

It has been found that if the protein content should be below approximately 3% by weight of the mixture that there is not sufficient protein present to obtain the gel characteristics which are desired upon heating the mixture. Also, if the protein content is higher than approximately 15% by weight of the mixture, the protein would result in the gel becoming too tough or rubbery upon heating such that it could not be subsequently cooled and mechanically worked to obtain the desired consistency or texture for the final pudding product. Some of these ingredients normally found in puddings which may be utilized in the high-protein pudding include sugar, nonfat dry milk, vegetable oil, starch (modified tapioca), alginate, salt and other flavoring materials. After the pudding mixture has been prepared it is then subjected to a heating operation. This heating operation may consist of any well-known heating device and preferably would be either a steam infusor or a Jet Cooker which are commercially available. In either the steam infusor or the Jet Cooker, the pudding mixture is subjected to direct contact with pressurized steam so as to heat the mixture to a temperature of between 280° and 300°F.

During the heating of this mixture, a gel is formed to transform the aqueous slurry into a semisolid state by heating the mixture to a temperature above 250°F and holding the material for a predetermined period of time, the pudding being aseptically produced. After the material has been heated within the desired temperature range and the gel formation has occurred, the gel is transported to a standard commercially available swept surface heat exchanger. The gelled material within the heat exchanger is subjected to a cooling action by running cool water through the jackets of the heat exchanger and simultaneously subjected to mechanical working by the paddles or blades within the heat exchanger.

This cooling action is carried out for a sufficient period of time to cool the gelled product to a temperature of about 70° to 85°F. In addition, the paddles or blades within the heat exchanger are rotated with a sufficient speed to break up the gel structure so as to obtain the desired consistency or texture of the final pudding product. By altering the retention or cooling time that the gel is maintained in the heat exchanger as well as by altering the mechanical working to which the gel is subjected, it is possible to simply control the final texture or consistency of the pudding.

In particular, a hard texture pudding may be obtained by using a short cooling cycle and subjecting the mixture to slight mechanical working. Conversely, by retaining the gel for a long period of time for cooling purposes and by subjecting the gel to a more severe mechanical working, it is possible to produce a soft pudding texture or consistency. Normally, this cooling step will be carried out for a period of between 1 and 4 min with the paddle unit within the heat exchanger running at approximately 100 rpm. By altering the jacket temperatures of the heat exchanger and by altering the revolutions per minute of the paddle unit within the heat exchanger, it is possible to obtain any desired texture or consistency for the pudding.

After the material has been subjected to the cooling and mechanical working, the product is placed in cans and sealed which results in an aseptically produced product. The pudding is then ready for final distribution. If the pudding is to be sold in a frozen condition such that it will subsequently be thawed for consumer use, it may be desirable to add a minor amount of starch-containing material such as modified tapioca starch. This minor amount of starch is utilized to give a freeze-thaw stability to the high-protein product.

If the pudding product is produced without starch and is subjected to a freeze-thaw cycle, the product has a tendency to form ice crystals and appear lumpy. However, by including the minor amount of starch, not to exceed approximately 2% by weight of the product, the water is apparently held by the starch such that ice crystals are not found within the product and a desirable appearing pudding product results even when subjected to a freeze-thaw cycle.

Example 1: A pudding mixture was prepared having the following ingredients.

	Percent
Water	66.0
Sucrose	14.0
Nonfat dry milk	5.0
Soy protein isolate	10.0
Vegetable oil	4.0
Salt	0.5
Alginate	0.3
Flavoring	0.2
	100.0

This material was then passed through a Jet Cooker where the mixture was heated to a temperature between 290° and 295°F. The mixture exiting from the Jet Cooker was in a semisolid state and had a gel structure. This gelled material was then passed through a swept surface heat exchanger where the material was cooled to a temperature of 75° to 80°F while simultaneously being subjected to the mechanical working of the paddles in the heat exchanger. The gelled material remained in the heat exchanger for about 3 min, then the pudding material was placed in a can. The high-protein pudding had a very pleasing texture and consistency which was substantially similar to starch pudding.

Example 2: The procedures of Example 1 were substantially repeated except that the pudding mixture had the following ingredients.

	Percent
Water	67.0
Sucrose	12.0
Nonfat dry milk	8.0
Soy protein isolate	3.0
Vegetable oil	9.0
Salt	0.5
Alginate	0.3
Flavoring	0.2
	100.0

PROTEIN-FORTIFIED PECTIN JELLIES

Normally, pectin-based jellies contain only one-tenth of 1% protein and thus constitute empty calories in that virtually no protein foodstuffs are included in the remaining 99.9%. Protein, when introduced into the pectin-jelled composition, would precipitate causing a segregation of components of the mixture resulting in a turbid, cloudy appearance in the final jelly product. The cause of the precipitation was found to be the excessive acidity of pectin-based jelly products which have a pH ranging from 2.6 to 3.5.

A protein-supplemented, pectin-based clear jelly product is made by *J.E. Hunter; U.S. Patent 3,749,588; July 31, 1973; assigned to The Procter & Gamble Company.* The product is based on the use of an acid-soluble protein isolate as a jelly protein supplement, the protein supplement does not precipitate and therefore the jelly product maintains its characteristic clear appearance. The range of protein concentration in the supplemented jelly product is 1 to 5% although the concentration may be as low as 0.3% or as great as 10% and may be varied to suit particular protein sources and jelly flavors.

The terms pectin or pectins designate a group of high-molecular-weight substances related to carbohydrates and found in varying quantities in fruits and plants. More specifically, the term pectin designates those water-soluble pectinic acids of varying methyl ester content and degree of neutralization that are capable of forming gels with sugar and acid under suitable conditions. Pectins are generally derived by dilute-acid extraction of the inner portion of the rind of citrus fruits, or of fruit pomaces, usually apple.

The pectin so produced may find ready use in the formation of jellies by combining with various fruit and fruit juices. More specifically, the pectin may be produced in the form of a mixture containing 5 to 8% of pectin together with a substantial amount of dispersant or filler, usually a carbohydrate such as glucose or dextrose.

The amount of pectin in a jelly will be dependent on the jelly grade, i.e., the jelling ability. This mixture commonly contains an acid in an amount sufficient to permit attainment in the jelly product of the desired pH. Citric acid is commonly used in an amount sufficient to give a product a pH of 2.6 to 3.5. The actual improvement in nutritional value due to the jelly protein supplement is extremely difficult to determine. However, one method that can be used to rate nutritional value is percentage protein in the final jelly product.

As stated previously, normally jelly contains at most a 0.1% protein concentration whereas this jelly product contains a 0.3 to 10% protein concentration. Therefore, at a minimum there is a threefold and up to a hundredfold increase in the final jelly product. Sources of the necessary acid soluble protein as a nutritional supplement include soybean, cottonseed, milk protein, egg protein, and other forms of animal and vegetable-derived protein. The preferred protein source is cottonseed.

The acid-soluble protein isolates may be obtained by means of several commercially known processes. By way of example, cottonseeds or soybeans are subdivided by means of grinding and then flaked in the presence of heat to a flake thickness of 0.007". The flaked product is then hexane-extracted resulting in flakes containing approximately 52% protein by weight and an oil product which is retained for other purposes. The flakes are then subjected to low-temperature desolventizing followed by treatment with dilute sodium hydroxide resulting in a dried product which can be used as animal feed and also resulting in a solution containing essentially all the carbohydrates and a small percentage of the protein. The remaining two thirds form an isoelectric precipitate containing about 95% by weight protein which can be put into solution and then reprecipitated at the desired pH, i.e., 2.6 to 3.5.

The final precipitate is in the form of an insoluble whey which can be discarded and a soluble whey which can be used in the jelly product following drying to remove excess water. This process is not unique to cottonseed, and may be used with other protein sources with minor variations in procedure. Actual laboratory experimentation done with Edi Pro N (soy-protein isolate) as vegetable protein source resulted in vegetable protein isolate that was 67% soluble at a pH of 3.0 and only 1.5% soluble at a pH of 4.0. Therefore, at normal jelly acidity, i.e., pH 3.0, two-thirds of the protein produced according to this process may be added to the jelly.

As mentioned above, animal protein such as Calcium Protolac can be used as the acid soluble protein isolate. Calcium Protolac is recovered from dairy cheese whey by precipitation with polyphosphate and neutralized with calcium hydroxide. One particular experiment resulted in 33.4% solubility of the protein isolate in acid at a pH of 3.0 and 27% solubility at pH 3.5. The same favorable conclusion may be drawn from this data as was drawn from the preceding data given with respect to a vegetable protein, i.e., that the protein isolate, because of its acid solubility within the pH range of 2.6 to 3.5, can be conveniently added to acidic jellies.

An alternative method of producing the protein isolate involves washing cottonseed meal with water at a pH of 7.0. This results in a first supernatant which is discarded and a precipitate (residue) which is treated with sodium hydroxide. The resulting residue and second supernatant have a pH of 10 to 12. The residue is discarded and the second supernatant, which contains 50 to 70% protein by weight, is treated with an edible acid, i.e., citric, hydrochloric (dilute), sulfuric (dilute), to lower the pH to 7.0. This results in a third supernatant which is discarded and a brownish precipitate which is 100% acid soluble at a pH of 3.0. This precipitate may be conveniently added to jellies at a pH of 3.0.

In the preparation of protein-supplemented jellies it is critical that the fruit juice base be divided into two portions. The sugar (sucrose) and pectin are added to the first portion in the presence of heat and stirring until the boiling point has been reached and maintained for at least one minute. The acid-soluble protein isolate is added to the second fruit juice portion without the addition of heat as proteins are denatured by heat which ultimately results in a significant loss in nutritional value.

Likewise, when the two fruit juice portions are combined to form the final jelly product it is critical that the first portion (heated to boiling point) be cool enough so that the protein isolate is not denatured. Denaturation of the protein isolate is undesirable because the highly ordered structure of the protein isolate would be replaced by a looser and more random structural configuration that reduces solubility, prevents crystallization, and decreases nutritional value of the acid-soluble protein isolate. Therefore, in order to prevent protein denaturation it is critical that the fruit juice-protein solution not be added until the fruit juice-sugar solution has cooled sufficiently to form a partial gel which will insure a temperature low enough to protect the molecular structure of the protein isolate.

Example: A product illustrating formation of a pectin-based, protein-supplemented, clear jelly was prepared by mixing 11.7 g of grape juice and 21.5 g of sugar (sucrose). The grape juice-sucrose mixture was heated to boiling with stirring. When stirring no longer reduced the boiling, 9.0 g of pectin was added and the mixture was boiled for 1 min. The boiled mixture was removed from the heat and allowed to cool to form a partial gel. Excess foam was skimmed off the resulting partial gel.

Ten grams of grape juice and 2.8 g of acid-soluble (at pH 3.0, 68% soluble) protein (cottonseed protein isolate prepared as previously described) are mixed together. The first grape juice-sucrose mixture is added to this second mixture and they are stirred together. The resulting combined mixture is poured into a container and allowed to cool. The jelly set to a firm gel on cooling. The jelly was rigid, readily unmoldable, and of excellent texture and clarity. In addition, the final jelly product contained 5.2% protein by weight, 16.3% pectin by weight, 39.1% sugar by weight, and 39.4% fruit juice by weight.

NUTRITIOUS MEAT-LIKE CHUNKS

The human derives psychological satisfaction in chewing. In addition to such psychological satisfaction, mastication is a necessary function for activity of the salivary glands and maintenance of adequate dentition. Most of the vegetable protein products are eaten in the form of mushes and gruels, or powdered supplements added to the staple diets which are usually high in carbohydrate, but deficient in protein. Some people, however, decline to accept such supplements, preferring their customary sources of food (rice, beans, the pastas, which provide carbohydrate but are deficient in protein), served in their traditional fashion.

It is known that an animal feed material can be made by subjecting solvent-extracted proteinaceous soybean meal, peanuts or other related means containing some moisture to pressure for a time and at an elevated temperature sufficient to convert the moisture in the meal into steam thereby debittering the meal, and to compact the meal particles by application of pressure from 2,000 to 5,000 psi. Pressure much below 2,000 psi tends to result in fairly loose pellets or particles of soybean meal rather than in the desired densely compacted mass; pressure above 5,000 psi may be used, but is generally not necessary and hence uneconomical.

Due to the extremely high compression to which the meal is subjected in conjunction with the removal of moisture by the action of the elevated temperature applied during compression, the meal particles are virtually fused or plasticized together in the form of a practically solid, very hard or horny and tough mass or cake, which can be fragmented into pieces or chunks of desired size. The mass also acquires a fiber-like consistency which is similar to the fibrous texture of flesh meat. As is evident, the chunks or pieces into which the mass is fragmented contain all the protein and other nutritional contents of the starting material. However, the physical characteristics of the chunks and or pieces make the same totally unsuitable for human alimentation. They are not palatable and cannot be chewed; in fact they are so hard that it is difficult even for animals to masticate the chunks.

H.A. Hoffman; U.S. Patent 3,620,755; November 16, 1971 treats the pieces with an edible oil prior to cooling with water to obtain uniform hydration. The function or action of the oleaginous material effects a coating on both external and internal surfaces thereby delaying hydration with cooking until the water has had sufficient time to make contact with all available surfaces to be cooked. As a result, cooking occurs uniformly and rapidly, produces finished cooked pieces which are uniformly softened and hence chewable, yet do not disintegrate into a mush or gruel but retain their general initial shape though enlarged two- to fourfold in volume. An oil-in-water emulsion instead of oil alone may also be used for coating the external and internal surfaces of the compacted pieces. The temperature of the emulsion must be maintained at a nonreacting range, such as a temperature between about

120°F and the temperature at which the emulsion still remains in the liquid stage. It has been found that the use of such emulsion improves in some instances the rapidity of the penetration of the oil into the pieces during the coating step thereby causing a more uniform hydration during subsequent boiling of the pieces. The pieces impregnated with the emulsion are permitted to dry before being boiled unless boiling immediately follows the treatment with emulsion.

It is possible to include in the cooking water nutritional and/or flavoring materials such as hydrolyzed vegetable protein, some salt and caramelized sugar to obtain chunks which were markedly palatable, readily chewable, and of noticeably attractive appearance, resembling chunks of roasted or braised beef and most important, equal or even superior in food value to average beef.

It was found that the dry pieces, although fat or oil-impregnated, when mixed with water in which was distributed hydrolyzed vegetable protein, salt, caramelized sugar, or any other supplementary nutritional materials, could be filled into cans; sealed hermetically; and heated under pressure (with simultaneous preservation by sterilization), to effect conversion into palatable chunks. Thus, a notably valuable food was made available that could be stored for prolonged periods of time (the same as other canned foods) and could be eaten either cold upon removal from the can, or upon heating depending upon the preference of the consumer. These palatable chunks may be mixed readily with high-carbohydrate staple foods to provide a diet of requisite protein levels in lieu of the supplementation with protein powders, gruels, etc., which many peoples refuse to use. The proportions of the components of the product are as follows.

Nonhydrated Product

	Suitable Range (parts)	Preferred (parts)
Toasted, debittered soy meal	50-75	64
Oleaginous material	12-30	23
To such product there is added:		
Hydrolyzed vegetable protein	7.0 parts	
Salt	1.3 parts	
Caramelized sugar	2.25 parts	
Sugar	1.3 parts	

When the hydrated product is canned, the proportions of the components of the canned product are as follows.

Hydrated Product

	Suitable Range (parts)	Preferred (parts)
Toasted, debittered soy meal	10-30	25
Oleaginous material	4-12	9
Water	35-80	61
To such product there is added:		
Hydrolyzed vegetable protein	3.0 parts	
Salt	0.5 parts	
Caramelized sugar	0.9 parts	
Sugar	0.5 parts	

Example 1: The pressed cake of disembittered (debittered) solvent-extracted soybean meal is broken into pieces. Although the pieces may vary in size, pieces varying in size from that of a hazelnut to a walnut are preferred. 100 kilos of the pieces are placed in a vessel to which is then added 36 kilos of liquefied 76°F coconut oil, held at a temperature of

120° to 180°F. The mass is tumbled until the pieces become completely impregnated by the coconut oil. The impregnated pieces have an almost dry appearance. Then there is added to the mass the following mixture: 12 kilos hydrolyzed vegetable protein, 2 kilos salt, 3 kilos caramelized sugar and 2 kilos sugar. The fat-impregnated pieces and the mixture are further tumbled until the mass acquires a uniform color, which is indicative of uniform distribution of the added material to the coated pieces. The mass is then added to 245 kilos of water and boiled for 30 min. The mass is then filled into cans according to conventional canning procedures, and the cans are sterilized.

Example 2: A dry formulation has the following composition: 64 kilos toasted disembittered (debittered) soy meal, 23 kilos 76°F coconut oil, 0.8 kilos hydrolyzed vegetable protein, 1.3 kilos salt, 2.25 kilos caramelized sugar and 1.3 kilos sugar. The foregoing ingredients are combined as described in Example 1, and packaged in containers of various sizes. The packaged product can be cooked with water as described above.

The amount of the oleaginous material and/or water, in any of the products above described, can be varied from that mentioned above, so as to produce a product of desired consistency. Thus, it is possible to produce a range of products extending from one virtually devoid of free liquid, to one which consists of the chunks in a gravy-like medium containing part of the vegetable protein hydrolysate, salt, caramelized sugar, etc. The quantity and the character of the medium depend upon the quantities of the fat and the water fractions.

The following table gives the analysis of the vegetable foods prepared in accordance with this process compared with beef.

	24.18% Protein	16.12% Protein	Dehydrated Base	Beef
Protein	24.18%	16.12%	40.30%	21.2%
Fat	9.48%	6.32%	15.80%	7.5%
Carbohydrates	14.83%	9.88%	24.71%	0
Mineral ash	7.10%	4.74%	11.84%	1.3%
Fiber	1.10%	0.73%	1.83%	0
Moisture	43.31%	62.21%	5.52%	70.0%
Weight of protein per unit weight of finished product shown	27.42 g/4 oz	27.42 g/6 oz	43.70 g/4 oz	24.04 g/4 oz
Calories per 100 g	237.71	158.48	395.28	152.30

ASCORBIC ACID IN CANDY

The process of *N.M.U. Somovilla; U.S. Patent 3,231,383; January 25, 1966* is directed to achieving a preparation that contains large quantities of l-ascorbic acid in caramel excipients in stable form with an agreeable excipient especially for children, which up to the present time has not been possible due to the risk of loss of the l-ascorbic acid during the caramelizing process when it is necessary to raise the temperature to 140°C. Apart from this difficulty, the incorporation of the l-ascorbic acid previous to or during the caramelizing process introduces a disagreeable taste to the caramel.

These difficulties concerning taste and loss due to temperature raising are eliminated by means of a process in which a calcic salt of l-ascorbic acid is used, which is much more stable than the l-ascorbic acid itself. This process moreover, allows the caramel to protect the l-ascorbic acid from oxidation and from the moist atmosphere, thus permitting the conservation of the medicament with all its vitaminic value indefinitely.

Example: Dissolve 7 kg of l-ascorbic acid in water and neutralize the solution with 2 kg of calcium carbonate; filter and complete the filtrate up to a volume of 50 l. Dissolve 100 kilograms of saccharose in the foregoing mentioned 50 l of dissolution, thus obtaining a solution of saccharose in calcium ascorbate which must be heated until the saccharose is completely dissolved at which moment, 20 kg of liquid glucose is added.

The heating is continued until the temperature reaches 130° to 140°C and when reached the heating is stopped. Traces of water are then eliminated by vacuum distillation. In this way a mass of caramelized product is obtained containing l-ascorbic acid in stable form, while the mass is endowed with an appetizing flavor and form. The amounts of calcium ascorbate cited are not limitative but may be widely varied, while any other salt of the l-ascorbic acid may be used instead of the calcium ascorbate.

As an alternative the l-ascorbic acid itself may be used, by first producing the caramelized mass at the temperature of 140°C necessary for the complete dissolution of the liquid saccharose whose integrant proportion varies between 3:1 and 4:1, allowing the mass to cool to 90° to 100°C and then adding the l-ascorbic acid in finely pulverized form, the acid being from 5 to 10% of the weight of the saccharose used, homogenizing its distribution throughout the mass and finally bestowing on it the desired taste and form.

FORTIFIED COATED SUGAR

The production of sugar having improved nutritional properties was developed by *J.F. Carter, Jr.; U.S. Patent 3,607,310; September 21, 1971.* Essential features of the process are dependent on the use of anticaking and odor-reducing compounds. The method is fully applicable to dry sugar irrespective of whether such sugar is provided in granular, powdered, flake, or similar dry form. In addition, such dry sugar, to be fortified, may be made from cane, beet, or corn sugar or any combination.

The method can be used to fortify liquid sugar; such method is fully applicable to liquid cane, beet, corn, or other similar liquid sugar product. Also, such liquid sugars may be used singly or in any desired combination, preferably within a range of 50° to 80° Brix. The final fortified sugar product whether in the dry or liquid form may contain any or all of the following nutrients as well as other water-soluble and fat-soluble vitamin, mineral, amino acid, and protein-type nutrients. The nutrients may appear from 10 mcg to 7 g of nutrients, per 100 g of final product dry sugar or per 100 g of bone dry sugar in solution in the case of liquid sugar. The following are the most desirable nutrients used in producing fortified dry sugar.

Coated thiamine mononitrate
Coated riboflavin
Niacin
Coated calcium pantothenate
Cobalamin
Vitamin A acetate
Vitamin D_2
Coated pyridoxine hydrochloride
Alpha-tocopheryl succinate
Vitamin K
Ascorbic acid
Monosodium phosphate
Monopotassium phosphate
Calcium lactate
Iron reduced by hydrogen (-325 mesh)
Potassium iodide
Sodium chloride

The following nutrients are the most desirable ones in producing fortified liquid sugar.

Thiamine hydrochloride
Riboflavin phosphate (sodium)
Niacinamide
Pantothenic acid
Pyridoxine hydrochloride
Cobalamin
Vitamin A palmitate
Vitamin D_2
Vitamin E
Ascorbic acid
Dibasic calcium phosphate
Dibasic potassium phosphate
Ferric citrate
Ferric gluconate
Potassium iodide
Sodium chloride

The method assures that the final product of fortified sugar does not have an objectionable color or odor. In the case of dry fortified sugar the nutrients previously listed are preferably coated with edible masking compounds, for example, edible stearic acid compounds. Such coated nutrients are available commercially.

In the case of vitamin nutrients the vitamins are coated so that the end vitamin product is within a weight range of 20 to 40%. To reduce and practically eliminate vitamin odor in fortified sugar, surface active absorption and/or adsorption compounds as silica gel, magnesium trisilicate, silicic acid and Cab-O-Sil M-5 are used. They serve as anticaking agents also.

To reduce the possibility of any hydroscopic compounds collecting moisture during processing of fortified dry sugar, the processing and packaging equipment must be maintained within a desirable range of 40 to 50% relative humidity at 75°F. Contamination of the fortified sugar is reduced by providing handling surfaces made of stainless steel, 61ST or 2S aluminum, or equivalent. Furthermore, it has been demonstrated that by transmitting a minimum amount of mechanical work to the sugar product being fortified, a product identical in appearance to regular unfortified sugar is produced. One series of method steps for producing dry fortified sugar (shown in Figure 10.2a) will be referred to as the multiple dilution method.

FIGURE 10.2: FORTIFIED COATED SUGAR

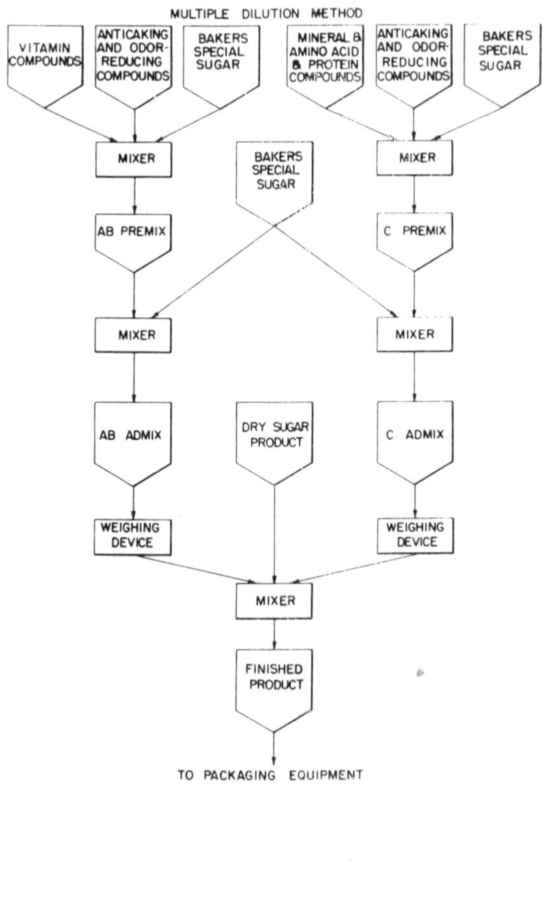

a.

(continued)

FIGURE 10.2: (continued)

b.

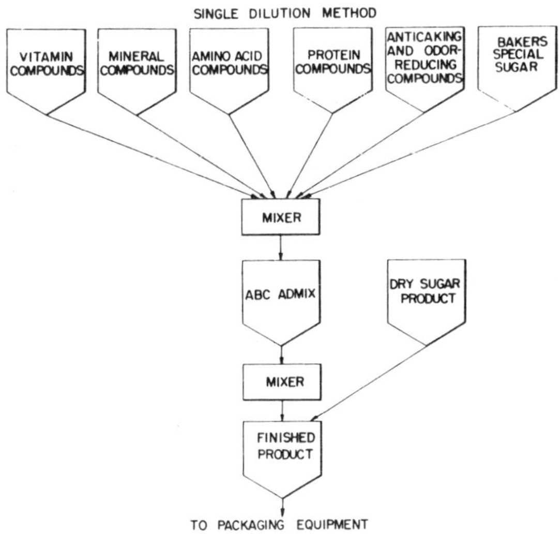

c.

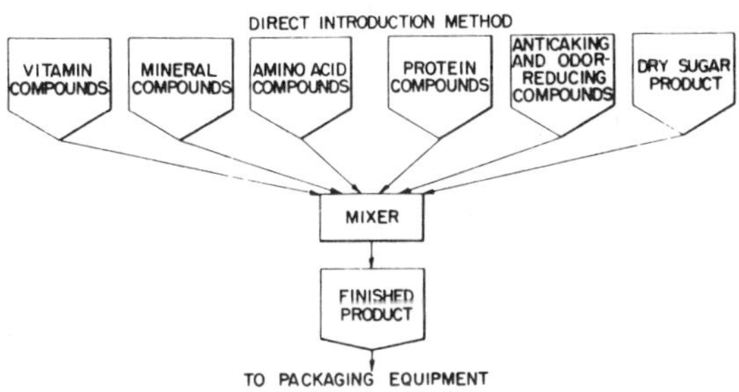

Source: J.F. Carter, Jr.; U.S. Patent 3,607,310; September 21, 1971

The method comprises preparing a first premix (which will be referred to as **AB** premix), preparing a second premix (which will be referred to as the **C** premix), preparing a first admix (referred to as **AB** admix) using the **AB** premix, preparing a second admix (referred to as **C** admix) using the **C** premix, and finally suitably mixing together the proper amounts of **AB** admix, **C** admix, and dry sugar product to produce the final fortified sugar product.

The **AB** premix is prepared by adding the proper amounts of vitamin-type compounds, anticaking and odor-reducing compounds, and a small amount of Bakers special sugar or equivalent to a blender of appropriate size and blending. The **AB** premix may range from 0.01 to 0.10%. The **C** premix is prepared by adding the proper amounts of mineral compounds, amino acid compounds, protein compounds, anticaking and odor-reducing compounds, and a small amount of Bakers special sugar or equivalent to a blender of appropriate size and blending.

It is not necessary to have compounds present from every group mentioned above in order to provide an acceptable **C** premix. **C** premix may range from 0.1 to 1.0% of the final sugar product. The **AB** premix and **C** premix are separately blended. In each case any suitable blender may be used although a twin shell tumble blender of the type having an intensifier bar is preferred. For both the **AB** premix and **C** premix, a blending time ranging from 1 to 10 min is generally sufficient for satisfactory results but optimum results are obtained when blending within a time range of 3 to 6 min.

The twin shell tumble blender has the appearance of being formed from an elongated right circular cylindrical body cut at an angle and joined together along the cut line to form a V shape. The axis of rotation is located at a point that provides equal torque loading throughout a full rotation. The twin shell blender combines the action of the inclined cylinder with the intermeshing action that occurs when the two inclined cylinders combine their flows and is the fastest of the precision blenders. The intensifier bar is an internal device to break up small agglomerates in the material being mixed.

Other types of blenders which have been successfully used for both **AB** premix and **C** premix comprise the ribbon blender and double cone blender using the same blending times as previously mentioned. The **AB** admix is prepared by adding the proper amount of **AB** premix and the proper amount of Bakers special sugar or the equivalent in a suitable blender and blending. The **C** admix is prepared by adding the proper amount of **C** premix and the proper amount of Bakers special sugar or equivalent in a suitable blender and blending. **C** admix may range from 3.0 to 5.0%.

The **AB** admix and **C** admix are separately blended. In each case a twin shell tumble blender is used either with or without intensifier bar. For both the **AB** admix and **C** admix, a blending time ranging from 3 to 30 min has been found sufficient, however, for optimum results a blending time of 10 to 20 min is used. The final fortified dry sugar product is obtained by placing the proper amount of **AB** admix, **C** admix, and dry sugar product in a suitable blender and suitably blending. The final blending action is accomplished using a continuous solids-solids blender. The optimum final blending time has been found to range between 1 and 3 min for batch blenders only.

The continuous solids-solids blender consists essentially of a number of twin shell tumble blenders joined together and inclined so that a preset amount of feed will flow through the blender to give the same blending results as the twin shell blender yet in a continuous manner. As in the case of a twin shell blender, an intensifier bar may be used if desired. A large variety of other blender types may be used for finally mixing the finished product comprising the ribbon blender, twin shell tumble blender, double cone blender, continuous double agitator mixer, and continuous mixing screw conveyor.

Another series of method steps for producing dry fortified sugar referred to as the single dilution method will be described (Figure 10.2b). It comprises preparing a first admix (which will be referred to as **ABC** admix), and finally mixing together the proper amount **ABC** admix and dry sugar product in a suitable blender and blending. The **ABC** admix is prepared by adding the proper amounts of vitamin-type compounds, mineral-type compounds, amino-acid-type compounds, protein-type compounds, anticaking and odor-reducing compounds and a small amount of Bakers special sugar or equivalent in a suitable blender and blending. **ABC** admix may range from 0.7 to 2.0% by weight.

The **ABC** admix is blended in a twin shell tumble blender with an intensifier bar. A blending time ranging from 10 to 20 min is acceptable. The final fortified product is prepared by adding together the proper amount of **ABC** admix and dry sugar product to a suitable blender and blending. The preferred blender is a continuous solids-solids blender. It may be preferred to mix the various constituents together by what will be referred to as the direct introduction method. A schematic flow diagram of the direct introduction method is shown in Figure 10.2c. In direct introduction the proper amounts of vitamin-type compounds, mineral-type compounds, amino-acid-type compounds, protein-type compounds, anticaking and odor-reducing compounds and dry sugar product are added to an appropriate size and type blender and blended. The blending action in the direct introduction method is accomplished in a continuous solids-solids blender having an intensifier bar. The blending time should be 2 to 6 min with larger times being applicable to batch blenders.

A process run for producing dry fortified sugar is given below as accomplished by the multiple dilution method (see Figure 10.2a). The process run presented is typical of dry sugar fortified with a maximum amount of nutrients. For the process run, **AB** premix in an amount sufficient for 20,000 lb of finished fortified dry sugar product is prepared by passing the following compounds through a 30-mesh screen and introducing such compounds into a twin shell stainless steel blender, for example, of the type previously described. The blending (mixing) action is accomplished for 5 min and the mixed product is held for the **AB** admix preparation.

Thiamine mononitrate (33% Rocoated)	360 g
Riboflavin (33% Rocoated)	408 g
Pyridoxine hydrochloride (33% Rocoated)	300 g
Silicic acid	180 g
Bakers special sugar	1,362 g
Cobalamin (vitamin B_{12})	0.2 mg
Calcium pantothenate	200 g
Niacin	1,250 g
Biotin	0.2 mg
Inositol	400 g
Ascorbic acid	408 g
para-Aminobenzoic acid	0.2 mg
Alpha-tocopherol acetate (1,210 IU/g)	0.3 mg
Vitamin D_2 (25,000 USP units/gram)	0.25 g
Vitamin A acetate (500,000 USP units/gram)	91 g
Glycine	900 g
Phenylalanine	450 g
Alanine	90 g
Leucine	90 g
Tryptophan	9 g
Methionine	90 g

C premix in an amount sufficient for 10,000 lb of final fortified sugar product is prepared by passing the following compounds through a 30-mesh screen into a stainless steel twin shell blender and blending for 5 min. The mixed product is then held for the **C** admix preparation.

Manganese sulfate	22 g
Potassium iodide	7.4 g
Magnesium trisilicate	454 g
Bakers special sugar	7.8 lb
Choline bitartrate	5 lb

AB admix in an amount sufficient for 20,000 lb of finished product is prepared by mixing 361.7 lb of Bakers special sugar and the **AB** premix previously prepared in a stainless steel twin shell blender for 15 min. The **AB** admix preparation is then preferably placed in 100-lb bags and suitably identified.

AC admix in an amount sufficient for 10,000 lb of final fortified sugar product is prepared by mixing the following together with the **C** premix in a stainless steel twin shell blender for 15 min. The **C** admix thus prepared is then placed in 100-lb bags and suitably identified.

	Pounds
Calcium lactate	108
Iron reduced by hydrogen (-325 mesh)	1.3
Monopotassium phosphate	40
Sodium chloride	10
Anhydrous magnesium sulfate	5
Monosodium phosphate	20
Bakers special sugar	201.7

The final fortified dry sugar product is prepared by mixing the following in a continuous solids-solids blender for 1½ min. The various materials are added in the sequence and quantities stated to produce 425 lb of the final fortified dry sugar. The mixing action is repeated as necessary to produce the desired quantities of fortified sugar. Of course, the final product is then transferred to packaging equipment and suitably packaged.

	Pounds
Regular granulated sugar	200
AB admix	4
C admix	8½
Regular granulated sugar	200
AB admix	4
C admix	8½

The maximum content of nutrients in the fortified dry sugar product based on distribution factors will be as tabulated below.

	Content/100 g of Finished Product
Vitamin B_1	1.0 mg
Vitamin B_2	1.2 mg
Vitamin B_6	1.0 mg
Vitamin B_{12}	Trace
Calcium pantothenate	2.0 mg
Niacin	11 mg
Biotin	Trace
Inositol	3.5 mg
Vitamin C	3.0 mg
para-Aminobenzoic acid	Trace
Vitamin E	Trace
Vitamin D_2	Trace
Vitamin A (USP units)	400
Glycine	8.0 mg
Phenylalanine	4.0 mg
Alanine	0.8 mg
Leucine	0.8 mg
Tryptophan	80 mcg
Methionine	0.8 mg
Choline bitartrate	40 mg
Manganese	0.12 mg
Iodine	0.1 mg
Magnesium	10.0 mg
Potassium	85 mg
Sodium	50 mg
Sulfur	10 mg
Silicon	3.0 mg
Calcium	110 mg
Iron	10 mg
Phosphorus	100 mg
Chlorine	55 mg

A similar process run could be done with a minimum amount of nutrients.

The mixing steps required to produce the improved fortified liquid sugar are as follows. The first series of method steps given will be referred to as the dry mix batch method. **AB** and **C** admix or **ABC** admix as described in the previous detailed description of the dry sugar disclosure are/is added to a given amount of liquid sugar product in a batch tank. The previously described most desirable nutrient forms for liquid sugar are used. The tank is agitated until the nutrients are dispersed and in addition to agitation the product may be recirculated by a suitable pump or a separate recirculation pump. The end product is pumped by a centrifugal pump to its usage point or to the inlet of a homogenizer.

The combination of nutrients and amounts are added in the same ranges shown in the dry section, but calculated on the basis of bone dry sugar product in the liquid. Another satisfactorily used technique for producing improved fortified liquid sugar is referred to as the dry admix batch method with homogenization. The procedure is the same as followed in the dry admix batch method with the addition of passing the final product through a homogenizer. A third satisfactory series of method steps for producing fortified liquid sugar will be referred to as the liquid admix batch method. This method may be used with or without homogenization.

The first step comprises preparation of liquid admix by adding together the proper amount of any or all of the vitamin-type compounds, mineral-type compounds, amino-acid type compounds and protein-type compounds along with the proper amount of liquid sugar in a blend tank and agitated. In addition to agitation the product may be recirculated by the discharge pump or a separate recirculation pump. The final brix of the liquid admix may or may not be the same as that of the final product. When the product is thoroughly mixed it is then pumped to a final usage point or through a homogenizer if more dispersion is needed. Homogenizer pressure may range from 200 to 1,200 psig. Liquid admix may be produced so that the level of added nutrients ranges from 0.1 to 3% by volume.

In some applications it may be desirable to prepare the liquid admix as described above leaving out the liquid sugar product. The liquid admix and liquid sugar product are then metered in proper proportion into a batch tank and agitated to insure proper nutrient distribution. In addition to agitation, the product may be recirculated by the discharge pump or a separate recirculation pump. The blended product is pumped by centrifugal pump to final usage or through a homogenizer if more nutrient dispersion is needed.

Another series of method steps used in producing fortified liquid sugar will be referred to as the chemical pump method. In the chemical pump method a large positive displacement pump pumping liquid sugar is matched with a small positive displacement pump pumping liquid admix. Both pumps feed a common header which in turn feeds a centrifugal pump. Both of such pumps are matched so that the proper amount of liquid admix is metered into the liquid sugar.

SURVIVAL FOOD PRODUCT

When a normal diet is not available, the food supplement of *R.L. Lindblad; U.S. Patent 3,058,828; October 16, 1962* can be used for survival or other emergency situations. The food supplement contains as basic elements, in carefully regulated critical proportions, animal protein, sugar, methylcellulose, nonfat dry milk substance and flavoring.

Protein is provided in a minimum amount to satisfy bodily protein needs and additionally at least 10% of bodily caloric needs. Sugar is provided for reason of its caloric content and in a dry form to prevent introduction of moisture into the mixture in quantities deleterious to other ingredients of the complete supplement. Methylcellulose is provided in an amount sufficient to provide a chewable substance in dry form and as a nutrient and primarily a filler in either wet or dry form. Additionally the hygroscopic nature of this substance aids in keeping the other ingredients in dry form and prevents caking.

The nonfat dry milk solids provide calcium as well as enriching the general mixture. Flavoring is supplied for palatability and additionally as a vehicle for combined sugar, salt, lecithin, color and certain vitamins. The flavoring can be processed cocoa which, gives the important additional advantage of providing bowel stimulation. A food supplement was prepared using the indicated ingredients and proportions.

	Percent
Animal protein (sodium caseinate), 90% protein, 4% ash, 1.5% fat, 1.3% Na, 0.2% inerts and 3% moisture at a particle size of 100% minus 120 (Tyler) mesh	10-16
Methylcellulose (chemically pure substance Methocel MC)	8-13
Granulated sugar, commercial grade	27-33
Chocolate flavor mix including sugar, cocoa processed with alkali, artificial and natural flavors, lecithin, salt and color (such as Baker's instant chocolate)	24-29
Nonfat dry milk solids	26-32

The relative proportions set forth above are important. Obviously, one may be increased in absolute quantity without detriment to the nutritional value of the mixture. However, the stated proportions cannot be materially changed on more than one ingredient without rendering the mixture unpalatable and seriously impairing its usefulness as a food supplement. Tests on actual subjects prove that while any single ingredient can be increased as much as about 50% without harmful effect no more than one ingredient can be increased beyond a maximum of 5% beyond the top of the range stated without destroying palatability.

Having mixed the ingredients in proper proportions, the mix is packaged into convenient-size packets for handling. It is best to provide packets containing one-half the daily adult nutritional requirements. Using the mix in the proportions set forth above, a packet containing the one-half daily requirement weighs only 37 g. Thus two packets, or a full day's needs would weigh only 74 g. Vitamin concentrates to supply established minimum daily requirements can either be mixed with the food supplement or taken separately. Such vitamins will not increase the weight of any single packet more than 1 to 2 g depending on the exact vitamin compounds utilized. The prepared mix is packaged in small polyethylene bags to protect it from attack by the elements.

In an actual test under severe conditions a food supplement was prepared containing 11.6% animal protein as described, 8.41% methylcellulose as described, 28.1% granulated sugar, 24.8% instant chocolate, and 27% nonfat dry milk solids. The supplement composition was put up in 37-g packets each of which was fortified as described below.

The following was added in the amounts indicated: vitamin A, 8,000 units; vitamin B_1 (thiamine HCl), 2 mg; vitamin B_2 (riboflavin), 4 mg; vitamin B_6 (pyridoxine HCl), 0.5 mg; vitamin B_{12} activity (cobalamin concentrate), 2 mcg; choline bitartrate, 20 mg; methionine, 20 mg; para-aminobenzoic acid, 10 mg; inositol, 6 mg; calcium pantothenate, 2 mg; niacinamide, 20 mg; vitamin C (ascorbic acid), 60 mg; vitamin D (irr ergost), 2,000 units; vitamin E (alpha-tocopherol acetate), 6 mg; calcium (bone meal), 170 mg; phosphorus (bone meal), 77 mg; iodine (potassium iodide), 0.2 mg; cobalt (cobalt sulfate), 0.2 mg; iron (ferric citrate), 20 mg; copper (copper sulfate), 1 mg; and fluorine (bone meal), 200 mcg.

After a thorough physical examination, the subject was parachuted into mountainous wilderness area with only 15 packets of the above food material, 15 caloric bars and equipment. He lived on the kit for 15 days during which time he traveled over 160 miles to a rendezvous point. The subject was immediately given a second supervised physical examination and the contents of the kit checked. The physical examination showed the subject to be in excellent condition both before and after the test and to have lost only 12 lb from a starting weight of 161 lb.

A check of the kit showed that the subject had used both the 15 food supplement packets and the 15 caloric bars. Aside from water he had partaken of no other food. Every food supplement packet was eaten dry rather than being dissolved in water.

COMPANY INDEX

The company names listed below are given exactly as they appear in the patents, despite name changes, mergers and acquisitions which have, at times, resulted in the revision of a company name.

INVENTOR INDEX

U.S. PATENT NUMBER INDEX

NOTICE

BREAKFAST CEREAL TECHNOLOGY 1974

by Roger Daniels

Food Technology Review No. 11

Today's breakfast cereal products are nutritive and have excellent eating qualities. This is especially true of the so-called ready-to-eat or cold cereal class.

While years ago emphasis was on vitamins and mineral enrichment, today it is on protein content, and high protein, ready-to-eat, or cook-in-the-bowl cereals are the public's favorites.

Ready-to-eat cereals can be produced in many forms and shapes such as flakes, crunchy pieces, small biscuits, granules, shreds and many more. Most of these must undergo a puffing operation. During puffing, relatively dense flakes and pieces are converted into lighter, more porous flakes and pieces by causing trapped moisture to expand very rapidly from the liquid state to the vapor phase. This affects the eating quality considerably, since well-puffed cereals are more crisp and tender. For this reason two chapters expound this most important operation in great detail.

In all, 102 processes for manufacturing these foods are described. Numbers in ( ) indicate the number of processes per topic. Chapter headings are given here, followed by examples of important subtitles.

1. **DOUGH COOKING ANDEXTRUDING (10)**
   General Mills Continuous Process
   General Foods Dough Processes
   Tempering Process
   Gelatinization Step
   Extrusion
2. **TREATMENT PRIOR TO PUFFING (10)**
   Benson Method
   Case Hardening
   Use of Pregelatinized Starch
   Use of High Amylose Starch
   Coating with Glycerides
3. **PUFFING PROCESSES (11)**
   Kikkoman Shoyu Continuous Processes
   Five General Mills Processes
   Increased Production from Puffing Guns
   Microwave Puffing
4. **FLAKING PROCESSES (8)**
   Treatment of Oat Groats
   Drying and Tempering
   Vollink Process
   Compressed Flakes
   Organosilicon Coatings
5. **SHAPING PROCESSES (9)**
   Reinhart Method
   Rippled Surfaces
   Extrusion Cooking
   Benson Apparatus
   Curled Flakes
6. **QUICK-COOKING OATS (9)**
   Tressler Processes
   Heat and Pressure
   Use of Polysaccharide Gums
   Nabisco Processes
   Addition of
   Cereal Hydrolyzates
7. **OTHER QUICK-COOKING CEREALS (13)**
   Farina
   Addition of Gums
   Wheat
   Cook-in-the-Bowl Cereals
   Grits
   Use of Emulsifiers
   Rice as Breakfast Food
8. **SUGAR COATINGS (9)**
   Melts and Syrups
   Fused Powdered Coatings
   Spun Sugar Filament
9. **ADDING FRUITS AND FLAVORS (13)**
   Fruit Incorporation
   Pellets for Fruit Filling
   Use of Release Agents
   Honey Graham Flavor
   Ice Cream Flavor
   Simulated Nut Product
   Multicolored Cereals
10. **NUTRITIVE ENRICHMENT (10)**
    Protein Fortification
    Proteolytic Enzymes for Soybean Cereals
    Fortified Puffed Corn
    Enriched Wheat Cereals
    Legumes in Cereals
    Protein Coatings
    Addition of Vitamins and Minerals
    Addition of Rennet
    Shelf-Stable French Toast

ISBN 0-8155-0519-1 290 pages

SHORTENINGS, MARGARINES, AND FOOD OILS 1974

by M. T. Gillies

Food Technology Review No. 10

Margarines and prepared shortenings are now widely accepted by the public due to their lower cost and absence of saturated fat.

Polyunsaturated fats are believed to be essential for growth and the maintenance of normal skin conditions, also to reduce blood cholesterol levels. Since the newer margarines contain polyunsaturated fats in liberal amounts, even better public acceptance and increased sales can be predicted with certainty.

All margarines and many shortenings may be considered as emulsions. Without proper additives or interesterification of the constituent oils their texture can break down from excessive freezing or heating, making them unsalable because of granular, lumpy appearance.

The processing of oils and fats, plus antispattering agents, crystallization inhibitors, flavors and colors, has grown into a highly developed technology, mostly because of the physico-chemical aspects of the raw materials and of the finished products.

How to prepare and apply these essential food products on a commercial scale is the subject of this book, illustrated by processes from 174 late U.S. patents.

A partial and condensed table of contents follows. Numbers in parentheses indicate the number of processes per topic. Chapter headings are given, followed by examples of important subtitles.

ISBN 0-8155-0517-5

333 pages

DEHYDRATION OF NATURAL AND SIMULATED DAIRY PRODUCTS 1974

by M. T. Gillies

Food Technology Review No. 15

Commercial production of dried dairy products began early in this century and was expanded considerably during World War II, when the U.S. government bought huge quantities of dried milk and eggs for the armed services and for lend-lease programs.

This book begins with an overview of patented dehydration processes which are suitable for a broad range of dairy products. It then deals with the special problems of dehydrating whole milk and combinations of fats plus milk proteins. Many processes on the manufacture of nonfat dry milk also discuss agglomeration techniques to improve this product's wettability and dispersion characteristics upon rehydration.

Numerous miscellaneous dairy foods capable of dehydration are given extensive treatment and the last section discusses the drying of simulated dairy products for many purposes, including coffee whiteners, dessert mixes, whipped toppings, etc. These products are commonly made from edible caseinates plus fats and carbohydrates.

A partial and condensed table of contents follows. Numbers in parentheses indicate the number of processes per topic. Chapter headings are given, followed by examples of important subtitles. 173 distinctly different processes are described.

ISBN 0-8155-0539-6 328 pages

COMPRESSED FOOD BARS 1974

by M. T. Gillies

Food Technology Review No. 18

Reducing the weight and bulk of military food rations has been the concern of supply officers since the days of the Roman legions. Today the providing of high grade, compact rations for sustaining the life of astronauts is an important aspect of NASA research.

This also benefits hikers, campers, hunters, explorers and backpackers in general, all of whom are willing to pay extra for food that weighs less and can be carried or stored for a long time. Since water is a rather heavy substance, weighing 8.3 pounds a gallon, it must be removed by some form of dehydration, such as freezedrying. Mechanical compression into bars or oblong blocks is usually carried out afterwards.

There are food bars that are normally eaten without rehydration. Then there are bars or blocks for use in organized camps and dining halls where time and equipment permit slow rehydration, and finally bars for stress situations that can be rehydrated in a few minutes or can be eaten as is, if need be.

In this book results of government and industrial research are offered to the alert food processor who wants to enter this definitely growing market. Potential volume savings can be 5 to 1 with meat products and as high as 16 to 1 with vegetables.

A partial and condensed table of contents follows here. Chapter headings are given, followed by examples of important subtitles.

ISBN 0-8155-0547-7 **116 pages**

VEGETABLE PROTEIN PROCESSING 1974

by L. P. Hanson

Food Technology Review No. 16

Vegetable proteins offer a great potential as direct food for human consumption. Food shortages are on the increase, particularly in India and the recent drought zones in Africa. Proteins are essential to growth and good health, and the nutritive values and chemical compositions of basic proteins are identical regardless of their provenance. Meat proteins, the traditional choice of most Americans, are becoming more and more expensive to produce.

The subject coverage of this book has been aimed at the processing of agriculturally produced vegetable proteins to make them usable as food products.

While grain proteins from wheat, rye, corn and other *Gramineae* are acceptable to the Western palate without much ado, proteins obtained from soybeans, cottonseed, rapeseed, and the like, require processing to provide foodstuffs having organoleptic properties acceptable for human consumption. Sometimes toxic ingredients have to be removed from the plant material before it can be used in animal feeds or in human food products.

Over 220 processes based on U.S. patents and government reports. A partial and condensed table of contents follows here. Numbers in parentheses indicate the number of processes per topic. Chapter headings are given, followed by examples of important subtitles.

A most timely book furnishing detailed process information on a rapidly developing phase of food processing. Of utmost value to food manufacturers including the meat processing industry.

ISBN 0-8155-0546-9 **306 pages**

FOOD ADDITIVES TO EXTEND SHELF LIFE 1974

by Nicholas D. Pintauro

Food Technology Review No. 17

Aside from freezing, canning and sophisticated methods of packaging, food is preserved by dehydration, salting, sugaring, smoking, curing, and certain types of fermentation.

A newer effective approach toward prevention of spoilage is by the use of chemical additives other than sugar, salt, vinegar, and spices.

Food additives, as defined by the National Academy of Sciences are those relatively nontoxic chemicals that may be incorporated into foodstuffs during the growing, processing, or storing periods. Every chemical added must serve one or more of these general purposes: Improve or maintain nutritional value, enhance quality, increase consumer acceptability, and facilitate preparation. In modern applications food additives are combined with established, classical methods of food preservation to maximize stability for extended shelf life. There is a great demand for additives to prevent or retard food deterioration. These additives include antioxidants, antibacterial agents, mold inhibitors, color stabilizers, anticaking agents, antibrowning agents, cloud stabilizers, metal scavengers, enzyme inhibitors.

This book describes over 140 processes involving the newest technology available in the U.S. patent literature using food additives. A partial and condensed table of contents follows here. Chapter headings are given, followed by examples of important subtitles.

ISBN 0-8155-0548-0 **402 pages**

BAKERY PRODUCTS 1975

—Yeast Leavened—

by D. J. De Renzo

Food Technology Review No. 20

Bakery products vary greatly in appearance and taste, yet they can be classified readily by the way they are leavened. In the processes described in this volume this is always done with yeast. Its single fungal cells ferment simple sugars and produce alcohol, thereby evolving carbon dioxide which is entrapped by the gelatinous gluten of the dough.

Commercial baking processes generally require flour, water, salt, milk, sugar, eggs, shortening, yeast and yeast foods (such as ammonium chloride and potassium bromate), additional enzymes, mold inhibitors, flavors and enrichment ingredients (thiamine, riboflavin, niacin, iron) as required by law for white bread.

How this is done economically in practical ways with the aid of additives and special procedures is the subject of this book. More than 200 patent-based processes are described.

A partial and condensed table of contents follows here. Numbers in parentheses indicate the number of processes per topic. Chapter headings are given, followed by examples of important subtitles.

ISBN 0-8155-0559-0 **456 pages**

PREPARED SNACK FOODS 1973

by M. Gutcho

Food Technology Review No. 2

The April 24, 1972 issue of the "Chemical Marketing Reporter" places the value of the U.S. snack industry at $2 billion a year. It is therefore not surprising that many companies are working on improving existing processes or on developing new snack foods altogether.

The dictionary defines a snack as "food eaten between meals," which implies that snacks purchased in a supermarket or other food store should already be prepared, ready to eat. The present book therefore deals to a large extent with what is called "crisps" in the food processing industry, i.e., potato chips, corn chips, popcorn, expanded products, nuts and nut products, and french fried potatoes. The preparation of additional crisps, such as onion rings and apple crisps, is described under miscellaneous snack products.

151 patent-based processes. A partial and condensed table of contents follows. Numbers in ( ) indicate the number of patents per topic. Chapter headings are given, followed by examples of important subtitles.

ISBN 0-8155-0474-8 289 pages